Digital Twin and Industry 4.0

Advances in Digital Twin Computing and Sensor Networks

Digital Twin and Industry 4.0

Concepts, Frameworks, and Applications

Abid Haleem

Department of Mechanical Engineering, Jamia Millia Islamia, New Delhi, India

Mohd Javaid

Department of Mechanical Engineering, Jamia Millia Islamia, New Delhi, India

Series Editor

Tuan Anh Nguyen

ELSEVIER

Elsevier
Radarweg 29, PO Box 211, 1000 AE Amsterdam, Netherlands
125 London Wall, London EC2Y 5AS, United Kingdom
50 Hampshire Street, 5th Floor, Cambridge, MA 02139, United States

ISBN: 978-0-443-54019-6

For Information on all Elsevier publications
visit our website at https://www.elsevier.com/books-and-journals

Publisher: Mara Conner
Acquisitions Editor: Craig Smith
Editorial Project Manager: Asmi Bharati
Production Project Manager: Sathyavani Deiva Sigamani
Cover Designer: Bakyalakshmi S.

Typeset by MPS Limited, Chennai, India

Contents

About the authors

Dr. Abid Haleem is a professor of mechanical engineering at Jamia Millia Islamia (JMI), India. After completing his PhD in systems from IIT Delhi, he started an MBA program at the JMI, heading mechanical engineering. Establishing the Internal Quality Assurance Cell of the university as the first honorary director, he helped the university achieve the best levels of national and international rankings. He has been an honorary visiting professor at IIT Delhi and taught engineering and management students. In a career spanning more than 39 years, his teaching and research focused on engineering management, including Industry 4.0, additive manufacturing, healthcare management, innovation management, supply chain management, halal, sustainability, and related areas. Under his supervision, more than 36 scholars have received their PhD degrees. He has guided hundreds of master's students on dissertations in engineering or management.

Today, with more than 380 publications in Scopus-indexed journals and many more in other journals, Dr. Haleem has been listed among the top global scientists (Elsevier and Stanford top 2% lists) since 2019. Presently, he is holding editorial positions with indexed journals in management, engineering, and healthcare. With professional experience, he has undertaken consultancy work for DFID, the Indian government, etc. He worked as an independent director on the board of Telecommunications Consultants India Ltd. He has been a member of the Ad hoc Task Force (ATF) of PMG of the Cabinet Secretariat of the Government of India. He serves on the advisory/research boards of various institutions and is involved in multiple institutional activities related to policy planning, administration, and accreditation.

Dr. Mohd Javaid is an associate professor of mechanical engineering at the Faculty of Engineering, JMI (a central university by an Act of Parliament) in New Delhi, India. He graduated in mechanical engineering from the Government College of Engineering and Technology, Jammu, India; received his master's in industrial and production engineering from NIT Kurukshetra, India; and obtained his PhD in the area of additive manufacturing from JMI, New Delhi, India. He has more than 13 years of teaching and research experience in the upcoming areas of engineering. He has taught undergraduate and postgraduate students, and two scholars have received PhD degrees under his supervision. Dr. Javaid has published more than 220 research papers in SCI/Scopus/ESCI/Web of Science-indexed journals. Presently, his Google Scholar h-index is 93, with more than 50,000 citations. Dr. Mohd Javaid has been recognised in Stanford University's list of the top 2% of scientists since 2021. His teaching and research interests are in Industry 4.0, additive manufacturing, healthcare technologies, and sustainability.

Preface

The world's industrial landscape is changing dramatically, with the Fourth Industrial Revolution, also known as Industry 4.0, provides a new phase of technological development. However, we do not find any book that covers all these technologies and is comprehensible to undergraduate students and others who are not aware of what the new, revolutionary technologies are all about. This book covers almost all the technologies used to implement the Industry 4.0 environment.

Here, we aim to provide an organised understanding of how digital twins and other Industry 4.0 technologies enhance innovation, efficiency, sustainability and competitiveness in the era of smart manufacturing by integrating theoretical underpinnings with real-world industrial applications.

Digital twin is being treated at the core of this revolution. With a focus on the crucial role of the Digital Twin, *Digital Twin and Industry 4.0: Concepts, Frameworks, and Applications* aims to provide a clear, organised and thorough introduction to Industry 4.0 technologies and the digital twin's role within them. Intelligent decision-making based on precise virtual models is all made possible by it. Industry 4.0 combines advanced manufacturing systems, intelligent industrial networks and sophisticated information technology. It is a strategy that integrates data and strategic information at all organisational levels to improve efficiency, quality, flexibility and sustainability. Industry 4.0 places greater emphasis on interconnection, intelligence and digital transformation than previous revolutions, which were mainly concerned with mechanisation or automation. The digital twin is a strategy framework that combines data, simulation, analytics and communication technologies to produce a synchronised digital replica of physical systems.

This book caters to a diverse audience, starting with graduate students in engineering, computer science, operations management and information systems, who will find the conceptual and technical depth they require for a basic understanding. Those who want to update themselves about the basic concepts, applications and framework would benefit. Policymakers and researchers will find a forward-looking basis for these technologies, helping them make informed decisions. The text is well augmented with simple figures, making it easy to understand.

This book is divided into 18 chapters; the first chapter covers the basics of digital twin and Industry 4.0, and the remaining 17 chapters address the enabling technologies for Industry 4.0. This book's main goal is to raise undergraduate students' awareness of the digital technologies that characterise Industry 4.0, with a specific focus on the digital twin. It is structured to provide a progressive and cohesive reading experience. It methodically explores each enabling technology, elucidating its principles, applications, advantages, challenges and its unique relationship with digital twin frameworks.

Each chapter is written to be both self-contained and contextually connected to the volume's broader themes. Accordingly, this book covers almost the complete gamut of Industry 4.0 technologies. Coordinated use of several technical platforms is necessary for the successful implementation of Industry 4.0. Cyber-physical systems (CPS), artificial intelligence, Internet of Things, additive manufacturing (AM), big data analytics, cloud computing, robotics, 5G communication, blockchain, sensors, machine vision, virtual reality (VR), holography, drones, robotics and strong cybersecurity frameworks are some of these. Each of these technologies makes a distinct contribution to the Industry 4.0 ecosystem, but when they work together under digital twin's integrative architecture, their full potential is reached. These disparate technologies are connected via the digital twin.

The authors hope that this book will serve as both a rigorous academic reference and a practical guide for those navigating the complex terrain of Industry 4.0 adoption. We hope that readers will find in this book a thorough introduction to the futuristic technologies shaping manufacturing; that researchers will discover frameworks and analyses that stimulate new avenues of inquiry; and that industry practitioners will gain actionable insights to support smarter, more resilient and more competitive operations.

Students can utilise this information as a springboard to investigate more complex research subjects, conduct case studies and apply Industry 4.0 solutions in practical applications. In addition to comprehending individual technologies, students also need to understand how various technologies work together in integrated digital ecosystems. This book aims to close the gap between basic engineering education and sophisticated industry operations.

Additionally, this book presents cutting-edge automation, security and visualisation technologies that support the use of digital twins in contemporary industries. While cybersecurity frameworks ensure secure and reliable operations, immersive techniques such as VR and holographic visualisation enhance interaction with digital models. Robots, drones, CPS, machine vision, 3D scanning, holography and other emerging applications demonstrate how digital twins enable intelligent monitoring, predictive analysis, simulation and optimal decision-making in complex situations.

The pace of technological change in the past decade, Industry 4.0, is no longer a futurist's forecast; it is the operating reality of competitive enterprises worldwide. The literature on these technologies is, at present, richly diverse and definitions vary across researchers. We have adopted a definition that is convenient for readers to understand. The application and framework are also diverse, and a middle path has been chosen.

Prof. Abid Haleem and Dr. Mohd Javaid have published numerous research articles and book chapters on Industry 4.0. Their work has been widely acknowledged, read and cited, and they are conducting research in Industry 4.0, AM and healthcare technologies. Dr. Abid Haleem has more than 39 years of teaching, research and corporate experience and has been listed among the top 2% of global scientists (Elsevier & Stanford) from 2019 to date; similarly, Dr. Javaid has been listed since 2020. Dr. Javaid completed his PhD in AM under the supervision of Prof. Haleem and now, in close collaboration while working in the same department, is making a perfect blend for writing a book with a similar understanding.

The authors sincerely thank their parents for their blessings and unceasing support, which have been the cornerstone of their academic and professional careers. We also like to express our gratitude to the editorial team and publisher for their cooperation and advice in making this effort a success. This publication was made possible by their contributions.

We sincerely thank our family, teachers, mentors, researchers and colleagues. We are indebted to the community of researchers, designers and teachers whose work has informed this book's text. We have cited them so that readers can access those papers and expand their horizons to the latest research and development. This will help them even undertake the latest research and implementations. We want to help humanity harness these technologies to make digital twins and Industry 4.0 a reality and a benefit. We understand that this text is a reliable guide and motivates the readers to explore further. Finally, in the age of digital twins and Industry 4.0, this book will serve as both an academic reference and an inspiration for creativity, research and technological improvement.

1

Introduction to digital twin and Industry 4.0

1.1 Digital twin

Digital twin technology facilitates the replication of tangible objects or assets in a virtual environment. Compared to using actual items, this helps make development, testing, and maintenance quicker and lower cost. In recent years, digital twins have become increasingly popular for a wide range of applications in Industry 4.0. Digital twin technology's simulation capabilities can be expanded to include demand forecasting, actionable insights, and real-time tracking across supply chains. Thereby, appropriate choices and risk reduction can be greatly aided by obtaining a thorough understanding of the entire supply chain [1–3].

Digital twins can be used to replicate various habitats and situations with great precision and to monitor workplace conditions in real time. This ensures that companies are doing their part to protect their workers by enabling safety hazards to be thoroughly evaluated before workers are placed in potentially dangerous situations. Many important functions, including monitoring, control, and diagnostics, can be carried out remotely using the digitisation of operations and maintenance through a digital twin. This eliminates the financial and geographic constraints on uptime and productivity by enabling the decentralisation of numerous operational tasks [4–6]. Chapter 2 provides an overview of digital twins and their role in Industry 4.0.

1.2 Brief about Industry 4.0

The fourth industrial revolution, commonly known as Industry 4.0, integrates sophisticated information and advanced manufacturing technologies with intelligent industrial systems. In this paradigm shift, new technologies are introduced to allow designers and developers to focus on improving product design, development, and maintenance. Industry 4.0 is a strategy that integrates data and strategic information by improving efficiency and quality. Industry 4.0 requires various technologies, such as artificial intelligence (AI), the Internet of Things (IoT), additive manufacturing (AM), robotics, and others, to drive advancements across many areas of the manufacturing and service sectors [7,8].

Industry 4.0 is reshaping manufacturing and incorporating digitisation technologies. It uses technological advancements to consolidate factory operations. Process automation, linked machinery, and manufacturing units are promoted as part of Industry 4.0. End-to-end digitisation, human resources for supervision, an internally connected network, programmable logic controller system, sensors, actuators, supervisory control and data acquisition, communication protocols, cloud connectivity, etc., are all significant features of Industry 4.0 [9–11].

Digital Twin and Industry 4.0. DOI: https://doi.org/10.1016/B978-0-443-54019-6.00011-X

Industry 4.0 refers to a new way of producing goods that efficiently and successfully employs modern industrial techniques and advanced manufacturing technologies. A smart factory is enabled by intelligent production machines that help the manufacturing sector embrace Industry 4.0. Smart manufacturing enhances production system performance by improving efficiency and decision-making. It manages large amounts of manufacturing data with a wide range of products, high velocity, and high volume [12,13]. The primary technologies of Industry 4.0 and their significant roles are discussed in this book and shown in Fig. 1.1.

Intelligent manufacturing promotes product efficiency, quality, and mass customisation by increasing manufacturing flexibility. It meets the demands of producing a customised product in less time and with higher quality. It will generate new ideas in the quest to realise Industry 4.0's much-anticipated goals. There is a strong desire for collaboration and improved cross-cultural understanding to deliver an intelligent, world-class solution. Industry 4.0 technologies send and receive crucial information, such as factory data and product information, via the internet. The intelligent manufacturing system enables enterprises to conduct research and development [14,15].

By utilising digital manufacturing, particularly AM, Industry 4.0 helps improve worker conditions, address environmental concerns, and enhance safety. Smart technologies, robotics, and cloud computing enable higher-quality products at lower cost. Machines connected to the internet via IoT analyse data and digitise supply chains. It improves the production system's productivity. Industry 4.0 relies on a cloud platform to store and retrieve

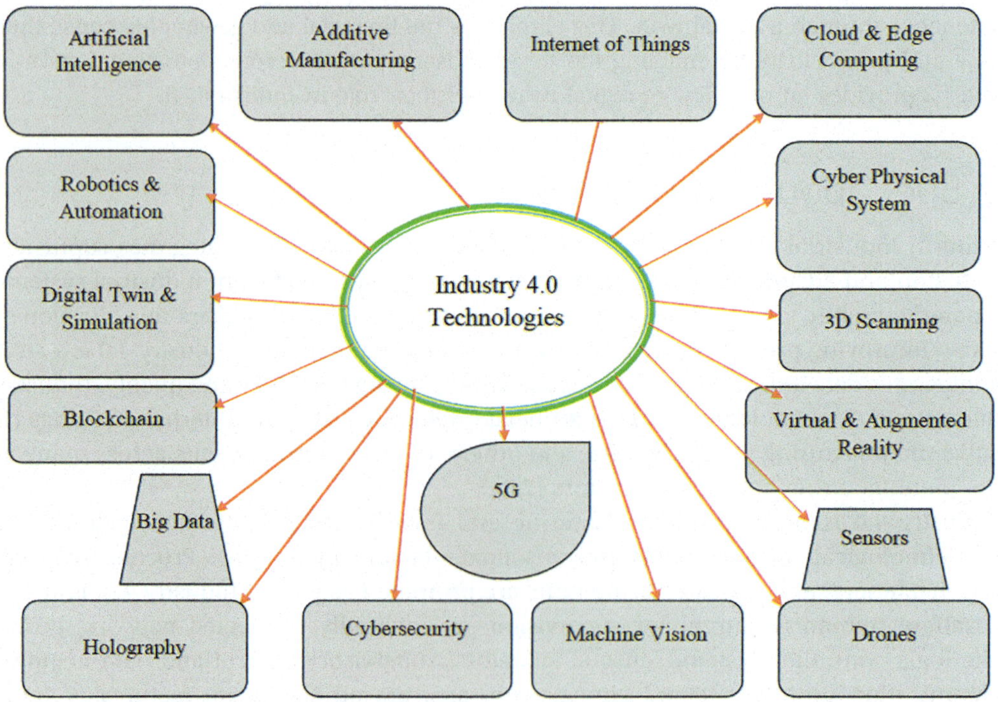

FIGURE 1.1 Major technologies of Industry 4.0.

data to perform critical tasks in real time. It improves data sharing across the entire manufacturing system and analyses customer data, market trends, and linkages [16,17].

1.3 First four industrial revolutions

From modest businesses to smart factories, the Industrial Revolution began with the production of goods. It alters culture and way of life to provide comfort. As a result, new manufacturing methods were invented to create various adjustments to save human energy. It also takes advantage of new modes of transportation to improve people's lives. People used to travel by horse or boat. Still, the Industrial Revolution introduced new modes of transportation, such as aeroplanes, automobiles, railroads, steamboats, and others, enabling people to move around the world [18–24]. Fig. 1.2 shows a brief overview of first four industrial revolutions:

1.3.1 First industrial revolution (Industry 1.0)

At the end of the 18th century (1760–1840), the first industrial revolution began. It incorporated machinery into the manufacturing process to reduce human labour. This first revolution saw improvements in manual production and the introduction of steam-powered machines to increase comfort. This made some advancement in the textile and agriculture industries. Then small firms expanded from serving a few customers to large organisations with owners, managers, and employees serving many more customers as production efficiency and scale improved. Industrial 1.0 starts an industry culture that prioritises quality, efficiency, and scalability.

1.3.2 Second industrial revolution (Industry 2.0)

Between 1870 and 1914, the Industrial Revolution began with the integration of railroads and telegraphs into industries. It was helpful for regular part manufacture, inspection, and quality

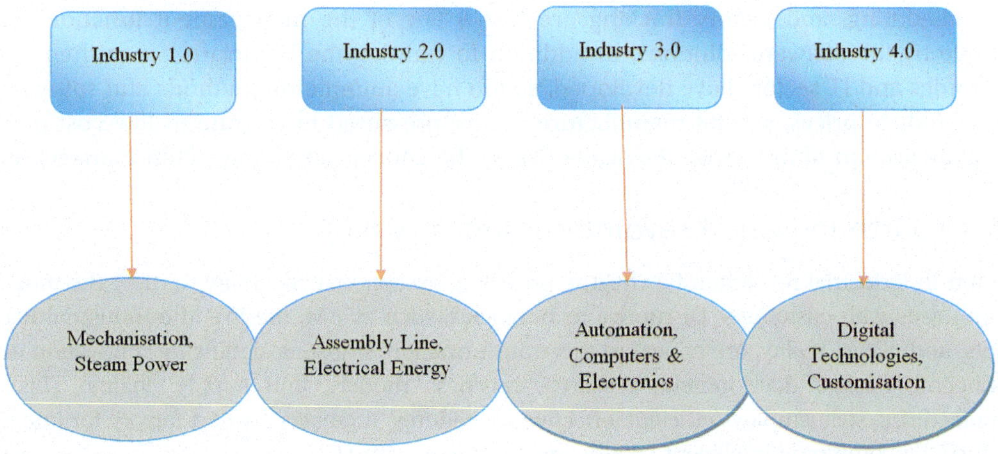

FIGURE 1.2 First four industrial revolutions.

control. It boosts production speed, but has limited design and manufacturing flexibility. It is a mass manufacturing system. This incorporates specific novel chemistry and related concepts into the manufacturing process. An important contributing component to this revolution was the invention of electrically powered machines. Electrical machines were less expensive to run and maintain than water- and steam-based ones, which were inefficient and resource-hungry. Mass production became the norm once the first assembly line was created during this time, greatly streamlining the mass manufacturing process.

During this time, the industrial culture that was developed in Industry 2.0 was transformed into management programs aimed at increasing the production facilities' efficiency. The underlying processes were optimised by a variety of production management techniques, including labour division, just-in-time manufacturing, and lean manufacturing concepts, which increased productivity and quality.

1.3.3 Third industrial revolution (Industry 3.0)

From 1950 to 1995, this third industrial revolution emerged, introducing quality, speed, delivery, safety, and flexibility through robots and programmable, flexible automation. It incorporated digital technologies with manufacturing systems, such as three-dimensional printers, computerised numerical control machines, and robots that are monitored using computers. It is a network that connects various software and technologies. It contributes to substantial advancements in computers, communication, and information technology (IT). This revolution ushered in the transition from analogue to digital systems. By combining several production technologies, this revolution in automation enables a new degree of flexibility and customisation in manufacturing. It improves the computer-aided design and manufacturing system, making the product design and development process.

The software development business was fuelled by the need for software systems to enable electronic devices because of the integration of electronic hardware into production processes. Enterprise resource planning, shipping logistics, inventory management, product flow scheduling, and factory tracking are just a few of the management functions made possible by the software solutions, in addition to running the hardware. Since then, as the electronics and IT sectors have developed, so too have automation methods and software. To further reduce prices, several manufacturers were pressured to migrate to low-cost nations. The globalisation of industrial sites gave rise to the concept of supply chain management.

1.3.4 Fourth industrial revolution (Industry 4.0)

The fourth industrial revolution has begun, providing exceptional capability to meet the industry's future needs with customers. Disruptive technologies, such as AM, the IoT, the Industrial IoT, AI, robots, and virtual reality, are critical to this transformation. It fosters significant innovation across production, design, development, research, business models, and supply chains. This new manufacturing style employs efficient, effective procedures. It creates a smart factory for intelligent products by combining intelligent production machines with IT.

Additionally, it brought about paradigm shifts in production, as common industrial operations began to integrate the virtual and real worlds. This distinction has been eroded by cyber-physical systems (CPSs), leading to several technological advancements across the business. CPSs make it possible for machines to interact more intelligently with one another across almost all geographic and physical barriers. Industry 4.0 makes machines smarter by using CPSs to communicate, assess, and guide intelligent actions across many processes. To suggest preventive and corrective measures, these smart gadgets are able to continuously monitor, detect, and foresee errors.

CPPs also make it possible for industry to be remotely monitored, controlled, and virtualised, which raises the bar for production. It enables more effective administration by integrating infrastructure, personnel, processes, and equipment into a single networked loop. Rapid technological advancements at reduced costs will happen as the cost of technology steepens. Industries are still implementing new techniques, and Industry 4.0 is still in its infancy. Industries must swiftly and economically implement new techniques to remain profitable and relevant. Industry 4.0 is here to stay for the next 10 years at the very least.

Utilising the concepts of quantum physics, such as superposition and entanglement, quantum computing is a sophisticated computational paradigm that can handle complicated problems that are much beyond the scope of traditional computers. Quantum computing has the potential to be revolutionary in the age of Industry 5.0, which prioritises sustainability, human-centric innovation, and intelligent human-machine collaboration. This technology helps create more individualised, effective, and ecologically conscious industrial ecosystems that are in line with Industry 5.0 principles by resolving extremely complex industrial problems.

1.4 Innovative requirements to be met by Industry 4.0

The concept of Industry 4.0 has evolved from the steam era to the 'Internet of Things', and developing countries offer fertile ground for this revolution as old industrial behemoths slow down. Its primary focus is to move forward from mass production towards mass customisation. This fourth generation of the industrial revolution has become a new reality in the manufacturing sector, bringing fresh concepts and ensuring competitiveness. It creates successful operational adjustments in both design and production processes. Thus to get a significant market share, industries must execute innovation swiftly. Companies launch new products with customised features at lower time and cost [25–28]. The following are the essential requirements that the adoption of Industry 4.0 fulfils:

- Extensive product customisation to meet a specific consumer demand.
- Required appropriate control mechanisms for modification.
- Automatically tracks the relevant goods by utilising the appropriate sensor system.
- Required sensor-based smart components.
- Detection of physical data and determination of related information.
- Data gathering and analysis to see if the product is suitable for the consumer.

- Ability to communicate with a centralised information system.
- Tracks shipments and logistics so that all stakeholders are up to date on the progress of the goods.
- Ability to provide multifunctionality for a variety of users and uses.
- Products tailored to the specific requirements of each customer.
- It is automated to track the proper product using the suitable sensor system.
- Use sensor-based smart components to create a customised product.
- It uses proper quality control techniques for customisation.
- Reduce manufacturing and other costs.
- The customer requests updated details for a customised product.
- Improve the flexibility of the system.
- Data are collected and analysed to determine whether the product is appropriate for the consumer.
- Innovative concept development.
- Launching new items/products/services.

1.4.1 Contemporary areas of concern while adopting Industry 4.0

The significant areas of concern while adopting Industry 4.0 are as follows [29–32]:

1.4.2 High costs of implementation

Industry 4.0 technologies are very costly, and a well-educated workforce is required to operate them, further increasing costs.

1.4.3 High failure rate towards successful implementation

The absence of goal-setting guidelines is one of the most difficult parts of starting Industry 4.0 initiatives. Projects may stutter out due to conflicting goals because they are often cross-functional endeavours with numerous stakeholders. There is a higher failure rate for its successful implementations.

1.4.4 System and component-level security

There is a security concern because people, goods, and equipment are all online. This allows hackers to gain access to networks even while it improves our ability to access data through the cloud.

1.4.5 Demand for highly skilled workforce

Manufacturing still depends on people to complete tasks, much like the rest of the sector. However, the need for highly skilled individuals increases as more systems become digitally connected, inadvertently decreasing the need for low-skilled labour.

1.4.6 Market and industry disruption

Current solutions will be phased out when new technologies become available. Certain industries won't be able to withstand what Industry 4.0 has to offer.

1.5 Goals of Industry 4.0

The overarching goal is to integrate new digital technologies into the manufacturing process as efficiently and effectively as possible [33–36]. The primary goals of Industry 4.0 are as follows:

1.5.1 Digital communication

The phrases 'digital' and 'communication' are combined to form 'digital communication'. A discrete-time-varying signal is called a digital signal. The IoT allows machines, sensors, devices, and people to interact and connect. These are the signals we use to communicate in our daily lives. In general, these signals, such as sound signals, are analogue. When communication over long distances is required, analogue signals are transmitted over wires, utilising several successful transmission methods.

Data transmission for long-distance communication utilising analogue technologies suffers from distortion, delays, interferences, and other losses. Several approaches, such as signal digitisation and sampling, improve the efficiency, clarity, and precision of the transmission process, thereby addressing various issues. Today, digital communication is a widely utilised technique in electronics. It gives us access to video conferencing, digital meetings, and online education, among other things. With the internet and other forms of digital communication, data may travel large distances in a fraction of a second. It does not save just the money but also time and effort. It has also helped improve individuals' social, political, and economic status.

1.5.2 Information

Sensor data in the information system creates a virtual copy of a physical thing. Digital information is data created or produced for electronic systems and devices, including computers, displays, calculators, and communication devices, and stored on these devices or in the cloud. Around the turn of the 21st century, electronic digital became the most common way to store, read, copy, edit, and transport data. Digital data is information stored in binary form, consisting of ones and zeros arranged in sequences to represent words and images. It is feasible to produce and store large volumes of data.

Industry 4.0 is helpful for analogue and digital data. The difference between analogue and digital data is striking. Analogue data may be physically represented, including mistakes and 'noise'. Because digital data is mainly error-free, it can be duplicated virtually endlessly with few mistakes. When an error occurs, it is almost always due to substitution, which occurs when a rogue symbol replaces an existing one. Another advantage of digital data is that large files may be reduced into smaller, more manageable packets. It enables the delivery of enormous volumes of data quickly, with the data then decompressed at the point of use.

1.5.3 Problem-solving

Assist humans by visualising and aggregating information gathered to solve pressing problems and make quick decisions. In particular, digital problem-solving involves using the abilities, methods, and techniques needed to use new tools, resources, and interfaces in effective and adaptable ways to accomplish both personal and professional objectives, as well as to navigate online in commonplace contexts, such as the library. The primary advantages of Industry 4.0 are straightforward: greater knowledge leads to greater efficiency, which saves time and money. There are also several lesser-known advantages, such as overcoming workforce issues. Many organisations in the manufacturing industry are facing a labour crisis.

Over the past 10 years, new Industry 4.0 technologies—from cloud computing to mobile—have made considerable strides and are now prepared for deployment as integrated, commercially available industrial systems. It is essential for real-time data and results, which will help the sector reach new levels of lean performance. It is used in a variety of contexts and covers a broad spectrum of technology.

1.5.4 Automatic decision

Systems can make decisions to fulfil tasks as autonomously as possible. IoT sensors may be used with everything from lighting to heating, ventilation, and air conditioning systems to manufacturing machines. Industry 4.0 is built on the IoT. The advantages of linked equipment are enormous, as it enables improved decision-making across all company levels. Large amounts of information are referred to as 'big data', but when robots are in charge, things change significantly.

Even though Industry 4.0 began in manufacturing, it now applies to organisations worldwide. This is because automating historically manual procedures and practices, and providing real-time data, dramatically improve human decision-making, decentralise data, and empower individuals locally. Data about local events might help retailers modify product placement. Factory managers can easily track and optimise automated operations.

1.6 Conclusion

Digital twins in manufacturing enable companies to monitor physical assets in real time, gain predictive insights, and simulate operations by building dynamic digital copies of physical assets. With digital twin solutions, these features greatly improve maintenance plans and increase operational effectiveness. Industry 4.0 is evolving into a smart manufacturing system that focuses on customising product design and manufacture to meet the needs of individual customers. The purpose of Industry 4.0 is to enable machines to connect. It can transfer information that enables it to quickly resolve an urgent situation and decide whether to carry out the required task. The primary components were created to enhance the manufacturing system's performance.

Industry 4.0 includes several aspects that attract attention and improve the production environment. In the age of Industry 4.0, AM offers a lot. It plays a critical role in achieving the goals of Industry 4.0, from customised design to the manufacture of complex products and mass customisation. It reduces inventory costs by keeping product digital and using design software to generate improvements that help the smart manufacturing system achieve its objectives.

Industry 4.0 technologies have a significant environmental impact by reducing material waste and maximising workplace efficiency to complete the job. It supports design software and advances digital technologies by automatically manufacturing the product. Industry 4.0 helps industrial facilities by reducing reliance on human labour and transforming the conventional relationships among producers, suppliers, and customers. These can improve the production environment, enabling the learning of new designs in less time.

References

[1] M. Javaid, A. Haleem, R. Suman, Digital twin applications toward Industry 4.0: a review, Cognitive Robotics 3 (2023) 71–92. Available from https://doi.org/10.1016/j.cogr.2023.04.003.

[2] T.H.J. Uhlemann, C. Lehmann, R. Steinhilper, The digital twin: realizing the cyber-physical production system for industry 4.0, Procedia CIRP 61 (2017) 335–340. Available from https://doi.org/10.1016/j.procir.2016.11.152.

[3] A. Martínez-Gutiérrez, J. Díez-González, R. Ferrero-Guillén, P. Verde, R. Álvarez, H. Pérez, et al., Digital twin for automatic transportation in industry 4.0, Sensors 21 (10) (2021) 3344. Available from https://doi.org/10.3390/s21103344.

[4] M. Nagy, M. Figura, K. Valaskova, G. Lăzăroiu, Predictive maintenance algorithms, artificial intelligence digital twin technologies, and internet of robotic things in big data-driven industry 4.0 manufacturing systems, Mathematics 13 (6) (2025) 981. Available from https://doi.org/10.3390/math13060981.

[5] L. Judijanto, L. Qadriah, I.A. Prabowo, W. Widyatmoko, P.C. Sabila, Trends in digital twin technology for industry 4.0: a bibliometric study, The Eastasouth Journal of Information System and Computer Science 2 (2024) 92–104. Available from 10.58812/esiscs.v2i02.382.

[6] B. Elghomri, F. Messaoudi, Digital twins for sustainable and low-carbon supply chains in industry X.0: crossing the fragmented bridge between academic research and industrial adoption, Operations and Supply Chain Management: An International Journal 19 (2025) 13–29. Available from 10.31387/oscm0640504.

[7] C. Bai, H.A. Li, Y. Xiao, Industry 4.0 technologies: empirical impacts and decision framework, Production and Operations Management 35 (2026) 50–68. Available from https://doi.org/10.1111/poms.13813.

[8] P. Hines, F. Magnani, J. Mula, R. Sanchis, Toward lean industry 5.0: a human-centered model for integrating lean and industry 4.0 in an automotive supplier, International Journal of Production Research 64 (2026) 250–273. Available from https://doi.org/10.1080/00207543.2025.2551238.

[9] A. Raj, G. Dwivedi, A. Sharma, A.B.L. de Sousa Jabbour, S. Rajak, Barriers to the adoption of Industry 4.0 technologies in the manufacturing sector: an inter-country comparative perspective, International Journal of Production Economics 224 (2020) 107546. Available from https://doi.org/10.1016/j.ijpe.2019.107546.

[10] P. Reaidy, M. Alaeddini, A. Gunasekaran, O. Lavastre, M. Shahzad, Unveiling the impact of Industry 4.0 on supply chain performance: the mediating role of integration and visibility, Production Planning & Control 37 (2026) 45–66. Available from https://doi.org/10.1080/09537287.2024.2440454.

[11] I.O. Raji, E. Shevtshenko, T. Rossi, F. Strozzi, Industry 4.0 technologies as enablers of lean and agile supply chain strategies: an exploratory investigation, The International Journal of Logistics Management 32 (4) (2021) 1150–1189. Available from https://doi.org/10.1108/IJLM-04-2020-0157.

[12] A. Samuels, Examining the integration of artificial intelligence in supply chain management from industry 4.0 to 6.0: a systematic literature review, Frontiers in Artificial Intelligence 7 (2025) 1477044. Available from 10.3389/frai.2024.1477044.

[13] S. Dehghan, S. SattarpanahKarganroudi, S. Echchakoui, N. Barka, The integration of additive manufacturing into industry 4.0 and industry 5.0: a bibliometric analysis (trends, opportunities, and challenges), Machines 13 (1) (2025) 62. Available from https://doi.org/10.3390/machines13010062.

[14] A.G. Frank, T.J. Sturgeon, G.B. Benitez, G.A. Marodin, S.F. e Cunha, How lean and Industry 4.0 affect worker outcomes and operational performance: a quantitative assessment of competing models, International Journal of Production Economics 279 (2025) 109475. Available from https://doi.org/10.1016/j.ijpe.2024.109475.

[15] A. Zeqiri, A. Ben Youssef, T. &Maherzi Zahar, The role of digital tourism platforms in advancing sustainable development goals in the industry 4.0 era, Sustainability 17 (8) (2025) 3482. Available from https://doi.org/10.3390/su17083482.

[16] K. Nikiforidis, A. Kyrtsoglou, T. Vafeiadis, T. Kotsiopoulos, A. Nizamis, D., Ioannidis, et al., Enhancing transparency and trust in AI-powered manufacturing: a survey of explainable AI (XAI) applications in smart manufacturing in the era of Industry 4.0/5.0, ICT Express 11 (1) (2025) 135–148. Available from https://doi.org/10.1016/j.icte.2024.12.001.

[17] M. Terziyska, I. Ilieva, Z. Terziyski, N. &Komitov, A review of the transition from industry 4.0 to industry 5.0: unlocking the potential of TinyML in industrial IoT systems, Sci 8 (1) (2026) 10. Available from https://doi.org/10.3390/sci8010010.

[18] S.B. Rane, Y.A. Narvel, Leveraging the industry 4.0 technologies for improving agility of project procurement management processes, International Journal of System Assurance Engineering and Management 12 (2021) 1146–1172. Available from https://doi.org/10.1007/s13198-021-01331-4.

[19] M. Ghobakhloo, M. Iranmanesh, B. Foroughi, M.L. Tseng, D. Nikbin, A.A. Khanfar, Industry 4.0 digital transformation and opportunities for supply chain resilience: a comprehensive review and a strategic roadmap, Production Planning & Control 36 (1) (2025) 61–91. Available from https://doi.org/10.1080/09537287.2023.2252376.

[20] S.B. Rane, P.R. Potdar, S. Rane, Development of project risk management framework based on industry 4.0 technologies, Benchmarking: An International Journal 28 (5) (2021) 1451–1481. Available from https://doi.org/10.1108/BIJ-03-2019-0123.

[21] H. Birkel, J.M. Müller, Resilient by nature or technology? How industry 4.0 enhances supply chain resilience until 2035, Supply Chain Management: An International Journal 30 (3) (2025) 304–322. Available from https://doi.org/10.1108/SCM-07-2023-0340.

[22] U. Techanamurthy, M.S. Iqbal, Z. Abdul Rahim, Industry 4.0 readiness and strategic plan failures in SMEs: a comprehensive analysis, PLoS One 20 (5) (2025) e0324052. Available from https://doi.org/10.1371/journal.pone.0324052.

[23] M. Farrukh Shahzad, H. Liu, H. Zahid, Industry 4.0 technologies and sustainable performance: do green supply chain collaboration, circular economy practices, technological readiness and environmental dynamism matter? Journal of Manufacturing Technology Management 36 (1) (2025) 1–22. Available from https://doi.org/10.1108/JMTM-05-2024-0236.

[24] F. Rosin, P. Forget, S. Lamouri, R. Pellerin, Impacts of industry 4.0 technologies on lean principles, International Journal of Production Research 58 (3) (2020) 1–18. Available from 10.1080/00207543.2019.1672902.

[25] T. Khan, M.M.H. Emon, Supply chain performance in the age of industry 4.0: evidence from manufacturing sector, Brazilian Journal of Operations & Production Management 22 (1) (2025) 2434. Available from 10.14488/BJOPM.2434.2025.

[26] F. Rosin, P. Forget, S. Lamouri, R. Pellerin, Enhancing the decision-making process through industry 4.0 technologies, Sustainability 14 (1) (2022) 461. Available from https://doi.org/10.3390/su14010461.

[27] S. Sajid, A. Haleem, S. Bahl, M. Javaid, T. Goyal, M. Mittal, Data science applications for predictive maintenance and materials science in context to industry 4.0, Materials Today: Proceedings 45 (2021) 4898–4905. Available from https://doi.org/10.1016/j.matpr.2021.01.357.

[28] S. Wibowo, Global trends in production management and operational excellence: a comprehensive bibliometric analysis (2024-2026, Journal of Collaborative Industrial Management (JCIM) 1 (2025) 182–201.

[29] S.M. Sepasgozar, Digital technology utilisation decisions for facilitating the implementation of industry 4.0 technologies, Construction Innovation 21 (2020) 476–489.

[30] L. Antonazzo, G. Sansone, M. Orel, The impact of industry 4.0 and state incentives on firms' skills needs: an empirical investigation of the Italian manufacturing sector, R&D Management 56 (1) (2026) 102–115. Available from https://doi.org/10.1111/radm.70009.

[31] L. Silvestri, A. Forcina, V. Introna, A. Santolamazza, V. &Cesarotti, Maintenance transformation through industry 4.0 technologies: a systematic literature review, Computers in Industry 123 (2020) 103335.

[32] A. Sharma, An empirical investigation into the nexus between industry 4.0 adoption and financial performance: evidence from listed Indian manufacturing firms, Advances in Consumer Research 3 (1) (2026) 1037–1046. Available from https://doi.org/10.5281/zenodo.18311084.

[33] A.G. Frank, M. Thürer, M. Godinho Filho, G.A. Marodin, Beyond industry 4.0–integrating Lean, digital technologies and people, International Journal of Operations & Production Management 44 (2024) 1109–1126.

[34] A.E.H. Gabsi, Integrating artificial intelligence in industry 4.0: insights, challenges, and future prospects–a literature review, Annals of Operations Research (2024) 1–28. Available from https://doi.org/10.1007/s10479-024-06012-6.

[35] A. Bueno, M.L. da Rocha Azevedo, M. Godinho Filho, G.M. Ganga, F.L. Lizarelli, Industry 4.0 skills in industrial engineering courses: contributing to the role of universities toward sustainable development, IEEE Transactions on Engineering Management 71 (2024) 8369–8387.

[36] M. Ghobakhloo, M. Iranmanesh, M. Fathi, A. Rejeb, B. Foroughi, D. Nikbin, Beyond industry 4.0: a systematic review of industry 5.0 technologies and implications for social, environmental and economic sustainability, Asia-Pacific Journal of Business Administration 17 (2025) 889–914.

2

Digital twin for Industry 4.0

2.1 Introduction

Digital twins are real-time virtual representations of systems, processes, and other entities. A digital twin provides a digital representation of a factory's operations, activity on a communications network, or the flow of goods via a logistics system. Industry 4.0 effectively leverages digital twins [1–3]. Manufacturers can better understand how assets behave in real time by installing sensors on specific assets, collecting data, creating digital copies, and leveraging machine intelligence. They can make informed decisions with assurance, enhancing business success. This technique includes, at the very least, the specific identification of the real thing [4,5].

Smart components are connected to physical objects that use sensors to collect data on working conditions, real-time status, or position, serving as digitised duplicates of machinery, equipment, or physical locations equipped with sensors, also known as the virtual counterparts of physical assets. The digital twin may simulate the processes of physical objects and anticipate how they would behave under certain conditions. By simply analysing a product, process, or system's features, performance capabilities, and potential issues, this virtual-based design method may identify and establish the optimal practical efficiency level for the given situation [6–8].

Digital twins collect data, making it easy to identify patterns in past events, pinpoint underlying causes, and improve operational processes. The number of new use cases that can be created depends on its ability to gather data from and about a physical system or product. Using the digital twin, engineers may remotely troubleshoot machinery, reducing problem resolution times and improving accuracy. The digital twin enables utilising data to simulate various scenarios for the new equipment and identify areas where new plants can outperform current production systems [9–12].

2.2 Processing steps of digital twin

A digital twin is a virtual model that mimics the physical elements and behaviours of an Internet of Things (IoT) device over the course of its life. Sensors are used to gather real-time data from physical items in order to build a digital counterpart. Teams may be better able to understand and assess real-world systems after using that data to generate a digital replica. The true advantage of a digital twin is that it offers real-time data that could aid in reasoning, learning, and comprehending how systems and objects function. Many companies employ digital twin cloud services to model and simulate infrastructure assets because of the numerous advantages [13,14].

Digital Twin and Industry 4.0. DOI: https://doi.org/10.1016/B978-0-443-54019-6.00008-X

A digital twin is an application that combines to produce a nearly real-time digital reproduction of the physical environment and process. The objective of a digital twin is to identify unwanted departures from ideal conditions in several dimensions. Digital twins increase the effectiveness of systems and businesses, leading to better results. The digital twin's foundation comprises several dimensions of substantial, cumulative, in-the-moment, and real-world data measurements. These measurements may lead to a growing digital profile of the thing or process, which can then influence physical actions, such as altering product design or production methods [15–17].

Digital twins provide advantages beyond the design phase and into the product's actual use. The twin can also provide crucial insights into how an asset will respond to stresses and various situations. The growing adoption of digital twins to enhance customer experience and resource management across various corporate verticals has significantly boosted market revenue growth. A digital twin may replicate a particular procedure or an entire system in a virtual setting, allowing changes to various characteristics to be made using information from the actual setup [18,19]. Fig. 2.1 shows the method utilised by the digital twin in Industry 4.0.

A virtual twin is used to take input from the physical system. Software, data, and techniques provide a virtual representation of the part/system. This virtual representation of the part/system provides valuable information in line with Industry 4.0 requirements. A digital twin aims to replicate a company's or factory's physical assets. Whether it is a

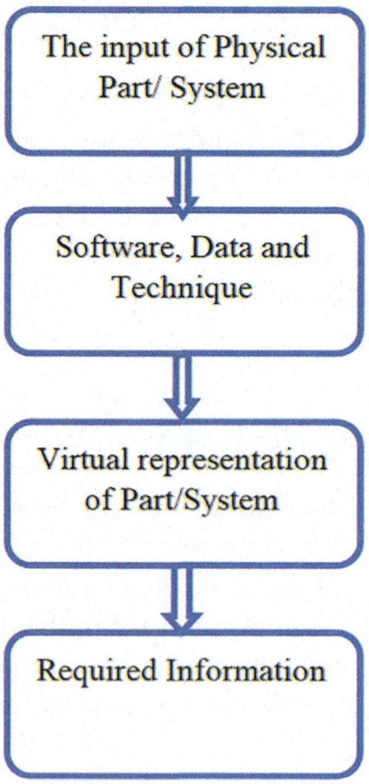

FIGURE 2.1 Processing steps of digital twin in Industry 4.0.

facility, a building, large machinery, expensive equipment, a system, or a structure, the digital twin brings tangible and intangible elements to a digital platform [20,21]. The digital twin must place strong emphasis on the data needed throughout the asset's life cycle. The uniqueness of the digital twin is that it provides businesses worldwide with a comprehensive solution.

2.3 Types of digital twin

There are four main types of the digital twin, which are as follows [22–26]:

2.3.1 Asset twins

Asset or product twins are virtual twin components that operate together. Asset twins are often used to narrow the study's scope to a few critical components, enabling synchronous performance mapping and pinpointing probable component failures.

2.3.2 Components twin

They speak about the tiniest virtual replicas of every system component. By putting them through dynamic high-range simulations, these virtual components help engineers understand the fundamental properties of each machinery component.

2.3.3 System twins

They are used to study how various assets fit together as fully functional components of a larger system and are also referred to as unit twins. They provide detailed insights into performance improvement and help simulate industrial chains.

2.3.4 Process twins

Process twins manage the highest degree of complexity in a simulated environment. They carefully virtualise each interconnected system, simulating the real-time interaction of its components, as in a factory.

2.4 Components of digital twin

A digital twin is a precise, continuously updated virtual model of a physical product or procedure that is used to test different situations, predict possible issues, and find areas for improvement. Unlike traditional computer-aided design and engineering models, a digital twin always has a unique, real-world counterpart, receives real-time data from it, and makes adjustments as needed to duplicate the origin throughout its existence [27–31]. Fig. 2.2 shows the main elements of a digital twin.

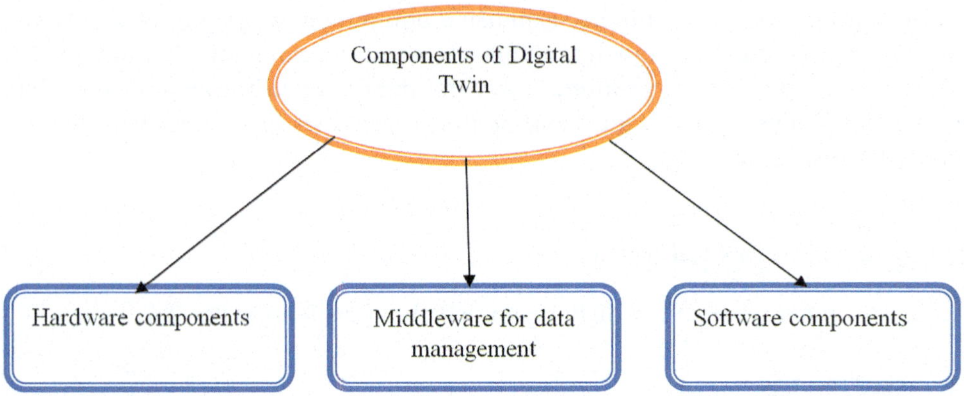

FIGURE 2.2 Major components of digital twin.

2.4.1 Hardware components

The main technology enabling the digital twin is the IoT sensors, which start the information flow between assets and their software representations. The hardware component also includes network devices, such as routers, edge servers, and IoT gateways, as well as actuators that translate digital information into mechanical motion.

2.4.2 Middleware for data management

A central repository that gathers information from numerous sources is an essential component. Connection, data integration, visualisation, processing, quality control, modelling, governance, and other associated tasks should ideally be handled by the middleware platform.

2.4.3 Software components

Because it converts unprocessed observations into useful business insights, the analytics engine is essential to digital twinning. The crucial elements of a digital twin include simulation software, design tools for modelling, and dashboards for real-time monitoring.

2.5 Advantages of digital twin

The significant advantages of digital twin are as follows [32–34]:

- The manufacturing sector is able to monitor a constant flow of real-time performance and consumption data.
- Data is collected and sent to the central repository about a physical object and its surroundings.
- The sector may use end-to-end asset or product life cycle data to generate digital threads.
- Data analysis and preparation for digital twin feeding.

- Industries support new product-as-a-service business models.
- The digital twin simulates the effects of changing environmental conditions.
- It drives developments in manufacturing, R&D, logistics, supply chain management, and services.
- Enhanced operational effectiveness and performance.
- Reducing resources and business costs.
- Remote monitoring and collaboration.
- Digital twin accelerates medical innovation.
- Supporting the development of new drugs and research on tumours.
- These technologies let doctors accurately maximise the effectiveness of patient-specific treatment plans.
- By creating a digital twin of a hospital, one can investigate potential modifications to operational strategy, capacity, staff, and care delivery models.

2.6 Roles of a digital twin for Industry 4.0

A digital twin provides manufacturers with a better understanding of their operations by accounting for many production-related factors. These consist of computerised maintenance, process optimisation, and equipment monitoring. Implementation success rates are increased through digital prototyping procedures employing augmented and virtual reality, artificial intelligence (AI), and machine learning (ML). With a digital twin, firms can improve processes, develop new products, and boost productivity. Modelling a production process enables manufacturers to assess how capacity can be increased or downtime reduced. Machines and equipment will employ self-optimisation to improve operations [35–38]. The significant roles of digital twin for Industry 4.0 are as follows:

2.6.1 Simplification of procedures

By providing insights into assets and production, digital twin technology detects bottlenecks, streamlines processes, and enables the creation of new products. If businesses have a thorough understanding of the condition and performance of their equipment, they can identify anomalies and deviations in their operations. Further rapid maintenance and spare component replenishment can be well-planned to decrease time-to-service and prevent expensive asset failures. Maintenance using digital twins may increase product longevity while providing a new service-based revenue stream. Virtual representations of products in operation provide in-depth information on use patterns, deterioration points, workload capacity, faults, etc. Digital twin technology enables robust simulations and virtual prototypes for feature testing based on empirical data.

2.6.2 Replicate the physical asset and activities

To replicate the physical asset and its activities, the digital twin application is often developed in the business's primary system language. Furthermore, the procedure for data management and interoperable networking may include standards and security measures. The digital twin may replicate a world that is more wealthy and less isolating. Additionally, digital representations of whole factories or production lines are made. This helps create new, better business opportunities that optimise all associated procedures. Undoubtedly, the idea of a digital twin is the next big thing in most business sectors. Predicting the current and future state of physical assets will benefit from evaluating their digital counterparts. A corporation might enhance customer service and get better, more accurate information about the success of its products by using a digital twin.

2.6.3 Opportunities for business success

The growth of Industry 4.0 offers many opportunities for business success in both services and manufacturing sectors. With its technological improvements, digital twin offers a virtual approach to system-based design, giving each piece of machinery or system a far more active role. By providing a precise digital replica of the machine, this technology helps operators recognise specific traits, performance, and potential flaws in the virtual simulation model. To generate real-time digital simulation models that update and respond to changes in their physical counterparts, the digital twin integrates information and technology, including AI, ML, and software analytics. The organisation benefits from a thorough digital footprint for product development, from design to deployment.

2.6.4 Improve product performance information

A digital twin enables excellent data-driven decision-making, enhanced customer service, and improved product performance information. Digital twins significantly impact how things are described. It decreased throughput times and improved production efficiency and excellence. Using this technology, a vehicle can be virtually recreated in the automotive industry. It records the vehicle's operational data and helps decompose the vehicle's overall operation and related features. Digital twin implementation can significantly improve the retail customer experience by creating virtual twins for each customer.

2.6.5 Transparency in manufacturing

Digital twin technology is one of the critical Industry 4.0 technologies that is now attainable. Using digital twins, the entire production process appears transparent. Engineers may analyse, explore, and evaluate existing assets, processes, and systems using virtual tools provided by this technology. Real-time updates are used to track its identical-duplicate state. This technology may be used in various situations, including product monitoring and throughout a product's life cycle. Operators may be forewarned of potential machine faults, downtime, and accidents by using this technology, which enables real-time monitoring of a physical plant via sensors connected to the entire setup. Industrial staff can improve machine performance, monitor device coordination, run diagnostics on virtual plants, and fix flaws without loss of production using real-time operational updates.

2.6.6 Provide a reliable information management solution

Digital twin uses sensors connected to the physical components of a production facility to provide a reliable information management solution. This mainly helps firms use digital and agile methods to monitor the past, present, and future performance of industrial machinery. Digital twin is designed to run continuously and reliably. It assesses tasks, learns from experience, and improves through cutting-edge technologies. The process produces data that is continuously analysed and stored in the cloud. This provides human labour with immediate access to critical business information.

2.6.7 Predict maintenance

The digital twin's capacity to provide organisations with predictive analysis sets it apart from other significant disruptors. The maintenance or operations team can predict how an asset will function over the long term and how it will impact overall company performance using digital twin. Industrial equipment failures and other business hazards are reduced as a result. A digital twin may be used by businesses to test and verify a product before it is ever used in the real world. A digital twin enables engineers to identify possible process flaws before the product is put into production by simulating the desired manufacturing process. It enhances risk analysis, expedites the production of new items, and boosts the manufacturing line's reliability. A virtual representation of an actual product may include financial information, such as labour and material costs.

2.6.8 Simulate intricate procedures

Digital twins are designed to simulate complex processes that interact with their surroundings in various ways. To track and analyse wear and tear while the asset is utilised in the field, this technology is widely used to simulate complex deployed assets. Digital twins might yield insights that could inform the design of future assets. Although there may be a digital twin of certain deployed assets, the manufacturing process is a particularly potent and appealing use case. The digital twin software performs real-time analysis of incoming data streams.

2.6.9 Innovation

Digital twins are widely used across several industries. By delivering accurate virtual representations of objects and simulations of operational processes, its solutions help various companies swiftly create innovation opportunities and enhance business processes and performance. Digital twins may be used to visualise basic and complex objects and processes, including product digital twins, component digital twins, and performance digital twins. The digital twin approach uses reconstruction models based on data from various risk indicators, operating conditions, and software settings to forecast maintenance failures. They contribute to cost savings, improved equipment performance, less downtime, and increased equipment longevity.

2.6.10 Create a virtual copy

Digital twin technology creates a virtual copy that can be tested and modified before a complex physical product is developed. Before producing the finished product, businesses may utilise digital twins to review and simulate each development step, identifying issues and potential failures. This helps development teams analyse how potential changes to the manufacturing process can affect production results and modify their manufacturing procedures to achieve the desired improvements. Because of its cutting-edge, reliable, practical, economical, and intelligent applications, the digital twin platform is being adopted across several sectors. One way in which creating a digital twin plays a role in the automotive industry is by collecting and analysing operational data from a vehicle to assess its status in real time and alert to product modifications.

2.6.11 Error detection ability

A digital twin technique creates a replica of the desired manufacturing process, allowing engineers to detect errors before the product is put into production. Engineers may disrupt the system to provoke unanticipated events, which can also be used to analyse the system's behaviour and to develop mitigating strategies. This additional capability enhances risk analysis, expedites the production of new items, and boosts the manufacturing line's reliability. Using digital twin technology, staff members might get training in dangerous real-world situations. Additionally, staff members need to be ready to work with equipment that is too costly or far away for them to have hands-on experience with. Users can remotely monitor and control system performance via a digital twin platform accessible from any location.

2.6.12 Speed up production

By using digital twin technologies, manufacturing may be made more accessible and efficient while also speeding up production. In addition to manufacturing and industry, the retail sector uses digital twin simulations to enhance the customer experience at every stage, from shopping centres to retail establishments. Industrial organisations must understand and use the technologies powering Industry 4.0 to be at the forefront of the new digital age. This technology allows data scientists, researchers, and engineers to mimic devices or develop assets before they are built. Digital twins are gaining significant attention as a way to better understand and predict a physical product's success before the real thing is constructed.

2.6.13 Data collection

Using this system, we may gather all data related to various assets, including manufacturing, operational, analytical, and physical data. Engineers can simulate, assess, anticipate, and avoid issues/upgrades in advance using a virtual model updated regularly by data streams. The digital twin concept in Industry 4.0 opens up many possibilities by combining data analytics, IoT, and AI fields. Digital twins are widely used across sectors, including telecommunications, engineering, and services, generating significant business value. We can graphically depict goods using digital twin technologies to increase production while lowering risk. This provides more insight into the production process.

2.6.14 Simulates mechanics, materials, and performance

A digital twin is an ongoing virtual representation that accurately simulates the mechanics and materials of a physical object or system. They may model a fulfilment centre operation to assess how people and robots interact before turning on certain robot features in virtual environments. Thousands of edge computers and other devices must receive updates from a deployment platform to implement AI from digital twins in the real world. An electric vehicle maker may use a car's digital twin to simulate software updates. Additionally, if the simulations show that software changes may boost the car's performance or fix a problem, those updates could be sent wirelessly to the actual vehicle.

2.6.15 Determine plant performance

Power plants can monitor sensors and other data that affect performance using a digital twin. The digital twin, when combined with AI and ML, provides insight into the optimal time to schedule maintenance work without disrupting production. Using the digital twin, brilliant facilities may promote a culture of preventive maintenance and reduce resource waste and equipment downtime. Digital twin for networks that simulate routing, security, automation, and monitoring for early installation. Additionally, they enhance current operations by reducing maintenance lead times through simulation analysis of network modification requests. Automation has also enhanced network operations capabilities.

2.6.16 Reduce design cost

The incorporation of digital twin technology has reduced the cost of the design, analytics, scheduling, and simulation services. Digital twin solutions are now available to small and medium-sized businesses for the resolution of difficult issues. This suggests that digital twin as a service is progressively emerging as a viable business option. Most businesses stand to gain from this technology, which has various applications. It enables companies to quickly and effectively test new concepts, whether they involve small or large products. Digital twins have as many uses as there are manufacturers that produce them. This technology will change as AI, image recognition, and other technologies advance, offering intriguing new applications for industrial production.

2.6.17 Identify potential issues

A digital twin can identify potential issues with its actual twin. Industry 4.0 offers several use cases and solutions for various issues. It is primarily propelled by improvements in software used in manufacturing plants. Digital twins are virtual representations of the essential resources in a given sector. The use of digital twins enables the collection of issues arising in the physical world, experimentation with data in Industry 4.0 asset representations, and the delivery of insights that can be applied in the real world as updated configurations. The facility may be coordinated and optimised using the data generated by the digital twin.

2.6.18 Increase self-awareness of the machine

An integrated digital twin, which would allow the machine to optimise performance for duty cycles, identify and fix noncatastrophic problems, and coordinate operations with other machines with little human contact, would be the cornerstone for enhancing the machine's self-awareness. A digital twin can be used in any industrial process to accomplish a number of goals, such as enhancing a component's performance in a particular process, lowering the consumption of raw materials, or quickly making operational decisions in response to outside influences. This digital matching component has become an essential technology for any business due to its capacity to resolve problems and enhance operational effectiveness. Digital twin technology provides businesses with a digital replica of their products, enabling them to identify physical defects in their machinery well in advance and take the necessary measures to fix them before they become problems.

2.6.19 Effective and user-friendly service

Services for digital twins have advanced, becoming more effective and user-friendly. Modern cloud-based digital twin services allow for increasingly complex digital simulations and models with autonomous learning capabilities. This technology provides more insightful, detailed data and intuitive dashboards that teams can utilise without the help of data experts. More and more sectors are adopting these models to improve performance, leading to better results as digital twin capabilities have advanced. Like conventional technologies, digital twin ecosystems can manage complex systems and processes.

2.6.20 Automate workflows and processes

Digitally enabled, smart factories may use technology to automate processes and operations. Smart factories' production performance and productivity may increase by integrating physical, human, and operational data. They can also anticipate and carry out repairs and monitor and manage inventory in real time. A smart factory can digitalise manufacturing processes and procedures, offering opportunities for automation. A digital twin helps learn about and assess a product's performance. With sophisticated algorithms and software programmes, the digital twin replicates data from the real world and analyses it to provide valuable insights about a product still in development. The digital twin is an excellent Industry 4.0 tool for evaluating new technologies, mimicking behaviour, and reducing risks. Digital twins enable the modelling of entire processes across all their phases. It could be directly applied to products during the design and manufacturing phases. The digital twin reflects the condition and operation of the power plants, enabling the modelling of their performance.

2.6.21 Observe systems and equipment

Through digital twins, businesses can proactively monitor their systems and equipment, schedule repairs before they break down, and boost production efficiency. Engineers may perform tests and inspections on a physical asset using simulations. Thus, sensors must first be put in the affected goods to collect real-time data on their status, operating conditions, physical location, and other factors to generate a digital twin. Digital twin technology helps businesses deliver a more critical customer experience by providing deep insights into client needs, upgrading current products, processes, and services, and reinventing organisations.

2.6.22 Save time on development

With a digital twin, manufacturing process errors are reduced. Additionally, it reduces development time and enables the identification of potential bottlenecks. To enhance and optimise logistical processes, the digital twin is also used to simulate material and workforce flows. They enable the simulation of numerous scenarios and the capture of all process states. These simulations provide decision-making information and help one choose the best course of action in each situation. Using these digital representations, potential issues during construction and subsequent vehicle operation may be avoided. To produce the exact virtual component of a specific physical product, engineers must gather and integrate data from various sources, including manufacturing statistics, product expertise, and analytics tools.

2.6.23 Real-time observation

Real-time observation of a manufacturing component, system, asset, or process is made possible via a digital twin. Manufacturers can gain a better understanding of what occurs on assembly lines and throughout the manufacturing process by improving monitoring capabilities. By utilising ML and input from skilled engineers, we can use the digital twin to detect issues before they occur and forecast their consequences. These projections include results under the current conditions, as well as results under other conditions. Using a digital twin, businesses can more accurately predict outcomes, innovate and build better products, handle physical challenges more quickly by recognising them, and ultimately better serve their customers.

2.6.24 Better product quality and traceability

Digital twin technology enables waste reduction, faster batch changeovers, improved product quality, and traceability. Integrating three-dimensional models with simulation and code emulating equipment control facilitates effective design and development. Additionally, having a digital twin enables virtual support and troubleshooting, easing the physical constraints of employing experienced engineers and creating the processes and data needed for the digital twin throughout the product life cycle, from asset creation to field use and maintenance in the real world. Process design is the first step in constructing a digital twin. Features added to the process design improve cost, time, or asset efficiency. These are often the underlying presumptions on which digital twin enhancements must be based.

2.6.25 Asset monitoring and maintenance optimisation

The digital twin can transform industrial processes by offering new approaches to cost reduction, asset monitoring, maintenance optimisation, downtime reduction, and the production of connected items. The digital twin approach is quickly gaining acceptance across manufacturing and other industries, such as construction. The ability to demonstrate the advantages and realise value from an organisation's investment in creating a digital twin may be a pressing issue for any organisation starting down the path to digitisation. The number of use cases and opportunities for establishing a digital twin and generating corporate value has substantially increased with the rise of more favourable storage.

2.7 Limitations of digital twin

The significant limitations of digital twins are as follows [15,39,40]:

- The security and privacy of data are one of the main problems.
- The system's use of intelligent software, which naturally has access to various resources and sensitive information about businesses, is vulnerable to hacker assaults and data theft.
- Internet access is necessary for technology to succeed.
- All supply chains will need to have a digital twin.
- A digital twin involves a high upfront cost in establishing it.
- The issues at hand include globalisation and contemporary production methods.
- When money is scarce, every business cannot afford the return on investment.
- Rather than 2D sketches, the concept of a digital twin is built on 3D CAD models.
- As the physical product evolves, managing all of this digital twin design data among vendors and partners will be challenging.

2.8 Future goals of digital twin for Industry 4.0

The future goals of digital twin for Industry 4.0 are as follows [40,41]:

- Future businesses will be able to more thoroughly assess and anticipate possible implementation challenges at the outset of the design phase using digital technology.
- Thus to prevent downtime, issues will be solved beforehand, or early warning signals will be given.
- With the improved operational and production potential provided by digital twins, it will be possible to evaluate the installation of advanced technologies in the future to create a fully functional smart factory.

- It helps in the analysis of industrial machine states and imparts knowledge about how future machine states will be.
- The connectivity of all factory floor operations will be made possible by the digital twin before the deployment of Industrial Internet of Things (IIoT) physical assets.
- The digital twin will expand on that intuition by showing how the issue is reflected in the data, helping investigative teams find issues with assets more quickly, and reducing the issue in the future.
- The digital twin will streamline a complex process throughout the product development cycle.
- The production machinery will be able to produce a distinctive product variant without the need for costly retooling, using the digital twin's exact product data transmission.

2.9 Conclusion

A virtual version of a physical system operating independently is known as a digital twin. This digital copy is a ‘twin’ of the information embedded in the physical system, much as the essential components are. With the development of the IoT, a digital twin may continuously collect sensor data and communicate with its physical counterpart throughout the system's lifespan. With the aid of digital twins, the entire manufacturing process can be represented from the smallest component to the largest facility. This technology is changing how industrial businesses operate by reducing costs, managing assets, and minimising downtime caused by equipment failures.

As a digital twin is an artificially created image of a real-world product, procedure, or service, it can replicate processes to gather information on how they will function, in addition to tangible assets. These programmes may increase output by utilising IoT, AI, and software analytics. A technician on-site and a remote expert may examine the same data and discuss what to do in an ideal maintenance scenario created by the digital twin. The industrial, retail, and automotive sectors have adopted digital monitoring and control technologies. A digital twin of a facility process might be created to examine how that system or process functions.

A digital twin is an abstraction or digital representation of a physical object, such as a device or machine. They are already extensively utilised in many different fields and applications. It makes use of a cloud-based IIoT platform and comes with remote assistance, specific operator training, and monitoring tools. In order to develop digital twins, high-fidelity virtual representations of the real world must be created. These representations must be able to change in real time as the physical environment does. Digital twins are being used or will be used by IoT-enabled organisations. Parts twinning helps engineers and developers better comprehend the mechanical, physical, and intellectual properties of a particular part in the context of the overall product. The digital twin seems to be becoming as widespread and fulfils various requirements of Industry 4.0.

References

[1] R.S. Kenett, J. Bortman, The digital twin in Industry 4.0: a wide-angle perspective, Quality and Reliability Engineering International 38 (2022), pp. 1357–1366.

[2] M. Javaid, A. Haleem, R. Suman, Digital twin applications toward Industry 4.0: a review, Cognitive Robotics 3 (2023) 71–92.

[3] S. Aheleroff, X. Xu, R.Y. Zhong, Y. Lu, Digital twin as a service (DTaaS) in Industry 4.0: an architecture reference model, Advanced Engineering Informatics 47 (2021) 101225.

[4] T.H.J. Uhlemann, C. Lehmann, R. Steinhilper, The digital twin: realizing the cyber-physical production system for Industry 4.0, Procedia Cirp 61 (2017) 335–340.

[5] A. Martínez-Gutiérrez, J. Díez-González, R. Ferrero-Guillén, P. Verde, R. Álvarez, H. Perez, Digital twin for automatic transportation in Industry 4.0, Sensors 21 (10) (2021) 3344. Available from https://doi.org/10.3390/s21103344.

[6] Q. Qi, F. Tao, Digital twin and big data towards smart manufacturing and Industry 4.0: 360-degree comparison, IEEE Access 6 (2018) 3585–3593.

[7] A.L. Hananto, A. Tirta, S.G. Herawan, M. Idris, M.E.M. Soudagar, D.W. Djamari, et al., Digital twin and 3D digital twin: concepts, applications, and challenges in Industry 4.0 for digital twin, Computers 13 (4) (2024) 100. Available from https://doi.org/10.3390/computers13040100.

[8] D. Guerra-Zubiaga, V. Kuts, K. Mahmood, A. Bondar, N. Nasajpour-Esfahani, T. Otto, An approach to develop a digital twin for industry 4.0 systems: manufacturing automation case studies, International Journal of Computer Integrated Manufacturing 34 (9) (2021) 933–949. Available from https://doi.org/10.1080/0951192X.2021.1946857.

[9] R. Rolle, V. Martucci, E. Godoy, Architecture for digital twin implementation focusing on industry 4.0, IEEE Latin America Transactions 18 (2020) 889–898.

[10] O. Bongomin, M.C. Mwape, N.S. Mpofu, B.K. Bahunde, R. Kidega, I.L. Mpungu, et al., Digital twin technology advancing industry 4.0 and industry 5.0 across sectors, Results in Engineering 26 (2025) 105583.

[11] K. Židek, J. Piteľ, M. Adámek, P. Lazorík, A. Hošovský, Digital twin of experimental smart manufacturing assembly system for industry 4.0 concept, Sustainability 12 (2020) 3658.

[12] V. Jovanovic, M. Kuzlu, U. Cali, D.H. Utku, F.O. Catak, S. Sarp, et al., Digital twin in iIndustry 4.0 and beyond applications, Digital Twin Driven Intelligent Systems and Emerging Metaverse, Springer Nature Singapore, Singapore, (2023), pp. 155–174.

[13] A. Gallala, A.A. Kumar, B. Hichri, P. Plapper, Digital twin for human–robot interactions by means of Industry 4.0 enabling technologies, Sensors 22 (2022) 4950.

[14] F. Tao, H. Zhang, A. Liu, A.Y. Nee, Digital twin in industry: state-of-the-art, IEEE Transactions on Industrial Informatics 15 (2018) 2405–2415.

[15] T.R. Murgod, S.M. Sundaram, U. Mahanthesha, P. Murugesan, A survey of digital twin for Industry 4.0: benefits, challenges and opportunities, SN Computer Science 5 (2023) 76.

[16] P. Bhambri, S. Rani, A. Khang, AI-driven digital twin and resource optimisation in Industry 4.0 Ecosystem, Intelligent Techniques for Predictive Data Analytics, IEEE, (2024), pp. 47–69. Available from https://doi.org/10.1002/9781394227990.ch3.

[17] A. Kampa, Modelling and simulation of a digital twin of a production system for Industry 4.0 with work-in-process synchronisation, Applied Sciences 13 (2023) 12261.

[18] S. Aheleroff, R.Y. Zhong, X. Xu, A digital twin reference for mass personalisation in Industry 4.0, Procedia Cirp 93 (2020) 228–233.

[19] T.Y. Melesse, V. Di Pasquale, S. Riemma, Digital twin models in industrial operations: a systematic literature review, Procedia Manufacturing 42 (2020) 267–272.

[20] S.M. Sepasgozar, M. Ghobadi, S. Shirowzhan, D.J. Edwards, E. Delzendeh, Metrics development and modelling the mixed reality and digital twin adoption in the context of Industry 4.0, Engineering, Construction and Architectural Management 28 (2021) 1355–1376.

[21] M. Raza, P.M. Kumar, D.V. Hung, W. Davis, H. Nguyen, R. Trestian, A digital twin framework for Industry 4.0, enabling next-gen manufacturing, 2020 9th International Conference on Industrial Technology and Management (ICITM), Oxford, UK, IEEE, (2020), pp. 73–77. Available from https://doi.org/10.1109/ICITM48982.2020.9080395.

[22] M. Ramakrishnan, 2022. How Digital Twin Solutions Affect Business: The Next Digital Transformation: https://emeritus.org/blog/technology-what-is-a-digital-twin/.

[23] A. Martínez-Gutiérrez, J. Díez-González, P. Verde, H. Perez, Convergence of virtual reality and digital twin technologies to enhance digital operators' training in Industry 4.0, International Journal of Human-Computer Studies 180 (2023) 103136.

[24] A. Gosavi, V.K. Le, Maintenance optimization in a digital twin for Industry 4.0, Annals of Operations Research 340 (2024) 245–269.

[25] D. Balderas, A. Ortiz, E. Méndez, P. Ponce, A. Molina, Empowering digital twin for industry 4.0 using metaheuristic optimization algorithms: case study PCB drilling optimization, The International Journal of Advanced Manufacturing Technology 113 (2021) 1295–1306.

[26] D. Preuveneers, W. Joosen, E. Ilie-Zudor, Robust digital twin compositions for Industry 4.0 smart manufacturing systems, 2018 IEEE 22nd International Enterprise Distributed Object Computing Workshop (EDOCW), Stockholm, Sweden, IEEE, (2018), pp. 69–78. Available from https://doi.org/10.1109/EDOCW.2018.00021.

[27] AltexSoft Editorial Team, 2025. Digital Twins: Technology, Use Cases, and Implementation Tips:- https://www.altexsoft.com/blog/digital-twins/.

[28] S. Attaran, M. Attaran, B.G. Celik, Digital twins and industrial internet of things: uncovering operational intelligence in Industry 4.0, Decision analytics journal 10 (2024) 100398.

[29] Assawaarayakul, C., Srisawat, W., Ayuthaya, S.D. N., Wattanasirichaigoon, S., 2019, December. Integrate digital twin to exist production system for Industry 4.0. In *2019, the 4th Technology Innovation Management and Engineering Science International Conference (TIMES-iCON)*, pp. 1-5. IEEE.

[30] F. Tao, Q. Qi, L. Wang, A.Y.C. Nee, Digital twins and cyber-physical systems toward smart manufacturing and industry 4.0: correlation and comparison, Engineering 5 (2019) 653–661.

[31] R. Wagner, B. Schleich, B. Haefner, A. Kuhnle, S. Wartzack, G. Lanza, Challenges and potentials of digital twins and Industry 4.0 in product design and production for high performance products, Procedia Cirp 84 (2019) 88–93.

[32] Wagner, C., Grothoff, J., Epple, U., Drath, R., Malakuti, S., Grüner, S., et al., 2017, September. The role of the Industry 4.0 asset administration shell and the digital twin during the life cycle of a plant. In *2017, the 22nd IEEE International Conference on Emerging Technologies and Factory Automation (ETFA)*, pp. 1–8. IEEE.

[33] I.C. Fantozzi, A. Santolamazza, G. Loy, M.M. Schiraldi, Digital twins: strategic guide to utilize digital twins to improve operational efficiency in Industry 4.0, Future Internet 17 (2025) 41.

[34] E. Negri, L. Fumagalli, M. Macchi, A review of the roles of digital twin in CPS-based production systems, Procedia Manufacturing 11 (2017) 939–948.

[35] H. Kaur, M. Bhatia, Scientometric analysis of digital twin in Industry 4.0, IEEE Internet of Things Journal 12 (2024) 1200–1221.

[36] M. Singh, R. Srivastava, E. Fuenmayor, V. Kuts, Y. Qiao, N. Murray, et al., Applications of digital twin across industries: a review, Applied Sciences 12 (2022) 5727.

[37] Sjarov M., Lechler, T., Fuchs, J., Brossog, M., Selmaier, A., Faltus, F., et al., 2020, September. The digital twin concept in industry–a review and systematization. In *2020 25th IEEE international conference on emerging technologies and factory automation (ETFA)*, Vol. 1, pp. 1789–1796. IEEE.

[38] R. Kerkeni, S. Khlif, A. Mhalla, K. Bouzrara, Digital twin applied to predictive maintenance for Industry 4.0, Journal of Nondestructive Evaluation, Diagnostics and Prognostics of Engineering Systems 7 (2024) 41008.

[39] J.S. Ortiz, E.K. Quishpe, G.X. Sailema, N.S. Guamán, Digital twin-based active learning for industrial process control and supervision in Industry 4.0, Sensors 25 (2025) 2076.

[40] S. Kosse, V. Betker, P. Hagedorn, M. König, T. Schmidt, A semantic digital twin for the dynamic scheduling of Industry 4.0-based production of precast concrete elements, Advanced Engineering Informatics 62 (2024) 102677.

[41] P. Bellavista, N. Bicocchi, M. Fogli, C. Giannelli, M. Mamei, M. Picone, Requirements and design patterns for adaptive, autonomous, and context-aware digital twins in Industry 4.0 digital factories, Computers in Industry 149 (2023) 103918.

3

Artificial intelligence for Industry 4.0

3.1 Introduction

Artificial intelligence (AI) is human-like intelligence displayed by computers, similar to natural intelligence, that aids in solving a wide range of issues. AI has a tremendous impact on the manufacturing sector, as it can perform various activities, mimicking human intelligence. AI is used to forecast product demand across time, geography, and socioeconomic dynamics, leveraging algorithms that account for macroeconomic cycles and weather patterns [1–3].

Industry 4.0 is heavily reliant on AI. AI automation systems that collect and analyse various types of data are used in production processes in smart industries, which depend on numerous devices interacting with one another [4,5]. Using sensors that monitor tooling efficiency and operating conditions, it aids in predictive equipment management. It provides pertinent information to support decision-making and alert people to possible issues. Industries will utilise AI to process the data provided by integrating IoT devices and connected machines into their equipment. Additionally, it enables companies to track every step of their operations and processes [6,7].

AI applications significantly improve manufacturing capabilities because they function across several business lines and levels, from staff planning to product design, maximising performance, improving product quality, and employee well-being. AI advances are at the heart of several developments, enabling robots to perform more complex computational tasks and make independent, real-time decisions based on environmental data in Industry 4.0. It entails juggling a variety of factors, including content types, production processes, financial constraints, and deadlines. Machine learning (ML) is an important part of AI that manages and assesses ideas and other critical duties, offering further insight into recent designs. In Industry 4.0, AI is being applied in factories to enable the predictive management of sensitive industrial systems to forecast asset breakdown [8–10].

Advanced AI techniques are applied to construct asset failure forecasts. AI algorithms are needed to notify manufacturing teams of evolving production errors that can lead to product quality issues. It can detect minor irregularities in machine behaviour, changes in raw materials, etc. AI algorithms inform manufacturing units of potential production flaws for quality management, leading to product quality issues. This technique is used in manufacturing to create a virtual version of a factory, product, or fundamental feature [11–14]. This represents real-time knowledge by using cameras, sensors, and other data-collection tools.

3.2 AI and its classifications

AI is one of humanity's most sophisticated and impressive creations. AI has brought advancements across industries that have led business executives and the general public to

Digital Twin and Industry 4.0. DOI: https://doi.org/10.1016/B978-0-443-54019-6.00002-9

believe that we are close to reaching the pinnacle of AI research and realising AI's full potential [15,16]. On the other hand, understanding the forms of AI that are already available and those that are potentially available will provide a clearer picture of current AI capabilities and the long road ahead for AI development. The degree to which an AI system can reproduce human skills is a key factor in distinguishing different kinds of AI, since AI research aims to make computers mimic human-like functioning [17–22]. The AI working process is shown in Fig. 3.1.

As a result, AI can be divided into a number of groups according to how different and effective it is in comparison to humans. In this concept, an AI will be considered more evolved if it can carry out more human-like tasks with similar degrees of proficiency. An AI with restricted capability and performance, on the other hand, will be regarded as a simpler, less sophisticated type. Based on this criterion, AI can be divided into two general categories [23,24]. One method is to classify AI and AI-enabled robots based on how similar their minds are to human minds and how they can 'think' and possibly 'feel' like people. Reactive

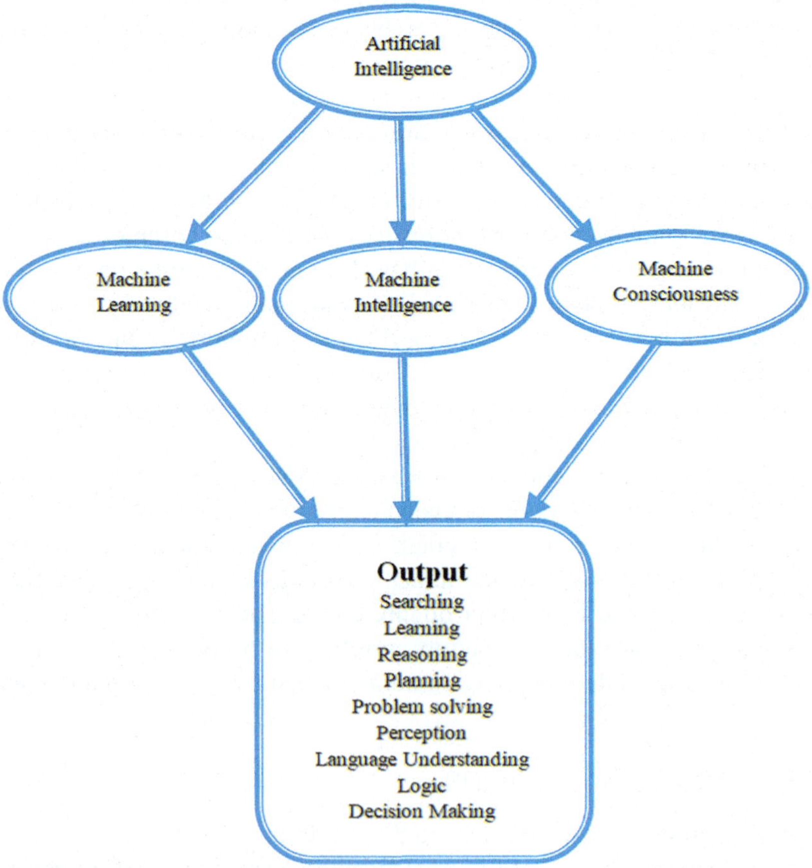

FIGURE 3.1 Working process of Artificial Intelligence.

machines, restricted memory machines, theory-of-mind systems, and self-aware AI are the four types of AI or AI-based systems that fall under this classification system [25–27].

3.2.1 Reactive machines

These AI systems are the most basic and have the fewest features. They mimic how the human mind reacts to different stimuli. These devices lack memory-based functionality. This implies that these robots are incapable of using past experiences to inform their present behaviour; in other words, they are unable to 'learn'. It is possible to programme these devices to react to a restricted set of inputs or combinations. They cannot rely on memory to make their processes better.

3.2.2 Limited memory

Computers with limited memory have the capabilities of fully reactive computers and can learn from past data to make decisions. This includes the majority of current AI applications that we are aware of. To build a reference model for handling future issues, massive volumes of data kept in memory are used to train all contemporary AI systems, including deep learning systems. For instance, hundreds of images and their labels are used to teach an AI that recognises images to name the objects it scans. When an AI scans an image, it uses the training photos as a guide to understand what's in the image. It then labels new images more accurately based on its 'learning experience'.

3.2.3 Theory of mind

The theory of mind is being actively developed by researchers studying AI systems. An AI at the theory-of-mind level will be able to recognise the desires, feelings, convictions, and thought processes of the entities it communicates with. Although leading AI researchers are concentrating on artificial emotional intelligence, which is already a growing industry, breakthroughs in other AI domains would also be necessary to achieve theory of mind AI. AI robots need to 'understand' people because they must acknowledge that they are unique individuals with minds that may have been influenced by a variety of experiences.

3.2.4 Self-aware

AI that has developed to the point that it resembles the human brain enough to be self-aware is known as self-aware AI. Developing this type of AI will always be the ultimate goal of all AI research. This type of AI will have its own emotions, desires, beliefs, and possibly objectives in addition to being able to identify and produce emotions in the people it interacts with. Furthermore, this is the type of AI that worries those who are sceptical of the technology. Self-awareness can hasten our civilisation's development, but it can also be disastrous. This is due to the possibility that, once self-aware, AI may embrace principles, such as self-preservation, which might directly or indirectly signal the end of humanity because such an entity could easily outsmart any human brain and come up with intricate plans to subjugate humans.

3.2.5 Artificial narrow intelligence

This type of AI includes all AI that is currently on the market, including the most sophisticated and capable AI ever created. AI systems that can do a single task on their own while maintaining human-like abilities are referred to as artificially narrow intelligence. Since these devices can only perform what they are made to do, their possibilities are quite limited. These systems are related to all reactive and limited memory AI, according to the previous classification.

3.2.6 Artificial general intelligence

The ability of an AI agent to learn, observe, comprehend, and behave similarly to a human is known as artificial general intelligence (AGI). These systems will drastically cut down on training time by independently developing a wide range of competencies and establishing connections and generalisations across areas. AI systems will be as capable as humans by imitating our multifunctional abilities.

3.2.7 Artificial super intelligence

Since AGI will be the most competitive kind of intelligence in the world, it will surely be the pinnacle of AI research. Artificial super intelligence (ASI) will be vastly better at everything it does and will imitate human cognition due to its significantly larger memory, quicker data processing and analysis, and superior decision-making abilities. AGI and ASI will converge, leading to the singularity. Although it is appealing to have such strong instruments at our disposal, these devices could endanger our existence or, at the very least, our way of life.

The state of our planet when more powerful forms of AI become accessible is hard to imagine. However, compared to where it is anticipated to go, AI development is still in its early stages, so there is still a long way to go. This implies that there is still time to ensure AI's safety and that it is a little too soon to worry about the singularity, according to those who are pessimistic about the technology's future. For those who are hopeful about the future of AI, the fact that we have only begun to explore the field makes it much more exciting.

3.3 Need for AI in Industry 4.0

The primary needs of AI in different aspects and significant ones are as follows [28–30]:

3.3.1 To automate data-driven learning and discovery

Data-driven learning and discovery are automated by AI. Instead of automating manual tasks, AI does frequent, high-volume automated tasks. It does so reliably and without fatigue.

3.3.2 To enhance intelligence

AI makes commonplace objects more intelligent. AI characteristics will improve a lot of currently utilised items. Combining automation, conversational platforms, bots, and intelligent robots with enormous amounts of data can improve many technologies. Improvements range from smart cameras and security intelligence to financial analysis for both homes and businesses.

3.3.3 To allow data to programme itself

AI adjusts by letting data use sophisticated learning algorithms to programme itself. AI searches for patterns and structure in data to help machines learn. An algorithm can learn how to play chess and determine what online product to suggest next.

3.3.4 To analyse data

AI evaluates more and deeper data, and they employ neural networks with several hidden layers. Building a fraud detection system with several layers of abstraction used to be difficult. All of that has changed with the advent of big data and computational power. Because deep learning models learn directly from training data, they need a lot of it.

3.3.5 To achieve precision

AI is made remarkably precise by deep neural networks. Every contact with Google and Alexa uses deep learning. The medical field can now more accurately identify cancer in medical images by using deep learning and object detection AI tools.

3.3.6 To achieve information

AI maximises the use of data. Data is an important resource for self-learning algorithms. The answers are in the data. Data may provide you with a competitive advantage because it is more important than ever.

3.4 Contemporary technologies used in AI

AI was coined in 1955 to refer to a brand-new area of computer science. AI is a rapidly developing field with high demand and significant impact on many facets of our daily lives. Internet giants and other startups are vying for their acquisition [31–35]. Fig. 3.2 shows the potent AI technologies, which are discussed in this part:

3.4.1 Natural language generation

Effective and efficient communication can be challenging, even for humans. In a similar vein, information processing by robots differs greatly from that of the human brain, which can be difficult and complex. A subfield of AI called 'natural language generation' helps computers express themselves as plainly as possible by turning text into data. It is widely used in customer service to produce market summaries and reports.

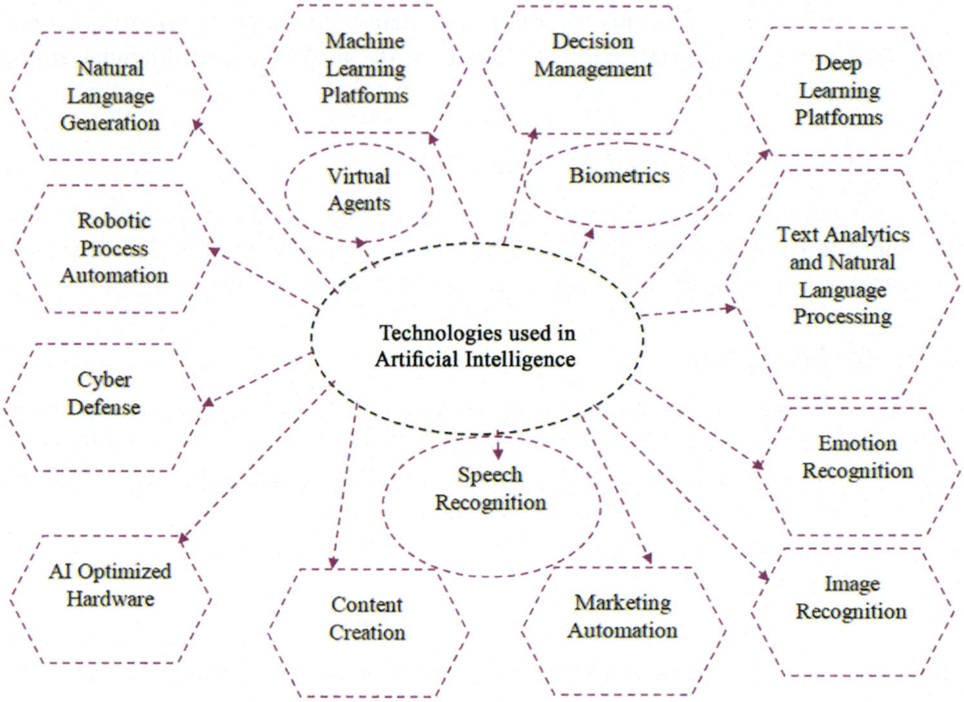

FIGURE 3.2 Contemporary technologies being used in Artificial Intelligence.

3.4.2 Virtual agents

A computer or piece of software that can interact with people in an efficient manner is called a virtual agent. AI is currently used as a smart home manager and in chatbots for customer service.

3.4.3 Decision management

Robots with AI can teach AI systems logic so that they are ready for tuning, maintenance, and training. In order to push and carry out automated decisions that add value to the business and increase its profitability, organisations are already integrating decision management into their apps.

3.4.4 Speech recognition

The process of translating human speech into a format that computer programmes can comprehend is known as voice recognition. Human language transcription and translation into usable formats are widespread and expanding quickly these days.

3.4.5 Machine learning platforms

A branch of AI is ML. Its objective is to develop novel approaches to computer learning and intelligence. Massive data, applications, development tools, algorithms, and application programming interfaces are all contributing to the growing popularity of ML platforms. They are frequently used for prediction and classification. Companies including Amazon, Google, H_2O AI, Microsoft, SAS, Skytree, Ad Text, and Fractal Analytics sell ML solutions.

3.4.6 AI optimised hardware

Because of improved graphics and central processing units, devices are being built and used to do AI-related tasks. An excellent example of this is the AI-optimised silicon chip, which can be integrated into any portable device. Businesses and organisations are therefore making significant investments in AI in order to accelerate the creation of the upcoming generation of apps. This technology service is offered by businesses including Alluviate, Google, IB, Cray, Intel, and Nvidia.

3.4.7 Text analytics and natural language processing

Natural language processing (NLP) studies the relationship between computers and human languages. ML examines the structure, meaning, and intent of sentences using text analytics. This method is frequently applied in fraud detection and security systems. Numerous apps and automated assistants employ NLP to produce unstructured data.

3.4.8 Deep learning platforms

Deep learning platforms are a subset of ML that analyses data and creates patterns for decision-making by using neural circuits that resemble those in the human brain. This innovative technology makes use of artificial neural networks. Among its applications are automated voice recognition, image recognition, and the prediction of anything detectable in the digital realm.

3.4.9 Robotic process automation

Robotic process automation mimics human labour to automate commercial processes. It is critical to keep in mind that AI is intended to supplement and improve human capabilities, not to replace them.

3.4.10 Biometrics

The study of identifying, quantifying, and evaluating the physical traits of the human body's composition, form, and behaviour is known as biometrics. It creates natural human-machine interactions using touch, images, words, and body language. Market research is its main application. This technology is offered by Affectiva, FaceFirst, Sensory, Agnitio, Synqera, and Tahzoo.

3.4.11 Content creation

Although individuals produce content for movies, advertising, blogs, and white papers, businesses do too. Wordsmith is a well-known programme created by Automated Insights that generates news articles using natural language understanding.

3.4.12 Emotion recognition

This kind of AI technology makes it possible to interpret and comprehend human emotions through strong image or audio processing. This device is often used by law enforcement during questioning. Three companies that use emotion recognition are Beyond Verbal, Viso, and Affectiva.

3.4.13 Cyber defense

Cyber defence is a computer defence mechanism that identifies, stops, and lessens threats and attacks against system infrastructure and data. Neural networks that can process input sequences and ML techniques can be used to develop learning technologies that can identify suspicious user behaviour and identify cyber threats.

3.4.14 Marketing automation

Departments and divisions in charge of marketing and sales have embraced AI and benefited from it. Campaign management, data integration, and automated consumer segmentation are examples of AI-enabled techniques that are commonly used. Adext AI has led the way in implementing marketing automation.

3.4.15 Image recognition

Image recognition is the technique of identifying a feature in a picture or video. It can significantly help with image searches, license plate recognition, illness diagnosis, and personality study. This technological service is offered by SenseTime, Clarifai, and GuGum.

3.5 Advantages of adopting AI

The significant advantages of AI are as follows [36–39]:

- AI provides businesses with a sophisticated degree of analysis that they may utilise to examine the results of their various components.
- AI database analysis can improve a facility's overall performance and quality of output.
- It enables robots or other intelligent equipment to detect anomalies and track parameters.
- It discovers, summarises, and analyses large amounts of data before sending it to other computers on a cloud-based network.
- It aids in the management of an ecosystem.
- AI helps programme developers and broadcasters discover which shows they can recommend to individual customers based on their behaviour.
- This technology has the potential to solve many of the industry's internal issues, ranging from skills shortages to decision-making complications, deployment challenges, and an abundance of knowledge.

- Using AI in manufacturing facilities allows businesses to alter their processes completely.
- AI can allow systems to monitor themselves to decrease downtime, maximise resource use, and predict malfunctions.
- It helps decision-makers in testing settings, improves asset efficiency, and avoids system failures.
- It assists businesses in appreciating their goods by allowing them to see how they operate in their production setting and in real-time by their employees.
- AI algorithms assist organisations in predicting market movements and optimising industrial supply networks.
- AI helps transform management from a reactive to a competitive position.
- AI algorithms calculate market demand by analysing position trends, socioeconomic and macroeconomic characteristics, environmental patterns, status policy, consumer behaviour, and other factors.

3.6 Significant areas where AI can be used for Industry 4.0 applications

AI improves a company's analytical power, allocating resources more effectively, generating more accurate predictions, and reducing stock costs. In Industry 4.0, there are several opportunities to use AI. This shows an extraordinarily efficient learning platform for various applications, including production process modelling and predictive quality analytics. AI identifies defects and reduces waste, thereby increasing revenue estimates. This saves a lot of time and money since the sensitive technology prevents waste from errors without the need for human intervention [40–43]. The following are the significant areas of AI in Industry 4.0:

3.6.1 Appropriate and adequate information

The human eye does not go unnoticed for many minutes in a manufacturing system. Advanced technologies, such as AI, may detect minuscule flaws on circuit boards that are invisible to the naked eye. As a result, in Industry 4.0, AI may provide relevant data and information. This is becoming more widespread in industries and provides longer-term safety.

3.6.2 Improve the production efficiency

Incorporating AI algorithms into industrial operations improves production efficiency. With highly accurate, robust robots, AI can swiftly replace visual inspections. Manufacturers may collaborate with specialists to develop appropriate, bespoke solutions to the AI problem in Industry 4.0. Industries grow more efficiently when AI is used. From concept to the production floor, this technology aims to revolutionise how items are produced and resources are processed, thereby improving production efficiency.

3.6.3 Monitoring the supply chain

Supply chain management is becoming more intelligent using advances in AI, machine vision, NLP, and language recognition. Many supply chain monitoring applications are possible with this technology. AI systems that use predictive analytics may also accurately estimate product demand. AI technology collects data from various sources and accurately estimates product demand. Furthermore, AI programmes may manage order records and install/uninstall new stock.

3.6.4 Aids in the completion of a typical job

AI is used to automate processes and execute regular operations. This technique enables avoiding and reducing human tasks to a negligible level. With AI, an industrial robot can track its accuracy and efficiency and plan to improve. The applications smartly cover assembly, welding, painting, stock testing, pick-up and placing, die moulding, boiling, glass manufacturing, and grinding. Some manufacturing robots are equipped with this technology, enabling precise movement in complex, unpredictable situations.

3.6.5 Demand forecasting and production planning

AI effectively manages production planning and demand forecasts, supporting resource planning. AI-driven demand prediction systems deliver more accurate results than traditional demand forecasting systems. Industries can better manage inventory levels, reducing the risk of cash-in-stock and cash-out scenarios. AI-driven technology will help organisations refine procedures to achieve sustainable output levels. Manufacturers can use AI-powered process tools to find and alleviate bottlenecks in organisational operations.

3.6.6 Management of production

One of the essential technologies for production, market, and inventory management is AI. Machines have the potential to be more powerful than humans. This is substantially faster than individuals in performing a variety of tasks. Without planning, AI-powered robots can easily perform repetitive tasks. AI improves production, customer service, quality, and market efficiency through smart management. The AI manufacturing system may help with process optimisation, lower operating costs, faster decision-making, higher quality, and improved customer experience.

3.6.7 Defect detection in a product

AI technology helps detect product defects. It examines items as they pass through the manufacturing process. This technique is used to detect problems in real time. Several products have the same flaw, which can be fixed in real time. This sensitive technology prevents waste caused by faults without requiring human intervention, saving significant time and resources. This can generate valuable data to help businesses develop creative, stable company strategies.

3.6.8 Improved monitoring and security

AI has become one of the most popular due to its improved monitoring and safety. Staff identification, thermal scanning, and employee contact monitoring for touch tracking and sanitation may all be made with this system. AI had aided in developing long-term protection measures before they occurred or in faster discovery of the underlying cause once they had occurred. This may lead to happier employees, safer work, and long-term employment.

3.6.9 Designing

AI plays a significant role in designing any items with generative architecture. It is an iterative design approach that feeds complete design information into AI algorithms. Many design requirements, such as processing techniques, product type, time restrictions, and budget constraints, may be easily covered by this technology. The algorithm will investigate each conceivable solution permutation and produce the best-suited output solutions by considering all these characteristics.

3.6.10 Assurance of high quality

Manufacturers recognise the relevance of AI in the dependable, timely detection and maintenance of the production line, and it is utilised to enhance computer vision for in-time detection of product faults for quality assurance. Continuous service is gaining traction to reduce downtime and improve product development. Furthermore, AI algorithms will alert production teams to emerging failures, such as minor equipment abnormalities and other issues, resulting in product quality issues.

3.7 Role of digital twin for AI

Digital twins provide a unique education to AI by creating virtual worlds that allow its algorithms to gain unprecedented knowledge about the real world. AI can also drive real-time process consistency if it has access to a digital counterpart. By making interfaces visible, the use of digital twins in the manufacturing industry improves product transparency. It facilitates dismantling silos and identifying weak points, quality issues, and other issues. This provides real-time monitoring of the entire manufacturing process, helping lower production costs, time, and waste [44–49]. The major roles of the digital twin for AI are as follows:

- It offers ongoing, historical, and real-time data to train, evaluate, and enhance AI algorithms.
- Provides a safe virtual environment for the training and validation of AI models.

- Provides precise data on system behaviour to improve the accuracy of forecasting and predictions made by AI.
- This creates a closed-loop system in which AI choices are continuously improved and tested in the digital twin.
- AI can discover early faults by detecting differences between virtual models and physical systems.
- Based on real-time system insights, the digital twin helps AI make data-driven, optimised judgements.
- This enables AI to monitor and modify systems to optimise and control them continuously.
- Offers a secure virtual environment where AI agents can learn by making mistakes without the dangers of the real world.
- This can continuously compare the performance of AI models to the behaviour of actual systems and detect drift.
- This allows AI systems to use real-time twin data to update and retrain dynamically.
- This platform enables AI-driven experimentation without affecting operations or tangible assets.
- This helps AI to track, forecast, and maximise system performance during its whole lifecycle.

3.8 Significant problem areas due to the implementation of AI

The major problem areas arising due to the implementation of AI are as follows [1,9,50]:

- AI will never be able to think out of the box.
- Integration issues and a lack of understanding of this technology.
- Compatibility with other systems and platforms, and usability.
- Consumer confidentiality.
- A possible lack of openness.
- Technical intricacy.
- It takes a lot of time and resources, which is quite expensive.
- The possibility of employment losses as a result of automation technologies.
- The necessity to repurpose or retrain personnel to keep them employed.
- Equitable sharing of machine-generated riches.
- It makes people lethargic and poses a risk to future generations.
- How human-machine contact affects human behaviour and thought.
- The requirement to address algorithmic bias in data stems from human bias.
- Fears over the security of potentially harmful AI systems.
- The requirement to protect against unforeseen consequences.
- AI necessitates the most expensive, cutting-edge hardware and software.

3.9 Future goals of AI

The primary upcoming goals of AI are as follows [7,51,52]:

- AI will supply essential data to corporate leaders, allowing them to create creative and stable business models.
- This technique will be highly beneficial in spotting patterns and occurrences that the average person cannot notice.
- AI will generate data that will help businesses to make sound, factual, and data-driven decisions.
- It will also aid in removing individual preconceptions from the equation and provide a complete overview of many aspects.
- AI and ML systems can collect data from various sources, identify growth, expansion, and even new market possibilities, and build new goods and services.
- AI will flourish, bringing new benefits that will aid in creating a wired, intelligent, and smart society.

3.10 Conclusion

In Industry 4.0, AI has been introduced to drive the enormous transformation of industrial businesses, leading to new business models and productivity gains. Industry 4.0 uses AI to automate the control of various manufacturing process phases entirely. Thus, to boost efficiency, this technology can share assembly lines with industrial operations. For asset and repair failure prediction, advanced AI systems are deployed. This technology has the potential to provide relevant viewpoints that contribute to manufacturing innovation.

Original equipment manufacturers use AI in manufacturing processes, enabling efficient operations in smart factories and advancing Industry 4.0. Many companies use AI to improve their strategy and automation platforms in their production processes. This makes it easy to respond to demand and move from the raw ingredients that start the manufacturing process to the finished product.

The usage of AI in Industry 4.0 is a trend that will drastically alter the industry. AI is one of the most exemplary rising technologies in manufacturing. Companies can incorporate high quality products using AI-enabled technologies. Furthermore, AI is the most advanced and cutting-edge technology for boosting overall operations by combining sensing device processes, machines, and data.

AI has a transformational impact across all industries. It is used in various settings, including healthcare, biological sciences, immobilisation, teaching, and manufacturing. Massive volumes of data will be converted into insightful predictions and insights by the AI revolution, which will encourage data-driven industries. AI is meant to completely transform how goods are manufactured and materials are handled, from the design and manufacturing floor to the supply and administrative chain. Furthermore, AI algorithms are increasingly being utilised to determine the quickest shipping route and simplify the distribution of goods to clients in diverse locations.

References

[1] Z. Jan, F. Ahamed, W. Mayer, N. Patel, G. Grossmann, M. Stumptner, et al., Artificial intelligence for industry 4.0: systematic review of applications, challenges, and opportunities, Expert Systems with Applications 216 (2023) 119456.

[2] J. Lee, H. Davari, J. Singh, V. Pandhare, Industrial artificial intelligence for industry 4.0-based manufacturing systems, Manufacturing Letters 18 (2018) 20–23.

[3] R.S. Peres, X. Jia, J. Lee, K. Sun, A.W. Colombo, J. Barata, Industrial artificial intelligence in industry 4.0: systematic review, challenges and outlook, IEEE Access 8 (2020) 220121–220139.

[4] I. Ahmed, G. Jeon, F. Piccialli, From artificial intelligence to explainable artificial intelligence in industry 4.0: a survey on what, how, and where, IEEE Transactions on Industrial Informatics 18 (2022) 5031–5042.

[5] H. Castro, E. Câmara, P. Ávila, M. Cruz-Cunha, L. Ferreira, Artificial intelligence models: a literature review addressing Industry 4.0 approach, Procedia Computer Science 239 (2024) 2369–2376.

[6] J. Ribeiro, R. Lima, T. Eckhardt, S. Paiva, Robotic process automation and artificial intelligence in industry 4.0–a literature review, Procedia Computer Science 181 (2021) 51–58.

[7] M. Sharabov, G. Tsochev, The use of artificial intelligence in industry 4.0, Problems of Engineering Cybernetics and Robotics 73 (10.7546) (2020).

[8] U. Murugesan, P. Subramanian, S. Srivastava, A. Dwivedi, A study of artificial intelligence impacts on human resource digitalization in industry 4.0, Decision Analytics Journal 7 (2023) 100249.

[9] A. Bécue, I. Praça, J. Gama, Artificial intelligence, cyber-threats and industry 4.0: challenges and opportunities, Artificial Intelligence Review 54 (2021) 3849–3886.

[10] N. Kraus, K. Kraus, O. Shtepa, M. Hryhorkiv, I. Kuzmuk, Artificial intelligence in established of industry 4.0, WSEAS Transactions on Business and Economics (2022) 1884–1900.

[11] A.E.H. Gabsi, Integrating artificial intelligence in industry 4.0: insights, challenges, and future prospects–a literature review, Annals of Operations Research (2024) 1–28.

[12] A.A.M. Al-Momani, Artificial intelligence and industry 4.0: catalysts for sustainable organizational development, Current and Future Trends on AI Applications, vol 1, Springer Nature Switzerland, Cham, (2025), pp. 231–245.

[13] M. Javaid, A. Haleem, R.P. Singh, R. Suman, Artificial intelligence applications for industry 4.0: a literature-based study, Journal of Industrial Integration and Management 7 (2022) 83–111.

[14] S. Mithas, Z.L. Chen, T.J. Saldanha, A. De Oliveira Silveira, How will emerging artificial intelligence and industry 4.0 technologies transform operations management? Production and Operations Management 31 (2022) 4475–4487.

[15] D. Mathew, N.C. Brintha, J.W. Jappes, Artificial intelligence-powered automation for industry 4.0, New Horizons for Industry 4.0 in Modern Business, Springer International Publishing, Cham, (2023), pp. 1–28.

[16] T. Ahmad, H. Zhu, D. Zhang, R. Tariq, A. Bassam, F. Ullah, et al., Energetics systems and artificial intelligence: applications of industry 4.0, Energy Reports 8 (2022) 334–361.

[17] Coursera Staff, (2025). 4 Types of AI: Getting to Know Artificial Intelligence:- https://www.coursera.org/in/articles/types-of-ai.

[18] IBM Data and AI Team, (2025). Understanding the different types of artificial intelligence:- https://www.ibm.com/think/topics/artificial-intelligence-types.

[19] Types of Artificial Intelligence (AI), (2026). https://www.geeksforgeeks.org/artificial-intelligence/types-of-artificial-intelligence/.

[20] A. Kalra, G. Tewari, Applications of artificial intelligence across industry 4.0, Machine Learning for Sustainable Manufacturing in Industry 4.0, CRC Press, (2023), pp. 23–43.

[21] S.K. Jagatheesaperumal, M. Rahouti, K. Ahmad, A. Al-Fuqaha, M. Guizani, The duo of artificial intelligence and big data for industry 4.0: applications, techniques, challenges, and future research directions, IEEE Internet of Things Journal 9 (2021) 12861–12885.

[22] A.A.H. Abdullah, F.A. Almaqtari, The impact of artificial intelligence and industry 4.0 on transforming accounting and auditing practices, Journal of Open Innovation: Technology, Market, and Complexity 10 (2024) 100218.

[23] P. Radanliev, D. De Roure, R. Nicolescu, M. Huth, O. Santos, Artificial intelligence and the Internet of Things in industry 4.0, CCF Transactions on Pervasive Computing and Interaction 3 (2021) 329–338.

[24] P.K. Ambadekar, S. Ambadekar, C.M. Choudhari, S.A. Patil, S.H. Gawande, Artificial intelligence and its relevance in mechanical engineering from an industry 4.0 perspective, Australian Journal of Mechanical Engineering 23 (2025) 110–130.

[25] N. Malik, S.N. Tripathi, A.K. Kar, S. Gupta, Impact of artificial intelligence on employees working in industry 4.0 led organizations, International Journal of Manpower 43 (2022) 334–354.

[26] G. Lampropoulos, Artificial intelligence, big data, and machine learning in industry 4.0, Encyclopedia of Data Science and Machine Learning, IGI Global Scientific Publishing, (2023), pp. 2101–2109.

[27] W. Chen, W. He, J. Shen, X. Tian, X. Wang, Systematic analysis of artificial intelligence in the era of industry 4.0, Journal of Management Analytics 10 (2023) 89–108.

[28] W.M.R.W. Ibrahim, R. Hassan, Recruitment trends in the era of industry 4.0 using artificial intelligence: pro and cons, Asian Journal of Research in Business and Management 1 (2019) 16–21.

[29] R. Rezazadegan, M. Sharifzadeh, Applications of artificial intelligence and big data in industry 4.0, Industry 4.0 Vision for Energy and Materials: Enabling Technologies and Case Studies (2022) 121–158.

[30] V.V. Febiandini, M.S. Sony, Analysis of public administration challenges in the development of artificial intelligence industry 4.0, IAIC Transactions on Sustainable Digital Innovation (ITSDI) 4 (2023) 164–168.

[31] Edureka, Top 15 Hot Artificial Intelligence Technologies, 2021. https://www.edureka.co/blog/top-15-hot-artificial-intelligence-technologies/.

[32] T. Babu, R.R. Nair, S. Kishore, Applications of artificial intelligence and machine learning in industry 4.0, Artificial Intelligence and Machine Learning for Industry 4.0 (2025) 107–143.

[33] S. Mukhopadhyay, R.K. Singh, T. Jain, Developing artificial intelligence enabled marketing 4.0 framework: an industry 4.0 perspective, Qualitative Market Research: An International Journal 27 (2024) 841–865.

[34] A. Chaurasia, B. Parashar, S. Kautish, Artificial intelligence and automation for industry 4.0, Computational Intelligence for Modern Business Systems: Emerging Applications and Strategies, Springer Nature Singapore, Singapore, (2023), pp. 357–373.

[35] S. Alvarez-Napagao, B. Ashmore, M. Barroso, C. Barrué, C. Beecks, F. Berns, et al., Knowledge project-concept, methodology and innovations for artificial intelligence in industry 4.0, 2021 IEEE 19th International Conference on Industrial Informatics (INDIN), IEEE, (2021), pp. 1–7.

[36] F. Delgado, S. Garrido, B.S. Bezerra, Barriers to visibility in supply chains: challenges and opportunities of artificial intelligence driven by industry 4.0 technologies, Sustainability 17 (2025) 2998.

[37] G. Nick, K. Zeleny, T. Kovács, T. Járvás, K. Pocsarovszky, A. Kő, Artificial intelligence enriched industry 4.0 readiness in manufacturing: the extended CCMS2. 0e maturity model, Production & Manufacturing Research 12 (2024) 2357683.

[38] M.A. Alohali, F.N. Al-Wesabi, A.M. Hilal, S. Goel, D. Gupta, A. Khanna, Artificial intelligence-enabled intrusion detection systems for cognitive cyber-physical systems in the industry 4.0 environment, Cognitive Neurodynamics 16 (2022) 1045–1057.

[39] G. Singh, G. Bhardwaj, S.V. Singh, N. Chaudhary, Artificial intelligence-led industry 4.0 applications for sustainable development, 2022 2nd International Conference on Innovative Practices in Technology and Management (ICIPTM), IEEE, (2022), pp. 339–343 vol. 2.

[40] A. Azizi, Applications of artificial intelligence techniques to enhance sustainability of industry 4.0: design of an artificial neural network model as dynamic behavior optimizer of robotic arms, Complexity 2020 (2020) 8564140.

[41] T.V.N. Rao, A. Gaddam, M. Kurni, K. Saritha, Reliance on artificial intelligence, machine learning and deep learning in the era of industry 4.0, Smart Healthcare System Design: Security and Privacy Aspects, Wiley, (2022), pp. 281–299. Available from https://doi.org/10.1002/9781119792253.ch12.

[42] S. Garg, N. Mahajan, J. Ghosh, Artificial intelligence and its impacts on industry 4.0, Industry 4.0 and the Digital Transformation of International Business, Springer, (2023), pp. 123–133. Available from 10.1007/978-981-19-7880-7_7.

[43] N. Nurainia, E.A. Apriadi, A literature review on the role of AI (artificial intelligence) in industry 4.0 transformation, International Journal of Technology and Computer Science 1 (2025) 25–37.

[44] K. Alexopoulos, N. Nikolakis, G. Chryssolouris, Digital twin-driven supervised machine learning for the development of artificial intelligence applications in manufacturing, International Journal of Computer Integrated Manufacturing 33 (2020) 429–439.

[45] V. Kharchenko, O. Illiashenko, O. Morozova, S. Sokolov, Combination of digital twin and artificial intelligence in manufacturing using industrial IoT, 2020 IEEE 11th International Conference on Dependable Systems, Services and Technologies (DESSERT), IEEE, (2020), pp. 196–201.

[46] V. Karkaria, Y.K. Tsai, Y.P. Chen, W. Chen, An optimisation-centric review on integrating artificial intelligence and digital twin technologies in manufacturing, Engineering Optimization 57 (2025) 161–207.

[47] K. Bartsch, A. Pettke, A. Hübert, J. Lakämper, F. Lange, On the digital twin application and the role of artificial intelligence in additive manufacturing: a systematic review, Journal of Physics: Materials 4 (2021) 032005.

[48] A.K. Tyagi, R. Richa, Smart manufacturing using the Internet of Things, artificial intelligence, and digital twin technology, Global Perspectives on Robotics and Autonomous Systems: Development and Applications, IGI Global Scientific Publishing, (2023), pp. 184–205.

[49] T. Kreuzer, P. Papapetrou, J. Zdravkovic, Artificial intelligence in digital twins—a systematic literature review, Data & Knowledge Engineering 151 (2024) 102304.

[50] A. TurandasjiPatil, B. Vidhale, A. Titarmare, Implementation of artificial intelligence in industry 4.0, future and its challenges: a comprehensive review, 2024 3rd International Conference for Innovation in Technology (INOCON), IEEE, (2024), pp. 1–5.

[51] Y. Li, J. Fan, X. Cheng, H. Jin, Y. Wang, F. Wei, et al., New revolution for quality control of TCM in industry 4.0: focus on artificial intelligence and bioinformatics—TrAC, Trends in Analytical Chemistry 181 (2024) 118023.

[52] D. Mhlanga, Artificial Intelligence in the industry 4.0, and its impact on poverty, innovation, infrastructure development, and the Sustainable Development Goals: lessons from emerging economies, Sustainability 13 (2021) 5788.

4

Additive manufacturing for Industry 4.0

4.1 Introduction to additive manufacturing

Additive manufacturing (AM) is the process of joining materials to make objects from 3D model data, usually layer upon layer. Facilitated through technological platforms, it is a digital process. Compared to other manufacturing processes, AM can offer several benefits in creating customised products and models. In contrast to subtractive manufacturing techniques, ASTM defines AM as the process of assembling materials to create items from 3D model data, often layer upon layer. The conventional subtractive manufacturing is a technological platform in which a part is produced by skilled model makers from 2D engineering drawings, and the final product is manufactured by subtracting material.

AM is an important technology that solves various requirements of Industry 4.0. Global manufacturing operations are benefiting from AM technologies that improve productivity, enable hybrid processes, and lower the cost of customised products. Software support, 3D printing, and 3D scanning are essential parts of the AM system. AM is a critical component for meeting the various requirements of Industry 4.0. Industry 4.0 is entering into a new market by focusing on customer delight through this technology. It has the potential to become an essential component in product invention and development [1–3].

With new technologies, Industry 4.0 is expected to change how businesses operate. AM demonstrates integrated technology by using 3D printing to transform a computer-aided design (CAD) file into a tangible object. By permitting digital production and mass customisation, Industry 4.0 will make production processes and supply networks more linked, adaptable, and efficient. This transformation may create additional opportunities for underdeveloped countries to address various concerns [4,5]. It offers a personalised product based on the customer's specifications simultaneously and consistently. AM provides significant inputs to smart production systems by using digital design and specialised software-based manufacturing technologies [6–8].

Customisation and digital production systems are the key challenges in contemporary Industry 4.0, which AM readily addresses. AM technologies use design software and digital inputs, even with advanced 3D scanners, to create flexible, intelligent products that meet worldwide market demand.

4.1.1 Some historical developments to the present

It started as 3D printing in the 1980s and has evolved from experimental prototyping to a driving force in modern manufacturing, enabling layer-by-layer fabrication of objects directly

Digital Twin and Industry 4.0. DOI: https://doi.org/10.1016/B978-0-443-54019-6.00005-4

from digital designs. In the early 1980s, Dr Hideo Kodama filed the first patent for AM, introducing the concept of photopolymer-based rapid prototyping (RP). However, the idea was not commercialised due to funding constraints. Then Chuck Hull's invention of stereolithography (SLA) in 1986 marked the first practical 3D printing process, laying the foundation for RP technology. Hull's SLA process used a UV laser to solidify layers of photopolymer resin, and he established 3D Systems to commercialise this invention. Selective laser sintering (SLS), developed by Carl Deckard in 1987, uses a laser to fuse powdered materials, further expanding the range of available materials for 3D printing. In 1989, Scott and Lisa Crump created fused deposition modelling (FDM), which remains the basis for many desktop and hobbyist 3D printers today. Commercialisation (1990s–2000s): Throughout the 1990s, additional AM technologies emerged, including solid ground curing and laminated object manufacturing (LOM). 3D bioprinting took its first steps in 1999, with the Wake Forest Institute producing the first 3D-printed tissue.

The early 2000s saw the democratisation of AM through the open-source RepRap project by Adrian Bowyer, enabling the public to build and share 3D printer designs. Then, in the late 2000s and 2010s, the rise of multimaterial and desktop printers occurred. The democratisation era began in 2011, when key patents for FDM and SLS expired. In 2014, NASA sent the first 3D printer to the International Space Station, enabling on-demand tool manufacturing in orbit. It developed continuous liquid interface production, which drastically increased the speed of resin printing. Then, in 2015, with extensive technological support, the medical and bioprinting breakthroughs began, with the launch of the first commercial bio-ink, making 3D bioprinting affordable for researchers. By the 2020s, AM became essential for patient-specific implants and dental aligners. With the AI era beginning in 2023, the factory emerged. The focus is now on AI-powered optimisation, real-time error-detection algorithms, and automated slicing of complex designs. In addition to meeting medical requirements, it also produces end-use metal parts for the aerospace and automotive sectors.

Modern industries are transitioning from mass production to a personalised manufacturing approach [9,10]. It is a direct manufacturing method that allows the product to be manufactured in a relatively short period. In the current industry, AM provides the necessary improvements to meet customers' unique needs [11,12] in much less time. Thus improving applicability in various engineering, medical, research, and related industries.

AM technologies are critical components of tomorrow's smart factories, leveraging software, technologies, materials, and procedures to deliver world-class outcomes. With just-in-time manufacturing, AM enables design, prototyping, and manufacturing in a quarter of the time, helping to reduce unnecessary inventory. AM speeds up product delivery by simplifying inventory management and transportation, and by assisting in fulfilling unique product needs based on client demand and market conditions. AM is utilised to validate product concepts, assess manufacturing scalability, estimate market demand, and identify supply chain needs [13–15]. With AI, AM is being developed to generate prototypes for a specific consumer in a shorter time, perhaps instantaneously, with high precision and optimised costs.

4.2 Technological platforms of additive manufacturing

AM allows the design and manufacturing of customised parts as per the design, and even as per the patient's specific requirements. There are four major areas through which AM is evolving, which are basically technological platforms, and the same are being provided below:

- Software platforms are a critical part of the process, including CAD software (e.g. Autodesk Fusion, SolidWorks) for design, and 'slicer' programs (e.g. Ultimaker Cura) that prepare the digital model and generate machine instructions (G-code) for the printer. Ansys, SolidWorks, Rhino, Mimics, 3D Slicer, 3D Doctor, Catia, Inspect Plus, and more than 200 software applications, both customised and generic.
- Hardware platform for various types of 3D printing technologies, 4D, 5D, bioprinting, hybrid, bioprinting, and other new developments.
- Digital modelling platform and data acquisition technologies includes CAD and inspection, scanning, magnetic resonance imaging (MRI), computed tomography (CT), coordinate measurement machine (CMM), CAD model, X-Ray, ultrasound, etc.
- Materials platform includes polymers, metals, ceramics, composites, gels, foams, and lots of newly developed materials.

These four major technological platforms, areas, and technologies are capable of manufacturing multimaterial jobs and even colour-complex jobs [16–18]. These colour 3D implants accurately depict the different sections of soft tissue, organs, and hard tissue, helping doctors and patients communicate clearly. Artificial 3D medical models can easily replace missing parts of a patient's body lost to disease or accident. These parts can perform their normal functions once properly implanted in the patient's body. It provides surgeons and doctors with innovative ideas and is helpful for student training and education.

Layer by layer, machines are instructed to deposit material in precise geometric shapes via CAD software or 3D object scanners. AM is the process of adding content to a product. However, material removal through milling, machining, carving, shaping, and other techniques is often necessary when producing an object using conventional procedures [19–21]. Using a laser or an electron beam, 3D metal printing machines spread metal powder layer by layer, melt it, and fuse it to create a product. The process is repeated until the component is finished. It provides multiple benefits as compared to conventional implant manufacturing methods. It enables the manufacture of complex metal parts to the required shape and size, thereby reducing the product's design and manufacturing time [22–24]. Fig. 4.1 shows the main components of AM.

4.2.1 Materials for additive manufacturing

4.2.1.1 Polymers (plastics and elastomers)

Polymers are the most prominent group in AM, categorised by their reaction to heat and light.

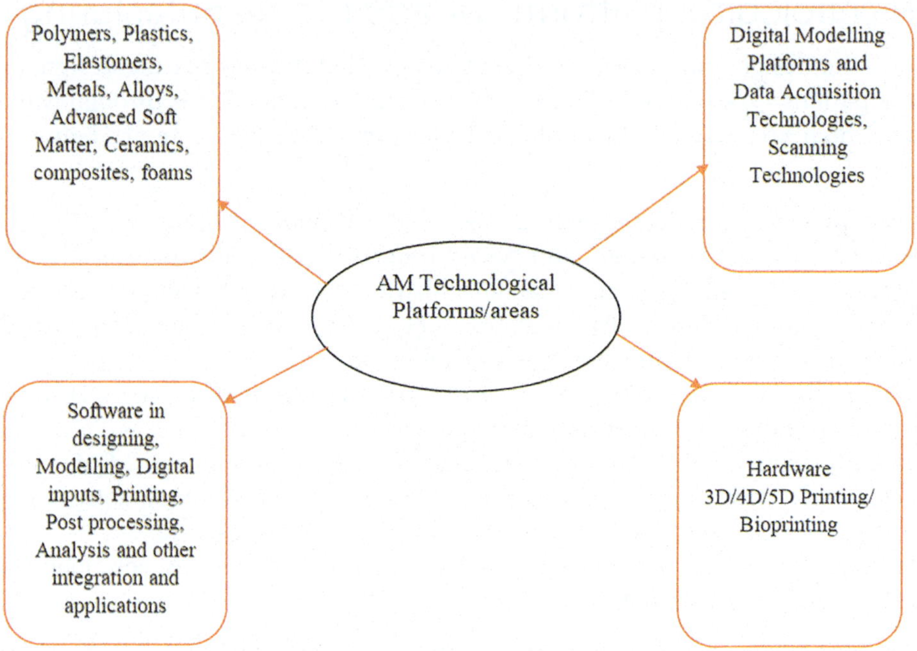

FIGURE 4.1 Four basic technological areas of additive manufacturing.

4.2.1.2 Thermoplastics

Materials that can be melted and solidified repeatedly, such as polylactic acid (PLA), acrylonitrile butadiene styrene (ABS), nylon (PA12), and high-performancepolyether ether ketone.

4.2.1.3 Thermosets (photopolymers)

Liquid resins that undergo a chemical reaction to become a permanent solid when exposed to UV light. Such as epoxies, acrylics, and flexible 'Rubber-like' resins.

4.2.1.4 Elastomers

Elastomers are polymers with high elasticity (stretchability), as used for making gaskets, wearable tech, and seals.

4.2.1.5 Metals and alloys

Metals are important for traditional components and, with new developments, due to their ability to create high-strength, lightweight parts, have extensive applications in the medical, aerospace, and automotive industries.

4.2.1.5.1 Traditional alloys

Stainless steel (316 L), aluminum (AlSi10Mg), and cobalt-chrome. Superalloys: nickel-based alloys like Inconel for extreme heat.

4.2.1.5.2 Reactive metals

Titanium (Ti6Al4V) is the standard for bone-integrated implants.

4.2.1.5.3 Advanced soft matter

Gels (hydrogels): Gels are mostly liquid; however, they behave like solids. In AM, these are primarily used in bioprinting, direct ink writing, etc.

Natural hydrogels: Natural hydrogels are largely made from biological sources (alginate, collagen, gelatin). These are used as bio-inks.

Synthetic hydrogels: Synthetic hydrogels are largely man-made polymers (PEG, PNIPAM) and are used for drug delivery systems, soft robotics, etc.

Aerosols and xerogels: Aerosols and xerogels are ultra-lightweight solids derived from gels in which the liquid is replaced with air, providing very good thermal insulation.

4.2.1.6 Architected and chemical foams

Foams are identified through their cellular structure, and there are two types of foams:

Syntactic and chemical foams: Syntactic and chemical foams are materials that contain gas-filled bubbles or blowing.

Lattice-based foams: Lattice-based foams, also called digital foams, are used to print a complex network of struts (lattices). It takes care of the squishiness of a part, such as in helmet liners and high-end running shoes.

4.2.1.7 Ceramics and composites

Technical ceramics: Technical ceramics, such as alumina and zirconia, are used to make parts for high-wear and high-heat industrial applications.

Fibre-reinforced composites: Fibre-reinforced composites are to achieve metal-like strength at a fraction of the weight. They are mostly plastics, supercharged with carbon fibre, Kevlar, or glass fibre.

4.2.2 Hardware as 3D printing

3D printing is an essential AM technology that creates a physical 3D object layer by layer from a CAD model. This printing technology was developed in 1980. The process of this technology is opposite to 'subtractive manufacturing processes', such as machining. In AM, instead of removing material from the workpiece, material is added layer by layer from a CAD input file in the standard triangulate language (STL) format. This technology enables the production of customised and complex-shaped products in less time and at lower cost [25,26].

The 3D printing technique was earlier called RP because it rapidly manufactures prototypes. Previously, creating a prototype was expensive. Now, with the 3D CAD file, a prototype/physical model can be quickly printed on the same machine. It is a new method for the customised manufacturing of end products with very few fixture requirements [27,28].

In the current scenario, 3D printing is an innovative technology that enables the rapid development of prototypes, new products, and devices at a lower cost. With technological developments, different printing technologies are available, capable of printing parts with different materials and combinations. It also enables the manufacture of complex objects from heterogeneous materials [29–31].

4.2.2.1 Steps used by 3D printing to print a part

3D printing is suitable for various purposes, including manufacturing prototypes, education and training, product design and development, modifying industrial products, medical and dental applications, architectural models, and research. This technology manufactures customised models at a low cost [32,33]. The significant steps involved in 3D printing technology are shown in Fig. 4.2.

- Importing a 3D digital CAD file into printing software.
- Setting files to print.
- The orientation of the part in the printing software.
- Printing of the part.
- Drying the part and emptying the built bed.
- Clean the part.
- Postprocessing of parts to improve the quality and strength.

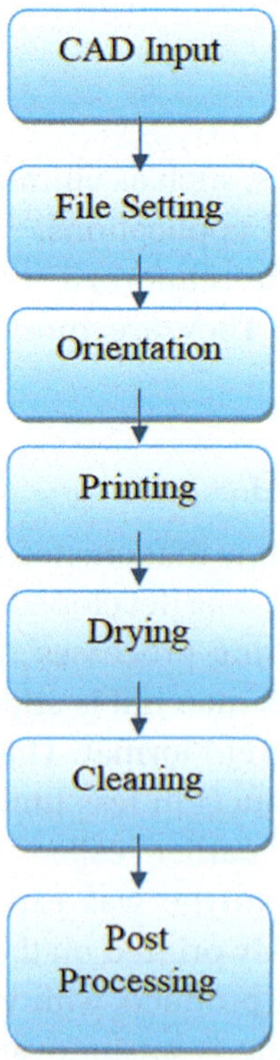

FIGURE 4.2 Major steps involved in 3D printing technology.

4.3 Types of 3D printing processes

3D printing has great potential to manufacture objects layer by layer. This technique covers various processes for creating a 3D solid object using a 3D digital model [34–38]. There are seven major types of 3D printing processes:

4.3.1 Vat polymerisation

In this process, a vat of liquid resin photopolymers is used to build a 3D solid object layer by layer. UV light is precisely directed through the mirror to create the desired product in less time and at a lower cost. This process does not require any support material during printing because the liquid forms an object. The steps used to build an object using this process are as follows:

- Prepare a build platform for a vat of liquid resin photopolymers.
- Layer by layer, cure the resin using UV light.
- Lower the built platform for additional layers as per layer thickness.
- Repetition of the process by adding the required number of layers to build a complete object.
- Finally, the vat is drained of resin, and the built object is removed safely.

4.3.2 Material extrusion

The raw material is fed to the machine by wire and extruded through a heated nozzle. This nozzle moves horizontally and vertically according to the shape of the required product. The layers of material are bonded together using chemical agents under proper temperature control. In the 3D printing extrusion process, support materials are also used to create complex-shaped products. The steps used to build an object using this process are as follows:

- Deposition of material using a nozzle as per the cross-sectional area of the product to be built.
- Adding layers on top of the previously built layer.
- Support material is also added during this process for a complex-shaped object.
- Fused the molten material to build a solid object.
- Remove the part and support material from the object.

4.3.3 Binder jetting

In the binding jetting process, the raw material is powder, and resin is used as the binding material. With the help of a roller, the powder is spread layer by layer, and the binder binds the powder sprayed from the print head. There is no requirement for support material in this process, as the powder serves as support. The steps used to build an object using this process are as follows:

- Spread powder material over the built platform using a roller.
- Deposits the binder on the powder with the help of a print head as per the requirement.
- Lower the platform as per the layer thickness of the object and repeat the step.

- Built the complete part.
- Remove the part and the powder from the surroundings of the object.

4.3.4 Directed energy deposition

In this process, the material is used in the form of metal, polymer, powder, or ceramic, and an electron-beam nozzle melts it to create the desired shape. In this process, the nozzle is mounted on a multiaxis arm, and molten material is deposited onto the specified surface. The specific axis of this nozzle is not fixed; thus it can be moved in multiple directions. The steps used to build an object using this process are as follows:

- Five-axis arm with a nozzle moves around the object to be built.
- Deposit of material on the entire surface of an object through the nozzle.
- Melting of material using a laser, electron beam, or plasma arc.
- The addition of material layer by layer solidifies to create the final object.
- Removal of part.

4.3.5 Powder bed fusion

An electron beam or laser fuses or melts powder material in an ultra-fine layer to create 3D objects. This method required a vacuum to create a functional part. In this process, a heated thermal print head fuses powder. The steps used to build an object using this process are as follows:

- Spread a layer of powder material on a built platform.
- Fuse powder using a laser/electron beam as per the cross-sectional area.
- Again, spread the new layer and repeat the step.
- Remove part from workpiece and powder during postprocessing.

4.3.6 Sheet lamination

A sheet of material is used to manufacture 3D physical models. In this process, a laser is used to cut sheets according to the product's design, and adhesive is used to join the sheets and manufacture the final product. This process requires relatively little energy, and different materials can be bound easily. The steps used to build an object using this process are as follows:

- The positioning of material on the cutting bed.
- Bound the sheet using adhesive
- Cutting of the material sheet using a laser, as per the size of the object.
- Adding material and repeating the step.
- Manufactured the final part and removed.

4.3.7 Material jetting

This jetting process is similar to a 2D inkjet printer. It allows the print head to move along the x-, y-, and z-axes to create a 3D object. Material is jetted onto a built plate as a drop and hardens layer by layer under UV light. It uses support material to create a complex object. In this process, waxes and polymers are commonly used for their ability. The steps used to build an object using this process are as follows:

- The positioning of the print head on the built platform.
- Deposition of material onto the surface in the form of droplets using a print head.
- Solidify the material droplet using UV light.
- Repeat the process and create another layer.
- Remove the built part and remove support material.

4.4 Types of 3D printing technologies

Various AM technologies are now available for printing colour 3D models, providing more precise information than scanning technologies. Major AM technologies used for printing of parts are as follows;

- Stereolithography (SLA).
- Selective laser sintering (SLS).
- Electron beam melting (EBM).
- Direct metal laser sintering (DMLS).
- Laminated object manufacturing (LOM).
- Fused deposition modelling (FDM).
- Sand binder jetting (SBJ).
- Inkjet 3D printing (IJP).
- Digital light processing (DLP).
- Polyjet 3D printing (PJP).
- Colour-jet-printing (JCP).
- Drop on demand (DOD).

4.4.1 Stereolithography

AM technology uses photochemical processes to create prototypes, models, parts, and patterns layer by layer. In this technology, ultraviolet (UV) lasers focus on the vat of photopolymer resin to manufacture 3D parts. The UV laser draws, as per the object's shape, on the photopolymer vat surface. It is a fast process, and almost all designs can be produced. SLA enhanced its application for the manufacturing of medical models, product development, and other applications.

4.4.2 Selective laser sintering

In this technology, an object is created using the fusion bed process. A CO_2 laser is used to sinter the raw material, which is in powder form. Here, the heat source is a laser, which melts the powder. Then, the second layer of powder is spread, and the laser sinters it. With repeated iterations of this process, the object is manufactured.

4.4.3 Electron beam melting

This technology uses high-energy electrons or beams to fuse metal powder particles. This beam fused and solidified the metal powder to a specific cross-sectional area. Products manufactured by this technology have relatively high strength.

4.4.4 Direct metal laser sintering

DMLS is a fascinating technology used to 3D-print metals such as titanium and aluminium. This advanced technology uses metal powder in a skinny printing layer. In this technology, metal powder is spread by a roller and fused at specific points using a laser beam as the heat source. The part manufactured using this technology has higher mechanical strength and greater accuracy.

4.4.5 Laminated object manufacturing

In this technology, 3D models are manufactured with material sheets. Here, sheets of material are used as raw material. The first sheet adheres to a heated roller, and the laser traces the prototype's desired dimensions. A laser removes material from the area, and the completed layer moves down. Sheets are joined using adhesive, with the fresh sheet material rolled into position. This process is repeated until the product is completed.

4.4.6 Fused deposition modelling

FDM is a commonly known AM method in which a product is manufactured layer by layer through material deposition. As it passes its glass transition temperature, the material is extruded through the nozzle (move in X, Y, and Z directions). This technology uses plastic filaments and material deposited layer by layer on the print bed. In this technology, the support material is also used to print complex-shaped products, with the number of layers required to form a part. Different materials are used in FDM, such as thermoplastics, pastes, metals, chocolates, and wood. This technology is cheap and efficient for daily life applications.

4.4.7 Sand binder jetting

This low-cost technology uses sand-binding jetting devices to produce 3D parts from the sand. The object is manufactured in full colour using plaster-based powder/silica sand with a liquid binding agent. In this process, first, the jets bind the agent, then print the head-jet colour to print the full-colour model. It repeats the process, layer by layer, until the part is manufactured. After manufacturing, the part is removed and cleaned with the unbound powder.

4.4.8 Inkjet 3D printing

This technology works like a 2D inkjet printer. Different fluids from a polymer solution are deposited layer by layer using a print head to create a 3D product. This technology can print a variety of materials at a lower cost and in less time.

4.4.9 Digital light processing

DLP uses a digital light projector to flash a single image of each layer. This image of each layer is composed of square pixels projected onto a digital screen. It achieves a faster print speed than SLA because the entire layer is exposed simultaneously.

4.4.10 Polyjet 3D printing

This technology uses two or more jetting heads: one for printing the part and one for support material. It sprayed photopolymer-like materials layer by layer and cured them with a UV lamp within the printer. This technology uses a gel-like support material that is easy to wash.

4.4.11 Colour-jet-printing

In this type of AM technology, the part is manufactured by spreading powder material in layers on the build platform using a roller. The adhesive (binder) is sprayed by the print head to bind the required powder, as specified in the CAD file. Again, the platform moved downward as per the layer thickness and repeated the step.

4.4.12 Drop on demand

This type of AM technology uses a pair of inkjets. It uses wax-like material as a built material with dissolvable support material. This printer uses a pointwise deposition method with a predetermined path for the jet material, manufacturing objects layer by layer. After creating each layer, this printer uses a fly cutter on the built area. The main application of this technology is to create a pattern for investment casting (lost-wax casting) and other mould-making applications.

4.5 Materials used in 3D printing

There are five primary forms of material by which parts are manufactured layer by layer using different 3D printing technologies, and they are as follows [39–43]:

- Molten material or semiliquid.
- Solidification of the photocurable resin.
- Cutting sheets of paper, metal, or plastic.
- Binding or fusing the granules of powder.
- Cell printing/living cells/bio-ink/scaffolds.

4.5.1 Molten material or semiliquid

In extrusion-based 3D printing, successive layers of an object are created by a nozzle that extrudes molten or semiliquid material. Usually, thermoplastics are used to build materials. These include ABS, bioplastic PLA, polycarbonate, and nylon. Whatever material is used, it is delivered to the print head in solid form, as a thin filament or strand, which is then heated and melted. FDM refers to this type of printing.

4.5.2 Solidification of the photocurable resin

The product is built using material solidification with light-sensitive polymers, which, when exposed to UV radiation, reproduce the 3D model. SLA refers to this type of form.

4.5.3 Cutting sheets of paper, metal, or plastics

The technology generates 3D models in these categories by adding layers of the defined sheet materials (metal, plastic, and paper). Firstly, a layer of sheet material is loaded; a laser (CO_2 laser) cuts the sheet as per the designed cross-section. After removing excess material from the sheet, another layer is applied over the previously manufactured layer. A laser tracing defines the following pattern based on the information in the STL file. Here, all layers are combined through adhesive. These steps are repeated further until a layered 3D model is generated. LOM refers to this type of printing.

4.5.4 Binding or fusing the powder granules

This process builds 3D models by spreading powder on a base, then selectively bonding it with a laser- or injection-based binder. This cycle is repeated many times. In these technologies, the powder is the core material, and an epoxy binder/laser is used to bind the powder. Project 3D printing and SLS are suitable examples.

4.5.5 Cell/tissue bio-ink/scaffolds printing

Because of its potential as an affordable method for creating customised human tissues and organs for transplantation or other medical applications, 3D bioprinting has been revolutionising regenerative medicine and tissue engineering. A 3D bioprinter requires bio-ink and other materials to produce cell/tissue bio-inks/scaffolds. Depending on the printed construct, different bioprinters may extrude layers or droplets of bio-ink. For instance, bio-ink droplets may be extruded from a bioprinter to produce a 3D-printed scaffold that can examine the impact of mechanical stress on cells and later be utilised to build tissue. Aqueous bio-inks are extruded into pellets or droplets to form scaffolds for tissue engineering.

4.6 Advantages of 3D printing

3D printing has certain advantages over conventional subtractive manufacturing. Some of the significant advantages of 3D printing are as follows [44–46]:

- Extensive customisation is possible at a lower cost.
- No requirement for fixture or tooling.
- Shorter setup time.

- Easily change the revised version of the product.
- Production enhancement on demand.
- Prints complete objects at one time.
- Lesser energy usage.
- Shorter build times—usually under 24 hours and often much shorter for small pieces.
- Reduced material waste during manufacturing.
- Helps create and support innovation.
- Provides increased industrial efficiency.
- Higher production and design flexibility.
- Parts having geometric complexity are produced without any extra cost of tooling.
- Extensive range of materials/multimaterials that can be used for different applications.

4.7 Limitations of 3D printing

Some of the main limitations of 3D printing are as follows [47–50]:

- The traditional manufacturing process is usually much faster for a mass-production system.
- Limited product size, that is, size depends upon the machine bed size.
- Per-piece cost can be high.
- Highly skilled human resources are required to operate these machines.
- Material changing options are limited.
- Printed products have a reduced surface quality.
- No break during the printing of a product; otherwise, the part will be damaged.
- Required postprocessing as it increases the final cost of the product.

4.8 Digital modelling platforms and data acquisition technologies scanning

3D scanning is an innovative technique for capturing the actual shape and size of an object as a 3D digital file using a specialised 3D scanner. This digital 3D data is saved in different design software formats and is editable. 3D printing technologies convert this data into a specific format before printing. This technology has extensive applications and is used gainfully as a disruptive innovation. It captures colour data and other information related to an object [51,52].

A 3D scanner is a device that converts the image of a physical object into digital form by acquiring data as a point cloud. These noncontact 3D scanners are used for various purposes, such as design modification, reverse engineering, prototyping, inspection, data storage, diagnosis, and product manufacturing, in combination with specialised software and an appropriate 3D printer. A camera and projector are used in this process, in which the projector projects light onto an object, and the camera captures the reflected image. These contactless 3D scanners now provide reliable measurements comparable to those of conventional CMMs that perform contact-type measurements [53–55]. 3D scanning technologies have broad applications in medicine, dentistry, education, design, inspection, and prototype manufacturing. The steps followed in 3D scanning are shown in Fig. 4.3.

In medicine, 3D scanning is used to achieve a high level of customisation, enabling a customised plaster to be designed and manufactured via 3D printing to provide patient comfort. It accurately shapes limbs and prosthetics and speeds up the therapeutic process. 3D scanning provides a perfect solution to improve the medical industry.

4.9 Different scanning technologies

Different scanning technologies are used to obtain information/data as discussed below [56,57]:

4.9.1 X-rays

This scanning technology captures medical images of dense structures, such as bones, using X-rays. Soft tissues, such as skin and organs, cannot absorb these rays. It is used in various medical treatments and diagnoses.

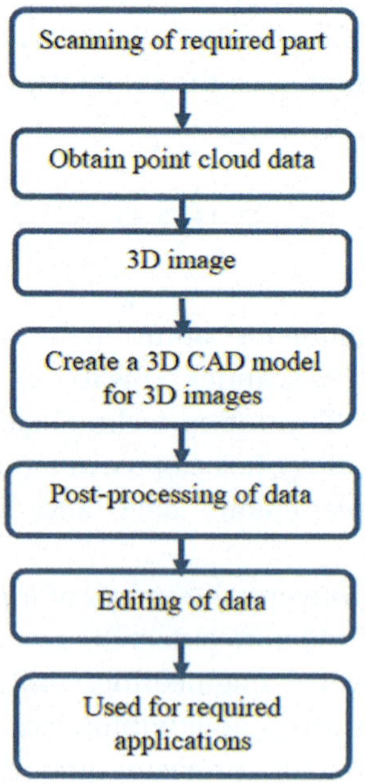

FIGURE 4.3 Major steps of 3D scanning.

4.9.2 Ultrasound

Using sound waves, a dedicated system captures an image of the inside of the body. The status of internal body organs, including pain, swelling, and infections, can be easily captured.

4.9.3 3D scanner

3D scanner captured 3D data as a point cloud/triangle mesh. It is an innovative scanning tool for designing and developing industrial products, medical models, implants, tissue, prosthetics, and anatomical models, as well as for their analysis. With this technology, 3D data can be stored digitally, further edited as per the required shape and size, and finally printed using 3D printing technologies.

4.9.4 Computed tomography scan

This technology easily captures data from a patient's internal organs. An X-ray beam captures detailed images of the spine, neck, bones, shoulder, eyes, heart, and other internal organs. This technology helps to check the status of defects in the body.

4.9.5 Magnetic resonance imaging scan

MRI technology captures detailed images of the bones, spinal cord, joints, organs, brain, heart, soft tissues, and liver using a strong magnetic field and radiofrequency pulses. It provides accurate information about normal and abnormal tissue using this technology. It is safe, painless, and quickly detects disease, providing information about bleeding, swelling, and brain tumours. This technology also provides information about the body's blood circulation.

Medical professionals use scans such as CT, MRI, X-rays, and ultrasound to provide valuable information about internal organs. CT, MRI, X-rays, and ultrasound scans provide information about internal body parts, whereas a 3D scanner provides information about the outer body surface. Thus to capture information about external body parts, 3D scanning technologies are well-suited, and today's products provide highly accurate measurements. It helps doctors/surgeons to enhance knowledge and benefits, and to minimise risks, for the best treatment of the patient.

4.10 Classification of 3D scanning

Literature classifies 3D Scanning products into the following categories [55, 58–60]:

4.10.1 Structured light

In the 3D noncontact scanning process, the object is measured by analysing the deformation of the light pattern on its surface.

4.10.2 Contact-based scanning

Contact-based scanning uses a touch probe to sample several points on the object's surface. This technique involves physical contact with the object's surface, which must be scanned.

4.10.3 Laser triangulation scanning

It projects a laser beam onto the object's surface and measures the deformation of the laser ray.

4.10.4 Laser pulse

This scanning method, also called time-of-flight scanning, is based on the time-of-flight of a laser beam. In this process, a laser beam is projected on the object's surface, and sensors collect the information.

4.10.5 Photogrammetry

Using computational geometry algorithms, 2D measurements of the object are captured and then used to create a 3D scan. This technology is simple, and the object is scanned from different angles. Here, the lens's focal length and distortion can be adjusted to create an accurate model.

4.11 Advantages of 3D scanning

3D scanning provides an accurate 3D image of the part. This 3D image can be visualised and used for analysis and decision-making. The light source is harmless to humans and accurately captures the required part. Various advantages of 3D scanning, as envisaged, are as follows [56,57,61]:

- Design an accurate model.
- Scan any surface, whether rough, smooth, dark, shiny, or any colour.
- Fast speed of scanning.
- Work in both stationary and portable cases.
- Obtain high accuracy by changing the lenses (field of view).
- Safe for the human being.
- Scan without any physical contact.
- Helpful in idea generation.
- Minimise time to design.
- Precise designing of a complex part.
- Identified improvement/changes.
- Simple for manipulation.
- Low risk of damaging any part.
- Low operational cost.

- Keep a record of historical sites in digital form.
- Design and development of tools, devices, and instruments.
- Improved learning and effective planning.
- Cost- and time-efficient, and improves the success rate.

4.12 Limitations of 3D scanning

3D scanning has the following limitations [51,55,62]:

- Provides information about the outer surface of the part.
- Not able to properly scan a transparent/reflective surface.
- The high cost of the scanner.
- Not able to provide material information during scanning.
- Equipment is accurate under standard environmental control.
- It does not provide information on holes, threads, and trapped surfaces.
- Does not provide information on moving objects.

4.12.1 Software (designing, modelling, digital inputs, printing, postprocessing, analysis)

Software is a set of instructions that tells a machine how to do specific tasks. This contrasts with hardware, which serves as the system's foundation and performs the actual work. High-level programming languages are used to write most software. Because they are closer to natural languages than machine ones, they are more accessible and efficient for programmers. Compilers, interpreters, or a mix of the two are used to translate high-level languages into machine language. Low-level assembly languages can also be used to write software, which strongly correlates with the computer's machine language instructions and is converted into machine language via an assembler [62 66].

The primary software used in AM is as follows:

There are extensive requirements for software in AM. They are required as software used in scanning and imaging, software for inspection, software in product modelling, analysis, simulation, and converting design files to printable files, software before making the actual printing, software supporting 3D/4D/5D/hybrid printing, software supporting postprocessing, software for machine calibration, and many more areas that need software. In addition, workflow and fleet management (MES/CMMS) systems are also essential.

As we are using AM in serial production, there is a requirement for software to manage multiple machines, and the process is complex. orders. Then we need manufacturing execution systems (MES)—platforms like Advanced Manufacturing or Authentise to track the 'digital thread' from order to delivery, managing schedules and quality control.

Maintenance and monitoring: Fabrico or OctoPrint monitors printer health in real-time, tracking 'print hours' to trigger automated maintenance for sensitive components like laser windows.

However, only a few software tools are listed for the 3D Scanning area, to give a sense of the requirements and the availability.

4.12.2 3D designing software

Major 3D designing software commercially available includes TinkerCAD, Fusion 360°, FreeCAD, BlocksCAD, Creo, CATIA, SolidWorks, AutoCAD, OpenSCAD, and Rhino.

4.12.3 3D scanning software

Major 3D scanning software available includes Clin 3D, Metashape, 3DF Zephyr, Inspect plus, RealityCapture, ReCap Pro, Mushroom, VisualSFM, Itseez3D, Scann3D, 3D Creator, and other support software.

3D printing software availability includes Cura, Prusa, Slic3r, 3DPrinterOS, MatterControl 2.0, SliceCrafter, ISL, OctoPrint, AstroPrint, 3D-Tool Free Viewer, Meshmixer, MeshLab, and Netfabb.

4.13 Advantages of adopting additive manufacturing

Significant advantages of AM are as follows [5,10,67]:

- Improved product design and development process.
- Enhanced customisation.
- Boost system productivity.
- New and easy design, lower design lead time.
- Provides chances for innovation.
- Increase the flexibility of production.
- Lower inventory costs.
- Contribute to the competitiveness of the industrial economy.
- Lowering total operational expenses.
- Lesser material Waste with efficient use of resources.
- Increased part dependability.

4.14 Roles of additive manufacturing for Industry 4.0

Industry 4.0 is an industrial revolution of smart, networked production becoming increasingly prevalent. This is useful for customisation, which increases innovation, efficiency, profitability, performance, and enhanced safety management. The primary goals of this revolution can be easily achieved with this technology. Automation is used in Industry 4.0 through the employment of AM systems [68–72]. The following are the critical roles of AM in Industry 4.0:

4.14.1 Prototyping

Prototypes/models may be built from CAD 3D files generated with various design software/3D scanners/CT/MRI, and printed using 3D printing technologies. This technology has a remarkable ability to bring items to market fast. AM prototypes may test concepts or processes before entering full production. It is a functional part of the intended design and can be easily manufactured.

4.14.2 Increase profit

Manufacturers benefit from value-added operations that increase the profitability of products and services. It motivates the organisation to perform effectively and maximise profit margins by improving facility utilisation. This enables software and technologies to accelerate the research and development of products, tools, and gadgets.

4.14.3 Higher level of customisation

In Industry 4.0, AM is widely used to produce personalised, customised products. It can quickly move toward digitisation and customisation using computer-aided manufacturing. This focuses on product innovation to improve efficiency and the product lifecycle. It enhances a new era of intelligent manufacturing that unlocks business growth opportunities. This enables businesses to meet every customer's expectations by manufacturing new, customised products on demand.

4.14.4 Product design and development

AM enables product design and development, resulting in better products in the market. This reduces potential risks before launching a new product in the market. AM meets market needs and increases profit by designing and manufacturing prototypes and the entire product. This technology has the potential to turn ideas into reality by bringing the product to market.

4.14.5 Reduce manufacturing time

Customised goods with the requisite strength are produced more quickly and at lower cost using this technology. Developing a conceptual model can accelerate the research and development process. Printing speed may be enhanced by increasing layer thickness. Any product can be manufactured with the required shape and material, given CAD input.

4.14.6 Digital inventory storage

In a traditional manufacturing system, inventory holding costs significantly increase the product's final cost. Using an AM platform, inventory is stored digitally as a 3D CAD file. Virtual inventory reduces supply chain costs. This eliminates the need for vendors, as the product is manufactured in a single step.

4.14.7 Recycling of material

Some powder-based 3D printing technologies can be implemented in Industry 4.0, where the powder is easily recycled. The product's final cost decreases due to lower raw-material losses. These technologies are environmentally sustainable.

4.14.8 Improves flexibility

Industry 4.0 will be popular due to its flexibility in product design and manufacturing. Now, design and manufacturing are more flexible thanks to AM technologies. Some 3D printing technologies also offer material flexibility, allowing the input materials to be changed. This improves the overall performance of the production system.

4.14.9 Improve the supply chain's efficiency

There is no need to pick a retailer who delivers goods to the consumer. This reduces supply chain costs for on-demand items in engineering, automotive, medical, and dental applications, since the product is delivered directly to the customer via AM.

4.14.10 Higher customer satisfaction

It has the potential to meet customer requirements more quickly. Customised products created by AM in less time strengthen the product's market reputation and improve customer satisfaction.

4.14.11 Lowering the risk

In Industry 4.0, AM reduces risk across all operations in global manufacturing enterprises. New products are manufactured quickly and can be tested before entering final production. It can work with customers and suppliers to strengthen relationships and satisfaction.

4.14.12 Improves accuracy

AM may be used to print metal, organic, and biocompatible materials with excellent accuracy and surface finish. AM uses a lot of powder material as input to enhance the accuracy of the results. The product is manufactured accurately as per the specified shape and size.

4.14.13 Enhance productivity

AM efficiently increases productivity by converting input into the required output through appropriate technology selection. It improves productivity by consuming less material, energy, and workforce. FDM 3D printing uses multiple nozzles to print multiple materials, increasing productivity. There is a lower chance of making mistakes during design and manufacturing.

4.15 Roles of digital twin for additive manufacturing

Using digital twins in the AM sector serves several purposes, including accelerating time-to-market and enabling more flexible manufacturing. Before design concepts are printed, they can be tested and improved using virtual prototypes and digital twins. It makes it easier to spot potential issues with the materials and the construction's structural soundness, allowing them to be addressed before the product is printed. This provides real-time monitoring, enabling the design to be optimised and modified based on actual data. The performance of 3D printed components is precisely measured using digital twinning. As a result, maintenance problems can be found at their source. To address any possible issues before they affect production, parts can be changed or made more resilient. This lowers maintenance expenses and downtime while increasing process reliability [73–76]. The major roles of digital twin for AM are as follows:

- Before the actual production, a digital twin simulates the layer-by-layer material deposition and builds a virtual model of the AM process.
- It virtually confirms the component's shape, structural strength, and dimensional accuracy.
- The digital twin examines and adjusts important process variables like laser power, layer thickness, scanning speed, and temperature, which helps to increase process efficiency, lower faults, and improve product quality.
- This tool forecasts the characteristics of materials, which guarantee that the finished part will have the appropriate mechanical properties
- It continuously tracks temperature, melt pool behaviour, and deposition accuracy by combining sensor data, which makes it possible to make quick corrections while printing.
- To guarantee dimensions and structural accuracy, real-time production data is compared with the original digital model, which improves overall product reliability and uniformity.
- Digital twins keep an eye on machine parts, including motors, lasers, and nozzles, that minimise downtime and avoid unplanned malfunctions.
- By anticipating thermal distortion, the digital twin adjusts the design in advance, which guarantees that the finished printed component will retain precise measurements.
- This examines how much energy is used during the AM process and makes recommendations for enhancements that help to enhance sustainability and lower operating expenses.
- The digital twin assesses material consumption, machine time, and operating costs.
- This establishes the ideal location and number of support structures needed for printing, which makes postprocessing easier and uses less additional material.
- Digital twins help with process optimisation, resource management, and build scheduling, which shortens the lead time for production and increases productivity.
- A digital twin forecasts the printed component's performance over time under actual operating conditions.
- This unifies logistical systems, production data, and design data on a single digital platform.
- It offers a secure online setting for operators to gain knowledge on equipment maintenance and troubleshooting.

4.16 Upcoming goals of additive manufacturing

The primary upcoming goals of AM are as follows [9,12,77]:

- In the future, AM will drive technical progress in design and manufacturing, as well as make a significant contribution to Industry 4.0's self-optimising production system.
- AM aids in reducing unused inventory and the fulfilment of customer- and market-specific demands.
- AM will quickly construct a customised prototype to fulfil any market demand.
- Industry 4.0 will enhance its progress in manufacturing through AM.
- Production, engineering, material, life cycle, and supply chain management will all benefit from AM.
- This technology will create a consumer-oriented, flexible market that can quickly and efficiently satisfy the unique requirements of customers.
- It will lead to advancements in research and development of any industrial part.
- With the aid of AM, Industry 4.0 will achieve a new manufacturing level of productivity and optimisation of the global industrial system.

4.17 Areas of concern with the adoption of additive manufacturing

The significant areas of concern with AM adoption are challenges for the researchers, industry, and we need research, development, and commercialisation here [77,78–80]:

- Materials options are limited.
- Build size restrictions.
- There will be no custom alloying.
- Postproduction and postprocessing are done.
- Job losses in the manufacturing sector, and upskilling is required.
- Need for better surface finish.
- Inaccuracies in the final part specifications.
- Requirement for skilled labour.
- Cost of a part for a mass manufacturing system.
- Generally, a lower strength of the part is printed.
- High cost of software, hardware, and materials.
- Slow process for a mass manufacturing system.
- Limited success in bioprinting.

4.18 Conclusion

By integrating 3D scanning, 3D printing, and various design, printing, and support software, AM is expected to contribute significantly to the implementation of Industry 4.0. It satisfies a variety of Industry 4.0 objectives, including personalisation, efficiency, speedy delivery, and waste reduction. AM has a solid capacity to create complex parts that meet the customer's specific needs while also reducing time and cost. It employs technologies that use various materials as inputs, including plastics, ceramics, powders, and metals. AM is a market-opening combination of flexibility and cost-effectiveness, improving design freedom and enabling new development opportunities.

Industry 4.0 will usher in a new era in manufacturing, introducing new manufacturing methods and creative design concepts. Using the AM system, it makes it easier to introduce new products with specific features at a reduced cost. Production can become more self-sufficient in producing individual components thanks to the application of AM in industry. Additionally, it reduces internal transportation and inventory costs. AM can help organisations to become more proactive, predictive, and responsive to some issues. It focuses on increasing production while lowering risk. Industries benefit from AM's greater flexibility across a wide range of product types. This revolution in manufacturing will improve flexibility, higher quality, productivity, and mass customisation through AM.

References

[1] U.M. Dilberoglu, B. Gharehpapagh, U. Yaman, M. Dolen, The role of additive manufacturing in the era of industry 4.0, Procedia Manufacturing 11 (2017) 545–554.

[2] J. Butt, Exploring the interrelationship between additive manufacturing and industry 4.0, Designs 4 (2020) 13.

[3] A. Haleem, M. Javaid, Additive manufacturing applications in industry 4.0: a review, Journal of Industrial Integration and Management 4 (2019) 1930001.

[4] M.E. Hernandez Korner, M.P. Lambán, J.A. Albajez, J. Santolaria, L.D.C. Ng Corrales, J. Royo, Systematic literature review: integration of additive manufacturing and industry 4.0, Metals 10 (2020) 1061.

[5] M. Mehrpouya, A. Dehghanghadikolaei, B. Fotovvati, A. Vosooghnia, S.S. Emamian, A. Gisario, The potential of additive manufacturing in the smart factory industrial 4.0: a review, Applied Sciences 9 (2019) 3865.

[6] M. Khorasani, J. Loy, A.H. Ghasemi, E. Sharabian, M. Leary, H. Mirafzal, et al., A review of industry 4.0 and additive manufacturing synergy, Rapid Prototyping Journal 28 (2022) 1462–1475.

[7] B. Bigliardi, E. Bottani, E. Gianatti, L. Monferdini, B. Pini, A. Petroni, Sustainable additive manufacturing in the context of industry 4.0: a literature review, Procedia Computer Science 232 (2024) 766–774.

[8] G. Prashar, H. Vasudev, D. Bhuddhi, Additive manufacturing: expanding 3D printing horizon in Industry 4.0, International Journal on Interactive Design and Manufacturing (IJIDeM) 17 (2023) 2221–2235.

[9] H. Gaub, Customisation of mass-produced parts by combining injection moulding and additive manufacturing with industry 4.0 technologies, Reinforced Plastics 60 (2016) 401–404.

[10] S. Dehghan, S. SattarpanahKarganroudi, C. Echchakoui, N. Barka, The integration of additive manufacturing into industry 4.0 and industry 5.0: a bibliometric analysis (trends, opportunities, and challenges), Machines 13 (2025) 62.

[11] A. Kumar, G. Singh, R.P. Singh, P.M. Pandey, Role of additive manufacturing in industry 4.0 for maintenance engineering, Applications and Challenges of Maintenance and Safety Engineering in Industry 4.0, IGI Global, (2020), pp. 235–254.

[12] J. Patalas-Maliszewska, M. Topczak, A new management approach based on additive manufacturing technologies and industry 4.0 requirements, Advances in Production Engineering & Management 16 (2021).

[13] H.G. Lemu, On the opportunities and limitations of additive manufacturing technology for the Industry 4.0 era, International Workshop of Advanced Manufacturing and Automation, Springer Singapore, Singapore, (2018), pp. 106–113 September.

[14] R. Godina, I. Ribeiro, F. Matos, B. Ferreira, H. Carvalho, et al., Impact assessment of additive manufacturing on sustainable business models in industry 4.0 context: sustainability, 2020, vol. 12, pp. 7066.

[15] M. Rodríguez-Martín, R. Domingo, J. Ribeiro, Mapping and prospective of additive manufacturing in the context of industry 4.0 and 5.0, Rapid Prototyping Journal 30 (2024) 1393–1410.

[16] A. Iqbal, G. Zhao, H. Suhaimi, N. He, G. Hussain, W. Zhao, Readiness of subtractive and additive manufacturing and their sustainable amalgamation from the perspective of industry 4.0: a comprehensive review, The International Journal of Advanced Manufacturing Technology 111 (2020) 2475–2498.

[17] B. Motyl, S. Filippi, Trends in engineering for additive manufacturing in the industry 4.0 era: a systematic literature review, International Journal on Interactive Design and Manufacturing (IJIDeM) 15 (2021) 103–106.

[18] C.F. Bănică, A. Sover, D.C. Anghel, Printing the future layer by layer: a comprehensive exploration of additive manufacturing in the era of industry 4.0, Applied Sciences 14 (2024) 9919.

[19] M. Ahmadi, D. Rahmatabadi, A. Karimi, M.H.A. Koohpayeh, R. Hashemi, The role of additive manufacturing in the age of sustainable manufacturing 4.0, Sustainable Manufacturing in Industry 4.0: Pathways and Practices, Springer Nature Singapore, Singapore, (2023), pp. 57–78.

[20] B.A. Aljabali, J. Shelton, S. Desai, Genetic algorithm-based data-driven process selection system for additive manufacturing in industry 4.0, Materials 17 (2024) 4544.

[21] K. Ishfaq, M.D.A. Khan, M.A.A. Khan, M.A. Mahmood, M.A. Maqsood, A correlation among industry 4.0, additive manufacturing, and topology optimisation: a state-of-the-art review, The International Journal of Advanced Manufacturing Technology 129 (2023) 3771–3797.

[22] D. Guo, S. Ling, H. Li, D. Ao, T. Zhang, Y. Rong, et al., A framework for personalised production based on digital twin, blockchain and additive manufacturing in the context of industry 4.0, 2020 IEEE 16th International Conference on Automation Science and Engineering (CASE), IEEE, (2020), pp. 1181–1186. August.

[23] V.A. Wankhede, S. Vinodh, Analysis of challenges of the wire-arc additive manufacturing process in the context of industry 4.0 using a graph theory approach, The International Journal of Advanced Manufacturing Technology 123 (2022) 1059–1078.

[24] N. Shahrubudin, T.C. Lee, R.J.P.M. Ramlan, An overview on 3D printing technology: technological, materials, and applications, Procedia Manufacturing 35 (2019) 1286–1296.

[25] G. Taormina, C. Sciancalepore, M. Messori, F. Bondioli, 3D printing processes for photocurable polymeric materials: technologies, materials, and future trends, Journal of Applied Biomaterials & Functional Materials 16 (2018) 151–160.

[26] A. Jandyal, I. Chaturvedi, I. Wazir, A. Raina, M.I.U. Haq, 3D printing–a review of processes, materials and applications in industry 4.0, Sustainable Operations and Computers 3 (2022) 33–42.

[27] I. Karakurt, L. Lin, 3D printing technologies: techniques, materials, and post-processing, Current Opinion in Chemical Engineering 28 (2020) 134–143.

[28] R. Ranjan, D. Kumar, M. Kundu, S.C. Moi, A critical review on classification of materials used in 3D printing process, Materials Today: Proceedings 61 (2022) 43–49.

[29] M. Ntousia, I. Fudos, 3D printing technologies & applications: an overview. In: Proceedings of the CAD 2020 Conference, Singapore, 2019 June, vol. 243, p. 248.

[30] J.Y. Lee, J. An, C.K. Chua, Fundamentals and applications of 3D printing for novel materials, Applied Materials Today 7 (2017) 120–133.

[31] M.M. Prabhakar, A.K. Saravanan, A.H. Lenin, K. Mayandi, P.S. Ramalingam, A short review on 3D printing methods, process parameters and materials, Materials Today: Proceedings 45 (2021) 6108–6114.

[32] A.E. Jakus, An introduction to 3D printing—past, present, and future promise, 3D Printing in Orthopaedic Surgery, Elsevier, (2019), pp. 1–15.

[33] P. Jagadeesh, M. Puttegowda, S.M. Rangappa, K. Alexey, S. Gorbatyuk, A. Khan, et al., A comprehensive review on 3D printing advancements in polymer composites: technologies, materials, and applications, The International Journal of Advanced Manufacturing Technology 121 (2022) 127–169.

[34] S. Park, W. Shou, L. Makatura, W. Matusik, K.K. Fu, 3D Printing of polymer composites: materials, processes, and applications, Matter 5 (2022) 43–76.

[35] S.F. Iftekar, A. Aabid, A. Amir, M. Baig, Advancements and limitations in 3D printing materials and technologies: a critical review, Polymers 15 (2023) 2519.

[36] D. Srinivasan, M. Meignanamoorthy, M. Ravichandran, V. Mohanavel, S.V. Alagarsamy, C. Chanakyan, et al., 3D printing manufacturing techniques, materials, and applications: an overview, Advances in Materials Science and Engineering 2021 (2021) 5756563.

[37] E. MacDonald, R. Wicker, Multiprocess 3D printing for increasing component functionality, Science 353 (6307) (2016) aaf2093.

[38] M. Jiménez, L. Romero, I.A. Domínguez, M.D.M. Espinosa, M. Domínguez, Additive manufacturing technologies: an overview about 3D printing methods and prospects, Complexity 2019 (2019) 9656938.

[39] A.M. Arefin, N.R. Khatri, N. Kulkarni, P.F. Egan, Polymer 3D printing review: materials, process, and design strategies for medical applications, Polymers 13 (2021) 1499.

[40] A. Jadhav, V.S. Jadhav, A review on 3D printing: an additive manufacturing technology, Materials Today: Proceedings 62 (2022) 2094–2099.

[41] T. Singh, S. Kumar, S. Sehgal, 3D Printing of engineering materials: a state of the art review, Materials Today: Proceedings 28 (2020) 1927–1931.

[42] T. Mateti, S. Jain, L.A. Shruthi, A. Laha, G. Thakur, An overview of the advances in the 3D printing technology, 3D Printing Technology for Water Treatment Applications (2023) 1–37.

[43] L.Y. Zhou, J. Fu, Y. He, A review of 3D printing technologies for soft polymer materials, Advanced Functional Materials 30 (2020) 2000187.

[44] Y. Song, Y. Ghafari, A. Asefnejad, D. &Toghraie, An overview of selective laser sintering 3D printing technology for biomedical and sports device applications: processes, materials, and applications, Optics & Laser Technology 171 (2024) 110459.

[45] J. Nyika, F.M. Mwema, R.M. Mahamood, E.T. Akinlabi, T.C. Jen, Advances in 3D printing materials processing, environmental impacts, and alleviation measures, Advances in Materials and Processing Technologies 8 (sup3) (2022) 1275–1285.

[46] M. Pagac, J. Hajnys, Q.P. Ma, L. Jancar, J. Jansa, P. Stefek, et al., A review of vat photopolymerisation technology: materials, applications, challenges, and future trends of 3D Printing, Polymers 13 (2021) 598.

[47] H. Zhai, X. Li, S. Yu, J. Wang, Y. Chang, J. Li, et al., Review on the 3D printing technology and application of magnetic materials: material-process-structure-application, Composites Part B: Engineering 298 (2025) 112387.

[48] Y. Zheng, X. Huang, J. Chen, K. Wu, J. Wang, X. Zhang, A review of conductive carbon materials for 3D printing: materials, technologies, properties, and applications, Materials 14 (2021) 3911.

[49] Y. Wang, T. Wu, G. Huang, State-of-the-art research progress and challenges in printing techniques, and potential applications of 3D Printing for advanced ceramic materials, Materials Today Communications 40 (2024) 110001.

[50] A. Haleem, M. Javaid, R.P. Singh, S. Rab, R. Suman, L. Kumar, et al., Exploring the potential of 3D scanning in industry 4.0: an overview, International Journal of Cognitive Computing in Engineering 3 (2022) 161–171.

[51] K. Bartol, D. Bojanić, T. Petković, T. Pribanić, A review of body measurement using 3D scanning, IEEE Access 9 (2021) 67281–67301.

[52] A. Kantaros, T. Ganetsos, F.I.T. Petrescu, Three-dimensional printing and 3D scanning: emerging technologies exhibiting high potential in the field of cultural heritage, Applied Sciences 13 (2023) 4777.

[53] M.A.B. Ebrahim, 3D laser scanners' techniques overview, International Journal of Scientific Research 4 (2015) 323–331.

[54] T.M. Onyia, I.A. Olarinoye, S.A. Jimoh, Advancements and challenges in 3D scanning: a comprehensive review of engineering applications, African Journal of Advances in Science and Technology Research 18 (2025) 191–206.

[55] K. Kamani, Types of 3D Scanning Technologies and 3D Scanners (2020). Available from https://www.linkedin.com/pulse/types-3d-scanning-technologies-scanners-karan-kamani/.

[56] N. Farahani, A. Braun, D. Jutt, T. Huffman, N. Reder, Z. Liu, et al., Three-dimensional imaging and scanning: current and future applications for pathology, Journal of Pathology Informatics 8 (2017) 36.

[57] Capture3D, Best uses of 3D scanning and its applications. Available from: https://www.capture3d.com/knowledge-center/blog/best-uses-of-3d-scanning-software.

[58] Printlab, The real value of 3D scanning and its applications. Available from: https://weareprintlab.com/blog/the-real-value-of-3d-scanning-and-its-applications.

[59] C.A. Hu, L. Kong, F. &Lv, Application of 3D laser scanning technology in engineering field, E3S Web of Conferences, EDP Sciences, (2021), p. 04014 vol. 233.

[60] K.T. Chang, J.R. Chang, J.K. Liu, Detection of pavement distresses using 3D laser scanning technology, Computing in Civil Engineering (2005) 1–11.

[61] R.M. Ruiz, M.T.M. Torres, P.S. Allegue, Comparative analysis between the main 3D scanning techniques: photogrammetry, terrestrial laser scanner, and structured light scanner in religious imagery: the case of the holy Christ of the Blood, ACM Journal on Computing and Cultural Heritage (JOCCH) 15 (2021) 1–23.

[62] A. Wiberg, J. Persson, J. &Ölvander, Design for additive manufacturing–a review of available design methods and software, Rapid Prototyping Journal 25 (2019) 1080–1094.

[63] L. Yi, S. Ehmsen, M. Glatt, J.C. Aurich, Modeling and software implementation of manufacturing costs in additive manufacturing, CIRP Journal of Manufacturing Science and Technology 33 (2021) 380–388.

[64] I. Gibson, D. Rosen, B. Stucker, M. Khorasani, Software for additive manufacturing, Additive Manufacturing Technologies, Springer International Publishing, Cham, (2020), pp. 491–524.

[65] B. Vayre, F. Vignat, F. Villeneuve, Designing for additive manufacturing, Procedia CIRP 3 (2012) 632–637.

[66] S. Anand, M.K. Satyarthi, Exploring the role of additive manufacturing in industry 4.0: a review of applications and advancements, Journal of Basic and Applied Engineering Research 10 (2023) 85–90.

[67] D. Patel, S. Vinodh, Analysis of technologies enabling additive manufacturing and industry 4.0 integration for SMEs using MCDM tool, International Journal of Process Management and Benchmarking 16 (2024) 281–295.

[68] M.A. Rahman, M.S. Shakur, M.S. Ahamed, S. Hasan, A.A. Rashid, M.A. Islam, et al., A cloud-based cyber-physical system with Industry 4.0: remote and digitised additive manufacturing, Automation 3 (2022) 400–425.

[69] J. Francis, L. Bian, Deep learning for distortion prediction in laser-based additive manufacturing using big data, Manufacturing Letters 20 (2019) 10–14.

[70] G. Schuh, M. Salmen, C. Kelzenberg, J. de Lange, Integration of tool making into agile product development using industry 4.0 technologies and additive manufacturing technologies, 2018, Portland

International Conference on Management of Engineering and Technology (PICMET), IEEE, (2018), pp. 1–9. August.

[71] V.V. Bhandarkar, B. Das, P. Tandon, Synergising industry 4.0 technology with additive manufacturing to enhance quality assurance and enable additive manufacturing as a service industry, Proceedings of the ASME 2024 International Mechanical Engineering Congress and Exposition. Volume 2: Advanced Manufacturing. Portland, Oregon, USA. November 17–21, 2024. V002T03A044, American Society of Mechanical Engineers, (2024). Available from https://doi.org/10.1115/IMECE2024-144967.

[72] L. Zhang, X. Chen, W. Zhou, T. Cheng, L. Chen, Z. Guo, et al., Digital twins for additive manufacturing: a state-of-the-art review, Applied Sciences 10 (2020) 8350.

[73] L. Jin, X. Zhai, K. Wang, K. Zhang, D. Wu, A. Nazir, et al., Big data, machine learning, and digital twin-assisted additive manufacturing: a review, Materials & Design 244 (2024) 113086.

[74] A. Phua, C.H.J. Davies, G.W. Delaney, A digital twin hierarchy for metal additive manufacturing, Computers in Industry 140 (2022) 103667.

[75] C. Liu, L. Le Roux, C. Körner, O. Tabaste, F. Lacan, S. Bigot, Digital twin-enabled collaborative data management for metal additive manufacturing systems, Journal of Manufacturing Systems 62 (2022) 857–874.

[76] I. Tartici, Z.M. Kilic, P. Bartolo, A systematic literature review: industry 4.0-based monitoring and control systems in additive manufacturing, Machines 11 (2023) 712.

[77] A. Charles, M. Bayat, A. Elkaseer, S. Scholz, Simulation in additive manufacturing and its implications for sustainable manufacturing in the era of industry 4.0, International Conference on Sustainable Design and Manufacturing, Springer Nature Singapore, Singapore, (2022), pp. 370–377 September.

[78] A. Suresh, R. Udendhran, G. Yamini, Internet of things and additive manufacturing: toward intelligent production systems in industry 4.0, Internet of Things for Industry 4.0: Design, Challenges and Solutions, Springer International Publishing, Cham, (2019), pp. 73–89.

5

Internet of Things for Industry 4.0

5.1 Introduction to Internet of Things

The Internet of Things (IoT), commonly known as IoT, is the network of interconnected devices that use the Internet. Depending on the usage and other peripherals connected to it, IoT can behave independently or dependently. This technology is a giant leap in Industry 4.0, enabling smart monitoring of the complete production system. Industry 4.0 relies heavily on the IoT [1–3]. It has a wide range of applications in the industrial and service industries for monitoring production systems. This technology offers new and inventive industrial opportunities by permitting increased performance. Robotic arms, synchronised robotics, industrial machinery, accuracy analysis, delivery logistics management, security, and other applications may all benefit from IoT. In Industry 4.0, this technology is cost-effective and improves communication levels [4,5].

IoT's main feature is the ability to collect and distribute data via internet-connected devices. It is linked to unique identifying numbers or codes that can be manipulated on devices like smartphones every day. The primary components of this technology are software, hardware, and network connectivity. In manufacturing, IoT has fostered disruptive innovation. Industries have observed increased efficiency in product manufacturing through the effective deployment of this technology [6–8]. For an industrial or consumer sector, the major parts of an IoT device are as follows:

- A computer gadget with an arm (Raspberry Pi, Arduino, or Beagle Board).
- Connecting ports/pins for multiinterface with various devices.
- Motors and actuators (steppers, brushless, etc.) with detectors (wireless, gyro, temp, water quality, pressure, chemical, IR, image, etc.).
- An adaptor or brick for power distribution.

5.1.1 Internet of Things and its components

The IoT is a collection of tiny gadgets connected to the Internet. The IoT extends the idea of the 'internet' beyond most people's use cases. These IoT devices collect and share data with other systems, apps, and devices; the data 'talks' to 'us' and other objects [9–11]. It is linked in a number of ways. IoT devices create sensory, biotelemetry, and other forms of data, from wearables to industrial sensors, making users accessible and linked to one another because of the Internet. Users may access their email or social media accounts online, allowing them to interact and exchange photos [12,13]. People have traditionally used gadgets to position, gather information, and convey messages from one location to another. Every industry will benefit from connecting its unique devices to the IoT to acquire the data needed to make better decisions about what to do and, more importantly, when to act [14–16].

Digital Twin and Industry 4.0. DOI: https://doi.org/10.1016/B978-0-443-54019-6.00018-2

Connecting a gadget to the Internet and gaining additional value from it will be straightforward. Individuals are more likely to be inventive because they incur lower failure costs. The IoT can include everything from spinning doors to subway turnstiles. While spraying crops with pesticides, small sensors may be mounted to tractors to identify which plants require treatment. Other instruments can be inserted into the soil to monitor its moisture and temperature. IoT-connected devices will continue to advance, allowing for increasingly complex applications [17–19].

Small computers will be able to do rudimentary computational tasks. Instead of sending a signal to a central server and waiting for a response, IoT devices can compute when a signal is detected and make decisions on the spot. Because less information is transferred back and forth between parties, security benefits have grown [20–24]. Fig. 5.1 shows the major components of the IoT system.

5.1.1.1 Sensors

Sensors initially gather information from their environment. This data could be anything from a live video to an actual temperature reading. We use the phrase 'sensors/devices' because many sensors may be combined or sensors may be a component of a device that performs functions beyond simple perception. Our smartphones can do a lot of things, but they also have a lot of sensors (camera, accelerometer, global positioning system (GPS), etc.).

5.1.1.2 Connectivity

After that, the data is sent to the cloud, but not until it finds a way to get there. Low-power wide-area networks, cellular, satellite, WiFi, Bluetooth, gateway/router connections, or direct Internet access. Each option compromises bandwidth, range, and power consumption. The IoT application determines the best connectivity option, but they all achieve the same objective: moving data to the cloud.

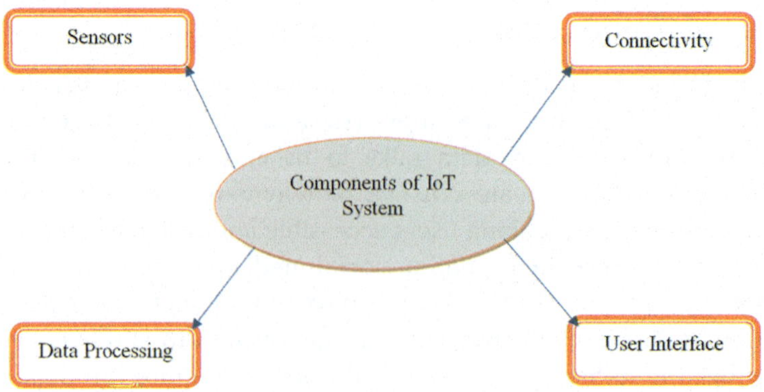

FIGURE 5.1 Significant components of the IoT system. *IoT*, Internet of Things.

5.1.1.3 Data processing

Once in the cloud, the data is processed by software. Making sure the temperature reading is within acceptable bounds may be all that is required. It might also be complicated, like computer vision-based object identification in a video.

5.1.1.4 User interface

After that, the data is somehow rendered usable for the final user. When the temperature in the company's cold storage is too high, a user may receive an SMS alert. An interface allows users to check the system beforehand. A customer can watch the video feeds on different properties using a web browser or a phone app. Depending on the IoT application, the user can be able to affect the system and carry out an action. Additionally, some actions are performed automatically. The system may use pre-programmed criteria to adjust the temperature rather than waiting.

5.2 Need for Internet of Things

The IoT is a computer concept that describes ordinary physical objects connecting to the Internet and recognising other devices. This would imply that common home goods would be connected with sophisticated technology to perform additional functions. IoT is sometimes known as 'smart' technology. It is now in its infancy, with particular gadgets such as smart TVs, smart refrigerators, and smart entry locks. Driverless automobiles and personal assistant technologies will become the norm sooner [25–27]. Sensors in these devices transmit data to a computer or software, enabling them to perform critical tasks. The number of devices connected to the Internet has increased due to their ease of use and high degree of automation. IoT is a vital technology that will only get better over time [28–31]. The major needs of IoT, through its comprehensive applications, are as follows:

5.2.1 Better decision-making

These gadgets may capture a vast quantity of data with the addition of sensors. In addition to the practical benefit of knowing which goods in an intelligent refrigerator are about to expire, this improved home appliance will also provide additional data on power usage, temperature, average time spent open, and more. As more data flows in, the firm behind the gadget may analyse significant trends and improve the device's functionality. Many businesses are recognising the value of this technology, leading to significant market expansion. The entire market value of IoT is, in reality, predetermined.

5.2.2 For tracking and monitoring

It not only helps a corporation track data, but it also benefits the user. These gadgets could monitor the current quality of household goods. Knowing the condition of belongings will help determine when to replace them without constantly examining quality.

5.2.3 To reduce workload

Having a gadget handle most of the job may save time and money. This decreases human labour significantly. It also leads to the development of gadgets that require less human involvement and can operate independently.

5.2.4 To increase efficiency by saving money and physical resources

It can also save money on the gadget in addition to saving time. This networked device may be used in a variety of ways. Machine-to-machine communication is encouraged by the IoT system, resulting in greater long-term efficiency for both the organisation and the user. Machine-to-machine communication is growing rapidly.

5.2.5 To improve the quality of life

This technology adds up to a higher quality of life. It is challenging to keep track of everything with so many gadgets being manufactured and new technologies being applied. This technology is fantastic for doing things as we like while a computer handles monotonous tasks. Regular exercisers can use wearable technology to monitor their heart rate, body temperature, and hydration levels to stay in shape and track their health.

5.3 Advantages of deploying Internet of Things

The major advantages of IoT are as follows [32–34]:

- Enhanced automation.
- Improved machine connection and production workflow.
- Intelligent item and device tracking.
- Constant monitoring of complete manufacturing processes.
- Lowering production costs.
- Better inventory control.
- Enhanced workplace safety and comprehensive safety.
- Increased operational effectiveness and productivity.
- Lowers the number of system faults.
- Improved reliability in forecasting.
- Improved quality assurance and control.
- Time-saving.

5.4 Roles of Internet of Things for implementing Industry 4.0

Using a virtual management system, IoT assists with real-time information. This technology detects any manufacturing fault. The data collected by this technology helps inform decision-making. IoT can remarkably foresee impending situations with a suitable monitoring system.

This technique is effectively employed in manufacturing to implement cost-effective production of essential parts. It intelligently manages all processes and ensures they are adequately supervised to preserve product quality [35–40]. The major roles of IoT in the effective implementation of Industry 4.0 are as follows:

5.4.1.1 Obtain appropriate information

This technology captures data in real time via internet-connected gadgets. It is used to reduce data collection time, energy use, and costs. IoT gathers information from a variety of commercial devices and other tracking systems. This is useful for gathering raw data that may be utilised for complicated analysis in Industry 4.0. It gains vital knowledge and makes informed production decisions going forward. The information gathered can be utilised to make long-term decisions.

5.4.1.2 Smart surveillance

IoT is used to smartly monitor various factors, such as humidity, temperature, and other environmental conditions, during production. It helps provide information in the event of a production system fault. This technology detects the damage location while maintaining proper manufacturing parameters. It is also used to monitor workers' health in real time. This technology provides supervisory control and can easily monitor the machining parameters.

5.4.1.3 Management of logistics in the warehouse

The IoT helps manage warehouses and inventory tracking. This technology is utilised in warehouse management on a day-to-day basis. It is used to boost warehouse production without the need for human monitoring. This supports warehouse decision-making and a smart warehouse management system. It intelligently and effectively monitors the storage platform and distributes labour to ensure a smooth production system.

5.4.1.4 Proper measurement of energy

It may measure energy and has applications in smart appliances, smart metres, renewable energy, and other energy-efficient technologies. It is helpful for intelligent monitoring, control, and communication, enabling better connectivity in the manufacturing plant. IoT is used in manufacturing plants for electricity generation, control, and proper power distribution. This assists in optimising the operation of the Industry 4.0 system.

5.4.1.5 Autonomous production system

IoT is being actively used to enable autonomous production. This technology is helping to develop self-driving cars. It is used to track the records and locations of integrated vehicles. It is being used to provide an innovative facility throughout the entire production process. This disruptive technology is becoming prominent for next-generation vehicles. Industry 4.0 improves the present service quality and supply security through IoT.

5.4.1.6 Management of materials

IoT appears to help manage all materials during ongoing production. This provides all the details regarding the inventory required for manufacturing. It can easily display real-time inventory updates with proper monitoring on a smartphone. This provides meaningful insights into the status, quality, location, and movement of materials.

5.4.1.7 Improve the working skills of workers

Through automation, the IoT improves worker skills and knowledge. It equips educators with the tools to succeed in their methods and performance. This technology aids improper management and provides training, education, and development. It enhances instructors' professional growth by providing them with better and more innovative ways. It provides individualised instruction to the employees through internet-connected devices.

5.4.1.8 Management of the supply chain

IoT is a revolutionary technology that uses the GPS to control and track product delivery. It can monitor product storage and improve the supply chain system. This technology is used to identify a specific location in a warehouse quickly. It allows for improved forecasting of traffic flow and product movement speed. This is useful for proper identification and management of certain items, as well as for route control.

5.4.1.9 Real-time monitoring of individual and system

IoT improves collaboration among employees and devices by enabling real-time monitoring. It helps develop analyses in line with Industry 4.0 criteria. This helps create a comfortable, productive space that improves indoor air quality. This technology helps build security.

5.4.1.10 Robotics with appropriate intelligence

The IoT is the best platform for connecting robots and sensors to the Internet. During manufacturing, intelligent devices monitor and control the entire process. IoT can utilise intelligent robots to perform specific tasks in complex situations. It improves manufacturing flexibility with fewer mistakes. IoT-enabled robots are employed to do specific tasks more efficiently and productively.

5.4.1.11 Development of new products

The IoT is employed in the product development process to help a company understand the challenges of continuous product development. It offers a more comprehensive solution through careful product design. This provides an innovative platform for product development. In Industry 4.0, this technology maximises user productivity by utilising internet-connected gadgets. This maintains the process by leveraging practical knowledge of existing goods to keep them up to date.

5.4.1.12 Simplify typical processes during manufacturing

This technology simplifies a typical manufacturing process. The IoT provides a better option for automated production by connecting all devices to the Internet. This technology uses cell phones, PCs, and cloud data storage for industrial automation. It can easily organise manufacturing data, allowing information to flow freely. Industry 4.0 provides a higher-quality and higher-performance computing platform.

5.4.1.13 Increase machine connectivity

IoT-based cloud systems increase connectivity with machines and devices, resulting in excessive communication. Due to its digital connectivity, it plays a critical role in Industry 4.0. This allows physical items to work together and communicate intelligently without human involvement. It gathers data from various sources and processes it using a cloud network.

5.4.1.14 Implementing a digital service environment

It is quickly expanding for various digital services that can be watched and managed remotely. IoT is used to build smart cities through various internet-connected devices. This technology enables smart parking and management. It provides improved opportunities to reduce pollution, improve traffic management, and keep citizens clean and safe. It can appropriately govern smart cities by utilising the central cloud platform.

5.5 Significant roles of digital twin for Internet of Things

Businesses can gain many advantages by integrating digital twins with IoT. The capacity to enhance maintenance and repair capabilities is among the biggest benefits. Businesses can precisely schedule preventive maintenance, track equipment performance, and spot potential issues before they arise with the help of real-time data from IoT sensors. The data can be received by the digital twin, which processes and analyses it in real time. Early problem detection enables companies to plan maintenance and repairs during scheduled downtime, minimising the impact on production schedules and reducing repair costs. Digital twins also enable remote monitoring and diagnostics, allowing technicians to fix faults and spot problems from any location in the world [41,42]. Businesses can use digital twins to build a virtual model of their production process and test it for errors or hazards in advance. This reduces the risk of costly, time-consuming production errors by helping engineers identify issues and develop mitigation strategies [43–45]. The significant roles of digital twins for IoT are as follows:

- It uses a virtual representation of physical assets to visualise real-time IoT sensor data.
- This makes it possible to remotely monitor IoT devices without being physically there.
- Helps with ongoing IoT data streams to forecast device failures.
- This helps to identify irregularities and strange activity in IoT networks.
- It consolidates information from several IoT devices into a single digital model.
- Models the behaviour of IoT systems in various operational scenarios.
- Improves security by spotting odd IoT network activity.
- Facilitates IoT system automation and intelligent control.
- Enhances the use of resources by all linked devices.
- Uses real-time IoT insights to help decision-making.
- Uses IoT data to monitor environmental conditions.
- Lowers maintenance and operating expenses.
- Improves client satisfaction with tailored IoT services.

5.6 Upcoming goals of Internet of Things

The primary upcoming goals of IoT are perceived as follows [46–49]:

- IoT will collect all detailed information in industries in the future to make production more efficient.
- It may be used to monitor and support the digital production system continually.
- With the use of smart sensing devices, IoT will become more efficient in Industry 4.0.
- This technology will successfully support all management procedures in the manufacturing line.
- Machines, inventories, and transportation systems will be easily monitored by using this technology.
- This technology will significantly influence industrial businesses' future appropriateness and connectedness.

5.7 Challenges with Internet of Things adoption

The significant challenges in adopting IoT are as follows [20,50–52]:

- The difficulty of storing information through internet-connected gadgets.
- When there is a lack of connectivity in a complex system, it might fail
- Internet dependency.
- A decrease in both mental and physical activities and loss of control of life.
- Improved data security requirements.
- Requires various hardware setups and connections.
- Possibility of software hacking.
- It requires more time and money when implemented.
- There are no worldwide standards or platform management.
- Leading to unemployment.

5.8 Conclusion

The IoT provides a new era in manufacturing innovation. Supporting a digitally regulated management system reduces risks and enhances overall performance. It can rapidly report associated difficulties during manufacturing using sensor applications. With adequate monitoring throughout product manufacture, IoT positively impacts Industry 4.0. IoT links many devices or equipment over the Internet in Industry 4.0. This captures a significant amount of data and shares it meaningfully, increasing automation in Industry 4.0. It increases overall manufacturing efficiency while lowering human mistake rates. This technology enables the manufacturing facility to operate autonomously with reduced human involvement.

The IoT has emerged as a powerful and popular technology for controlling and monitoring equipment usage. It creates a flexible, faster manufacturing line to meet customers' customised needs. This technology is essential for real-time inventory tracking and tracing. The firm may monitor the supply chain and available manufacturing resources with the aid of this technology. During production, the IoT removes the need for manual documentation. This reduces the risk of wasted organisational spending due to poor management. IoT will supply the backbone of production in Industry 4.0, completing a shift in the manufacturing landscape.

In Industry 4.0, this IoT networked platform effectively maintains worker safety and conveniently handles difficult situations. Powerful computing devices are utilised to improve research and development procedures. The application of this technology may meet the specific needs of industries. With increased machine connectivity, it plays a critical role in creating a smart factory. This technology boosts workplace safety, productivity, operational efficiency, and product tracking systems in Industry 4.0.

References

[1] M. Soori, B. Arezoo, R. &Dastres, Internet of Things for smart factories in industry 4.0: a review, Internet of Things and Cyber-Physical Systems 3 (2023) 192–204.

[2] G. Lampropoulos, K. Siakas, T. Anastasiadis, Internet of Things in the context of industry 4.0: an overview, International Journal of Entrepreneurial Knowledge 7 (1) (2019) 4–19. Available from https://doi.org/10.2478/ijek-2019-0001.

[3] M. Wollschlaeger, T. Sauter, J. Jasperneite, The future of industrial communication: automation networks in the era of the Internet of Things and industry 4.0, IEEE Industrial Electronics Magazine 11 (1) (2017) 17–27. Available from https://doi.org/10.1109/MIE.2017.2649104.

[4] P.K. Malik, R. Sharma, R. Singh, A. Gehlot, S.C. Satapathy, W.S. Alnumay, et al., Industrial Internet of Things and its applications in industry 4.0: state of the art, Computer Communications 166 (2021) 125–139. Available from https://doi.org/10.1016/j.comcom.2020.11.016.

[5] K.C. Rath, A. Khang, D. Roy, The role of Internet of Things (IoT) technology in industry 4.0 economy, Advanced IoT Technologies and Applications in the Industry 4.0 Digital Economy, CRC Press, Boca Raton, (2024), pp. 1–28. Available from https://doi.org/10.1201/9781003434269-1.

[6] M.S. Rahman, T. Ghosh, N.F. Aurna, M.S. Kaiser, M. Anannya, A.S. Hosen, Machine learning and Internet of things in industry 4.0: a review, Measurement: Sensors 28 (2023) 100822.

[7] C. Garrido-Hidalgo, T. Olivares, F.J. Ramírez, L. &Roda-Sánchez, An end-to-end Internet of Things solution for reverse supply chain management in industry 4.0, Computers in Industry 112 (2019) 103127.

[8] K. Ding, H. Shi, J. Hui, Y. Liu, B. Zhu, F. Zhang, et al., Smart steel bridge construction enabled by BIM and Internet of Things in Industry 4.0: a framework. in: 2018 IEEE 15th International Conference on Networking, Sensing and Control (ICNSC), 2018, pp. 1-5. IEEE.

[9] S. Attaran, M. Attaran, B.G. Celik, Digital twins and Industrial Internet of Things: uncovering operational intelligence in industry 4.0, Decision Analytics Journal 10 (2024) 100398.

[10] K. Witkowski, Internet of Things, big data, Industry 4.0–innovative solutions in logistics and supply chain management, Procedia Engineering 182 (2017) 763–769.

[11] I.H. Khan, M. Javaid, Role of Internet of Things (IoT) in adoption of industry 4.0, Journal of Industrial Integration and Management 7 (4) (2022) 515–533. Available from https://doi.org/10.1142/S2424862221500068.

[12] K. Marinova-Kostova, I. Kostov, Application of Internet of things in industry 4.0, Economics Ecology Socium 5 (2021) 49–58.

[13] J. Wan, S. Tang, Z. Shu, D. Li, S. Wang, M. Imran, et al., Software-defined industrial Internet of things in the context of industry 4.0, IEEE Sensors Journal 16 (2016) 7373–7380.

[14] M. Aazam, S. Zeadally, K.A. Harras, Deploying fog computing in industrial Internet of Things and industry 4.0, IEEE Transactions on Industrial Informatics 14 (2018) 4674–4682.

[15] R. Kumar, S. Rani, M.A. Awadh, Exploring the application sphere of the Internet of Things in industry 4.0: a review, bibliometric and content analysis, Sensors 22 (2022) 4276.

[16] E. Manavalan, K. Jayakrishna, A review of Internet of Things (IoT) embedded sustainable supply chain for industry 4.0 requirements, Computers &Industrial Engineering 127 (2019) 925–953.

[17] R. Rudenko, I.M. Pires, P. Oliveira, J. Barroso, A. Reis, A brief review of the Internet of Things, industry 4.0 and cybersecurity, Electronics 11 (2022) 1742.

[18] J. Branger, Z. Pang, From automated home to sustainable, healthy and manufacturing home: a new story enabled by the Internet-of-Things and Industry 4.0, Journal of Management Analytics 2 (2015) 314–332.

[19] J.I.R. Molano, J.M.C. Lovelle, C.E. Montenegro, J.J.R. Granados, R.G. Crespo, Metamodel for integration of Internet of Things, social networks, the cloud and industry 4.0, Journal of Ambient Intelligence and Humanized Computing 9 (3) (2018) 709–723. Available from https://doi.org/10.1007/s12652-017-0469-5.

[20] A.S. Gillis, What is the Internet of Things (IoT)? Available from: https://www.techtarget.com/iotagenda/definition/Internet-of-Things-IoT.

[21] M. Ghoreishi, A. Happonen, The case of fabric and textile industry: the emerging role of digitalisation, Internet-of-Things and industry 4.0 for circularity. In: Proceedings of Sixth International Congress on Information and Communication Technology: ICICT 2021, London, 2021, vol. 3, pp. 189–200. Singapore: Springer, Singapore.

[22] P. Radanliev, D. De Roure, K. Page, J.R. Nurse, R. Mantilla Montalvo, O. Santos, et al., Cyber risk at the edge: current and future trends on cyber risk analytics and artificial intelligence in the industrial Internet of Things and industry 4.0 supply chains, Cybersecurity 3 (2020) 13.

[23] How IoT Works – 4 Main Components of IoT System, Available from https://data-flair.training/blogs/how-iot-works/.

[24] N. Khan, W.D. Solvang, H. Yu, Industrial Internet of Things (IIoT) and other industry 4.0 technologies in spare parts warehousing in the oil and gas industry: a systematic literature review, Logistics 8 (2024) 16.

[25] A. Rani, V.K. Mishra, A.K. Mishra, V. Kumar, N.K. Pandey, N.K., Role and significance of Internet of Things (IoT) in industry 4.0. In: 2024 3rd International Conference on Power Electronics and IoT Applications in Renewable Energy and its Control (PARC), 2024, pp. 199-202. IEEE.

[26] A.Y. Alqahtani, S.M. Gupta, K. Nakashima, Warranty and maintenance analysis of sensor-embedded products using Internet of Things in industry 4.0, International Journal of Production Economics 208 (2019) 483–499.

[27] P. Radanliev, D. De Roure, J.R. Nurse, R. Nicolescu, M. Huth, S. Cannady, et al., Integration of cybersecurity frameworks, models and approaches for building design principles for the Internet of Things in industry 4.0. In: Living in the Internet of Things: Cybersecurity of the IoT-2018, 2018, pp. 1-6. IET.

[28] E. Ruiz Zúñiga, A. Syberfeldt, M. Urenda Moris, The Internet of Things, Factory of Things, and industry 4.0 in manufacturing: current and future implementations, Advances in Manufacturing Technology XXXI (2017) 221–226.

[29] B.B. Gupta, A. Tewari, I. Cvitić, D. Peraković, X. Chang, Artificial intelligence empowered email classifiers for Internet of Things-based systems in industry 4.0, Wireless Networks 28 (2022) 493–503.

[30] A. Lekan, A. Clinton, E. Stella, E. Moses, O. Biodun, Construction 4.0 application: industry 4.0, Internet of Things, and lean construction tools' application in the quality management system of residential building projects, Buildings 12 (2022) 1557.

[31] M. Majid, S. Habib, A.R. Javed, M. Rizwan, G. Srivastava, T.R. Gadekallu, et al., Applications of wireless sensor networks and Internet of things frameworks in the industry revolution 4.0: a systematic literature review, Sensors 22 (2022) 2087.

[32] L.M. Zawra, H.A. Mansour, A.T. Eldin, N.W. Messiha, Utilising the Internet of Things (IoT) technologies in the implementation of industry 4.0. In: International Conference on Advanced Intelligent Systems and Informatics, 2017, pp. 798-808. Cham: Springer International Publishing.

[33] V.C. Jadala, S.K. Pasupuletti, S.H. Raju, S. Kavitha, C.M.S. Bhaba, B. Sreedhar, Need of intenet of things, industrial IoT, industry 4.0 and integration of cloud for industrial revolution, 2021 Innovations in Power and Advanced Computing Technologies (i-PACT), IEEE, (2021), pp. 1–5.

[34] D. Mikołajewski, J.M. Czerniak, M. Piechowiak, K. Węgrzyn-Wolska, J. &Kacprzyk, The Internet of Things and AI-based optimisation within the industry 4.0 paradigm, Bulletin of the Polish Academy of Sciences, Technical Sciences (2024) e147346.

[35] M. Bayanati, Security and privacy analysis based on Internet of Things in the fourth industrial generation (Industry 4.0), International Journal of Innovation in Management, Economics and Social Sciences 3 (2023) 1–12.

[36] Z. Diao, F. Sun, Application of Internet of Things in smart factories under the background of Industry 4.0 and 5G communication technology, Mathematical Problems in Engineering 2022 (2022) 4417620.

[37] M. Faheem, R.A. Butt, Big datasets of optical-wireless cyber-physical systems for optimising manufacturing services in the Internet of Things-enabled industry 4.0, Data in Brief 42 (2022) 108026.

[38] C. Song, S. Liu, G. Han, P. Zeng, H. Yu, Q. Zheng, Edge-intelligence-based condition monitoring of beam pumping units under heavy noise in Industrial Internet of Things for industry 4.0, IEEE Internet of Things Journal 10 (2022) 3037–3046.

[39] G.B. Raja, Impact of Internet of Things, artificial intelligence, and blockchain technology in industry 4.0, Internet of Things, Artificial Intelligence and Blockchain Technology, Springer International Publishing, Cham, (2021), pp. 157–178.

[40] M. Jacoby, T. Usländer, Digital twin and Internet of Things—current standards landscape, Applied Sciences 10 (2020) 6519.

[41] A.R. Al-Ali, R. Gupta, T. Zaman Batool, T. Landolsi, F. Aloul, A. Al Nabulsi, Digital twin conceptual model within the context of the Internet of Things, Future Internet 12 (2020) 163.

[42] H. Xu, J. Wu, Q. Pan, X. Guan, M. &Guizani, A survey on digital twin for industrial Internet of Things: applications, technologies and tools, IEEE Communications Surveys & Tutorials 25 (2023) 2569–2598.

[43] Y. He, J. Guo, X. Zheng, From surveillance to digital twin: challenges and recent advances of signal processing for industrial Internet of Things, IEEE Signal Processing Magazine 35 (2018) 120–129 (*Gillis, A.S. What is the Internet of Things (IoT)?: https://www.techtarget.com/iotagenda/definition/Internet-of-Things-IoT*, no date).

[44] V. Souza, R. Cruz, W. Silva, S. Lins, V. Lucena, A digital twin architecture based on the industrial Internet of Things technologies. In: 2019 IEEE International Conference on Consumer Electronics (ICCE), 2019, pp. 1-2. IEEE.

[45] A.O. Adebayo, M.S. Chaubey, L.P. &Numbu, Industry 4.0: the fourth industrial revolution and how it relates to the application of Internet of Things (IoT), Journal of Multidisciplinary Engineering Science Studies (JMESS) 5 (2019) 2477–2482.

[46] S.N. Deshpande, R.M. Jogdand, A survey on Internet of Things (IoT), Industrial IoT (IIoT) and industry 4.0, International Journal of Computer Applications 175 (2020) 20–27.

[47] M. Faheem, G. Fizza, M.W. Ashraf, R.A. Butt, M.A. Ngadi, V.C. Gungor, Big Data acquired by Internet of Things-enabled industrial multichannel wireless sensors networks for active monitoring and control in the smart grid industry 4.0, Data in Brief 35 (2021) 106854.

[48] P. Matthyssens, Reconceptualising value innovation for industry 4.0 and the industrial Internet of Things, Journal of Business & Industrial Marketing 34 (2019) 1203–1209.

[49] A. Sadiq, M. Anwar, R.A. Butt, F. Masud, M.K. Shahzad, S. Naseem, et al., A review of phishing attacks and countermeasures for Internet of things-based smart business applications in industry 4.0, Human Behavior and Emerging Technologies 3 (2021) 854–864 (Gillis, A.S. What is the Internet of Things (IoT)?: https://www.techtarget.com/iotagenda/definition/Internet-of-Things-IoT, no date).

[50] V.R. Kebande, Industrial Internet of Things (IIoT) forensics: the forgotten concept in the race towards industry 4.0, Forensic Science International: Reports 5 (2022) 100257.

[51] K. Coatney, M. Poliak, Cognitive decision-making algorithms, Internet of Things smart devices, and sustainable organisational performance in industry 4.0-based manufacturing systems, Journal of Self-Governance and Management Economics 8 (2020) 9–18.

[52] R. Bavaresco, H. Arruda, E. Rocha, J. Barbosa, G.P. Li, Internet of Things and occupational well-being in industry 4.0: a systematic mapping study and taxonomy, Computers & Industrial Engineering 161 (2021) 107670.

6

Big data for Industry 4.0

6.1 Introduction to big data

Big data is an emerging technology for storing large amounts of data and has wide applications across different fields. It tracks and collects data in accordance with the system requirements. Big data has a vital role in driving industrial innovation. It uses powerful data analytics tools and devices to track all production operations. Big data's key goal in Industry 4.0 is to collect the right data to address relevant issues in production and other essential services [1–3]. Industry 4.0 smartly and flexibly transforms the production landscape by using this technology. Advanced technologies are being used to implement Industry 4.0, and big data can effectively manage all data to support its successful implementation [4,5].

Big data is a valuable source for forecasting and operational management based on market insights and statistics, making it useful for predictive analysis and operations. Big data will improve Industry 4.0 and is expected to play an essential role in its effective implementation [6,7]. This technology is employed in Industry 4.0 to address current challenges and enables the automation of all essential operations using vast amounts of recorded data. It is critical for identifying areas that require improvement. It tracks all inventory and indicates which parts are needed at any time [8,9].

The industry will generate a large volume of data for real-time manufacturing. As a result, it is necessary to eliminate flaws to integrate big data applications into the Industry 4.0 system, in which all machines and devices readily interact and communicate. It delivers excellent functionality in the industrial information infrastructure. It investigates the key drivers of Industry 4.0 to provide valuable, actionable data [10–12]. In the upcoming years, the manufacturing sector will improve production efficiency. Recently, software for industrial applications has been created to ensure significant data compliance. This technology has been applied in the automation industry to deliver effective solutions. This is a creative way to collect a large amount of data [13–14].

Collecting statistics or information from within the system is a crucial advantage of big data in Industry 4.0. Detection, diagnosis, and correction without involvement in the production line are achievable with this in-process monitoring during automation and manufacturing. Big data applications forecast failure or equipment fatigue based on past factory data [15,16]. It helps reduce costs, such as maintenance and downtime, among other things. This is beneficial to several businesses in terms of real-time production and data quality. This data storage platform provides a solid foundation and a wide range of data sources for extracting inclusionary cognition to improve efficiency and quality [17–20].

Digital Twin and Industry 4.0. DOI: https://doi.org/10.1016/B978-0-443-54019-6.00004-2

6.2 Big data and its classification

Data can be stored and transferred as electrical signals, or it can be recorded on mechanical, optical, or magnetic storage media. Data is defined as 'the quantities, characters, or symbols on which computer operations are performed'. Large amounts of data that are too big to process and analyse with conventional technology are referred to as 'big data'. Big data presents a significant analytical problem due to its exponential growth in volume [19–23]. Firstly, we are going to classify big data structuring in three ways:

- Structured data
- Unstructured data
- Semistructured data

Even though these three ideas are theoretically applicable at all analytics levels, they are essential with big data. Knowing where the raw data comes from and how it needs to be handled before analysis becomes even more important when working with big data sets. For the project to be successful, data extraction must be effective because there is so much of it. The data's structure is essential for figuring out how to use it and what insights it may provide. Each piece of data must go through an extract, transform, and load (ETL) process before it can be analysed. Data is collected, processed to make it readable by an application, and then stored for later use. Every data structure has a unique ETL procedure.

6.2.1 Structured data

Big, structured data is the most straightforward to work with. Its proportions are set by preset criteria, and it is well-organised. Quantitative information like age, billing, contact details, address, expenses, and debit/credit card numbers are crucial. Since structured data already consists of concrete numbers, it is much simpler for software to sift through and gather data. For structured data, schemas, roadmaps to particular data points are utilised. Each item's location and meaning are specified by these schemas. A payroll database contains information about an employee's name, pay rate, hours worked, and mode of payout.

Structured data is the most detailed data to analyse because it requires minimal preprocessing. While a proper query will require little interpretation or conversion, a user may need to purify and filter the data to only the most important parts. One of the main advantages of using structured data is the quicker integration of relational and business data. Making all sources compatible requires very little preparation because pertinent data dimensions are often determined and specific parts are in a uniform format.

The final product of the ETL process for structured data is kept in a data warehouse. The original data was collected with a specific analytics objective in mind. For that reason, these databases are meticulously arranged and filtered. Relational databases make it easy to query datasets. They allow users to access other data, study it on their own, or integrate it with their own data to offer more context. Relational database management systems use structured query language, a consistent language, to obtain data from a network of data sources and platforms. Scalable data processing is made possible by this standardisation.

6.2.2 Unstructured data

Not all data is organised and packed as neatly as structured data, and not all data has usage guidelines. Typically, only about 20% of the data is organised. Thus, all of the data that is accessible is unstructured. Since this data is not structured, we refer to it as unstructured data. Unstructured data is produced by practically all computer operations. In an analytical process, structured data saves time, whereas unstructured data requires time and effort to make it readable. For structured data, the ETL approach is really simple. At the transform stage, it is cleaned and validated before being loaded into a database.

The hardest part of analysing unstructured data is teaching an application to understand the information it obtains. Usually, this means turning it into some ordered data. This is not simple, and the details of how it is done differ depending on the format and the intended use of the analytics. Common methods include taxonomy-based content hierarchies, natural language processing (NLP), and text parsing. It frequently calls for a sophisticated algorithm that integrates contextualisation, interpretation, and scanning.

Because of its contextual component—combining internal data with external context increases its relevance—unstructured data is common in big data. With context, any model or analysis will be more accurate. In contrast to structured data, unstructured data is kept in data lakes, where all of its information is preserved in its unprocessed state. Warehouse data is limited to a particular schema.

6.2.3 Semistructured data

A combination of organised and unstructured data is known as semistructured data. This is typically unstructured data with associated metadata. Data can be added later as semantic tags, including time, location, device ID stamp, and email address. When paired with the appropriate datasets, semistructured data can be a useful tool since it fills the gap between structured and unstructured data. Machine learning and AI training can benefit from the use of metadata to link patterns. Semistructured data requires a defined schema. This could be both beneficial and difficult. Because the software needs to be informed of the meaning of each data item, it can be more difficult to manage. Metadata generation can organise queries on semistructured datasets, but it does not limit them.

6.3 Main components of big data

Conventional approaches cannot handle large, complex data. In order to evaluate, extract information, and have a deeper understanding of the data, we utilise big data. For large amounts of data, we look at volume, velocity, diversity, veracity, and value. Big data includes information gathered via social media. Big data makes it possible to analyse data trends and gain a deeper understanding of corporate and human behaviour. This makes processing more efficient and raises customer satisfaction. The data in big data may be time-related, natural, processed, structured, or unstructured [24–29]. The significant components of big data are as follows:

6.3.1 Machine learning

It is the science of teaching computers to learn on their own. In machine learning, a computer is supposed to use statistical models and algorithms to carry out predetermined tasks without direct guidance. Results from machine learning algorithms are derived from historical data. For instance, a number of smartphone applications now notify users of impending bill payments, give suggestions for savings plans, and summarise money and bills. One of these responsibilities is to read texts and emails.

6.3.2 Natural language processing

Spoken human language can be understood by a computer. These days, the most well-known examples that people may identify with are Google Home and Amazon Alexa. Both use several technologies, including NLP, to offer a virtual assistant experience. We are surrounded by NLP without realising it. It automatically fixes itself when we attempt to send an email without the attachment we mentioned in the email content, and these days it provides auto-suggestions to finish the email. When we attempt to send an email without the attachment that we mentioned in the text of the email, it immediately intimates us. This is a component of the backend NLP applications.

6.3.3 Business intelligence

In order for end users, including managers and corporate executives, to obtain actionable insights and make wise business decisions, business intelligence (BI) is a technology-driven approach or procedure for gathering insights through data analysis and presentation.

6.3.4 Cloud computing

Cloud computing may provide servers, storage, networking, software, databases, analytics, intelligence, and internet service to enable faster innovation, flexible resource allocation, and economies of scale.

6.3.5 Data sources

Significant data architecture begins with all company data sources. They are typically scattered and segregated across multiple tools before being centralised via data orchestration techniques and technologies in real-time sources, relational or conventional databases, and static files generated by applications.

6.3.6 Data storage

To process data, businesses need to store it. Often included is a data lake, which is a large, scalable unstructured database that can store enormous amounts of content in many formats.

6.3.7 Batch processing

Before beginning an ETL task to filter, consolidate, and get massive amounts of data ready for analysis, batch processing waits for a certain amount of raw data to be gathered. When data freshness is not a major concern, it is used. Large-scale data processing is common with open-source frameworks like Hadoop.

6.3.8 Real-time message ingestion

Real-time data necessitates rapid processing, which is why businesses need a system like Microsoft's Azure Event Hubs to consume real-time data sources and give maximum commercial value.

6.3.9 Stream processing

As soon as data enters the storage layer, usually within seconds of creation, it is processed using technologies like Apache Kafka.

6.3.10 Analytical datastore

BI tools are used to query and examine previously processed data in the data repository. For this purpose, prominent cloud data stores include Google BigQuery, Snowflake, and Amazon Redshift.

6.3.11 Analysis and reporting

To determine business value, teams must assess and comprehend data after consuming and processing data from various sources. This is accomplished by reporting and analytical tools that use embedded BI technology to produce helpful charts, graphs, and other visualisation forms. Therefore, the data may need to be interpreted by a data scientist.

6.4 Characteristics of big data

The significant characteristics of big data are provided through variety, volume, Variability and velocity; they are elaborated as follows [1,24,25,30–33]. Fig. 6.1 shows major characteristics of big data.

6.4.1 Volume

It is important to consider the size of the data in order to determine its value. Determining whether a given dataset falls under the big data category also depends on volume.

6.4.2 Variety

The term 'variety' describes the various types of data, depending on whether they are structured or unstructured. In the past, the majority of applications looked at data in

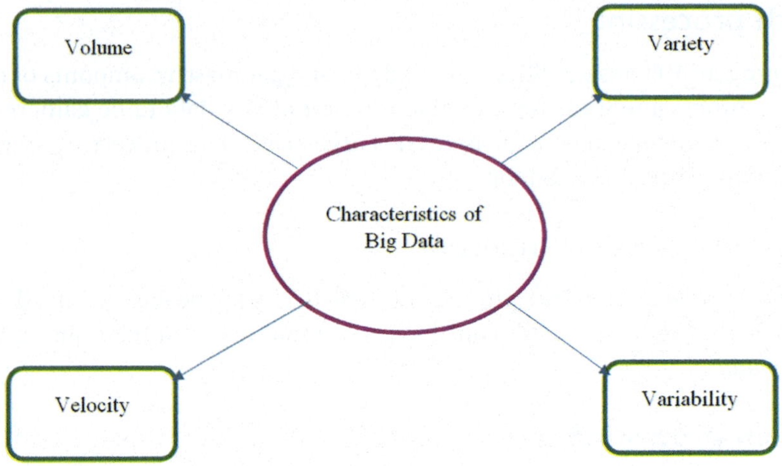

FIGURE 6.1 Major characteristics of big data.

databases and spreadsheets in rows and columns. Data does, however, come in a variety of formats, such as emails, pictures, movies, music, etc.

6.4.3 Velocity

Velocity is the rate at which data is produced. The speed at which data may be gathered and processed from a source determines its potential.

6.4.4 Variability

Data should be adaptable, which may require consistency and synchronisation without interfering with or impeding appropriate data administration and processing.

6.5 Best practices of big data

The primary best practices of big data are as follows [21,30,34]:

6.5.1 To align with the business vision

Big data architectures should prioritise users and be in line with the organisation's business objectives. This includes being aware of the organisation's objectives, key business drivers, enterprise architectural maturity level, and comprehensive data-architecture task standards.

6.5.2 To identify data sources

Identify and categorise structured and unstructured data sources so that system architects may normalise the data into a standard format.

6.5.3 Choose data services

They provide essential data sources and enable data integration and visualisation. They also speed up big data applications by providing access to data storage and enabling faster retrieval, processing, and analysis.

6.5.4 Design a user-friendly interface

A user-friendly interface is necessary for important data architecture. Users who make use of well-designed information architecture are more likely to locate what they need, for instance, through dashboards and exploration, and they also spend less time and effort looking for information.

6.5.5 Eliminate internal data silos

Information silos are ineffective and have inadequate data management since they store data in different places. After removing any barriers, clean and assess the data to make sure it is accurate and legitimate.

6.5.6 Account for differing data structures and formats

Build a solid data architecture that can manage information from both organised and unstructured sources. Get ready for data that is expanding, changing, and becoming more sophisticated.

6.5.7 Implement solid data governance

Implement a thorough data governance strategy that includes data security and compliance measures to build trust in the data used to make essential business decisions.

6.6 Advantages of using big data

The significant advantages of big data are as follows [35–37]:

- Store all inventory and vendor information.
- Maintain manufacturing-related quality data.
- Increase client engagement and happiness.
- Lower the cost of production.
- Assess the industry's risk.
- Keep track of all procedures performed during the production process.
- Enhances and improves the production process as a whole.
- Improve the efficiency of production.
- Assist in the monitoring of the whole production process.
- Ensuring that the production system is properly controlled.

- Maintain a database of all client information, including feedback.
- Error detection in the system and determining the source of failure.
- Provide possibilities for advancement.
- Allows the user to work with a wide range of data.

6.7 Roles of big data towards the Industry 4.0 environment

Extensive data systems store massive volumes of exchanged data in industrial and automation environments, leveraging IoT, cloud computing, cyber-physical systems, and AI. It may solve various practical problems in the manufacturing industry, such as improving security and achieving other objectives [38–40]. Big data stores all this in the cloud, transforming it into valuable, actionable information. In the specific context of Industry 4.0, this technology manages large volumes of data to enable improvements and innovation. Future preventive measures can be planned using the collected data, saving money on rework and waste [33,41–45]. The significant roles of big data for Industry 4.0 are as follows:

6.7.1 Keep all important data

Big data is beneficial for storing all necessary data to ensure the correct completion of industrial processes. This saved data is utilised to create new products. It continuously analyses all ongoing Industry 4.0 activities. It boosts data management efficiency by accurately and effectively forecasting trends and developing a dynamic pricing plan.

6.7.2 Gather information

Big data offers much promise in collecting information about industrial systems and analysing their performance. This technology can detect and correct any issue without interrupting the production process. It helps to grasp the progress of Industry 4.0's different new tools and approaches. It is also utilised for online marketing analytics and transaction records. It helps identify new ways and approaches to improve the production system.

6.7.3 Automate manufacturing

By leveraging massive, diverse, actionable data, this technology helps automate industries. Thus, to develop industrial big data applications, various software tools are employed. It invents new production processes that require less human involvement and attention. It improves the precision of industrial operations by automatically tracking all essential tools and resources throughout production. It creates automated manufacturing lines to increase production speed and quality. This technology can automatically detect faults and defects in the production process without causing any downtime.

6.7.4 Proper control of machines

Machines are correctly regulated using detailed information derived from big data. Manufacturing companies can improve operations and make better business decisions by utilising real-time shop floor data. It helps handle all supply chain-related tasks efficiently and effectively. Big data can enhance post-sale service. It assists in controlling all new manufacturing processes.

6.7.5 Manufacturing in the digital age

Big data is a valuable tool for improving the digital manufacturing system. All detailed information stored by this technology improves quality and efficiency. It provides a competitive advantage by solving various manufacturing challenges. It also plays a vital role in providing a better experience for new workers and employees. This provides a competitive advantage in solving various manufacturing challenges.

6.7.6 Failures forecasting

By analysing system data, big data can help forecast failures. It provides a highly personalised solution that includes improved predictive analysis and operations management. It boosts industrial efficiency, productivity, and failure prediction capacity. Using relevant recorded data to build a machine learning model to anticipate failure is beneficial. It keeps downtime costs from detracting from the production system's efficiency. It can quickly check the status where there is a significant risk of mistakes.

6.7.7 Well-informed decision-making

Big data is generated in various formats, at varying volumes, and at different speeds, enabling better decision-making. It is helpful when deciding on an overproduction system and aids in making the best decisions during complicated production processes. It is used to make well-informed, well-managed decisions about new activities.

6.7.8 Intelligent transportation system

This technology supports intelligent transportation system route planning during transportation. Proper route planning decreases fuel consumption and travel time. It also assists in the handling of all incoming materials and finished goods. Overall, it brings disruptive change to industries by effectively managing the transportation system.

6.7.9 Intelligent management of manufacturing systems

This technology can store significant amounts of data, enabling improved management during the production process. Big data enables improved personnel management in manufacturing, helping eliminate mistakes. It can be used for risk management and other relevant tasks. It tracks and manages all inventory to ensure the smooth flow process continues to develop. This supports testing and simulation of innovative production processes in Industry 4.0.

6.7.10 Detection of fraud

Big data may be used to detect fraud and improve environmental protection. Using real-time data, one can quickly analyse the pattern of any scam. It is also used to track client complaints. Insurance firms use this technology to address various insurance fraud issues. It helps detect and prevent health insurance fraud among workers and employees.

6.7.11 Assistance to the smart factory

Big data is a crucial component of Industry 4.0, providing vital assistance to smart factory management. It is a crucial tool that stores a wealth of data and other relevant information about the production system. In Industry 4.0, this technology intelligently manages all ongoing processes. It provides accurate, timely data that helps reduce product cycle time.

6.7.12 Track financial history

Big data tracks all financial records and estimates the cost to start production. This keeps track of all inventories and the final product. It helps determine the industry's turnover and profit. In the future, this technology may be utilised to improve production efficiency. It will prevent theft, fraud, and other threats to system productivity.

6.8 Roles of digital twin for big data

A digital twin is a virtual model that supports big data. Whenever there are changes to its physical counterpart, the virtual model must be updated accordingly. Proper data analysis and system monitoring would be made possible by the combination of the real and virtual worlds. Through simulations, future planning, and the creation of new opportunities, this would help resolve issues before they become serious and prevent downtime. During testing, the virtual model must behave similarly to the real thing [46,47]. They must also evaluate any information based on the advantages and disadvantages of the physical object, attempt to anticipate any unanticipated problems, and, if feasible, offer remedies [48–50]. The significant roles of digital twins for big data are as follows:

- Data from physical assets is synchronised with big data platforms in real time.
- Combining diverse datasets from several sources.
- Processing and ingesting data from streams at high speeds.
- Archiving large historical data for lifecycle analysis.
- Modelling predictive maintenance with extensive datasets.
- Operational optimisation through prescriptive analytics.

- Extensive simulation powered by real-time and historical data.
- Root cause analysis using correlation between multidimensional data.
- Forecasting demand and load with trend-based data models.
- Data-driven scenario analysis and risk modelling.
- High-quality analytics with datasets at production scale.
- Optimisation of resource utilisation by analytics of usage data.
- The ability to scale cloud-based analytics and storage.
- Compliance monitoring, traceability, and data governance.
- Large dataset visualisation and dashboarding in real time.

6.9 Upcoming goals of big data

The primary upcoming goals of big data are as follows [51–53]:

- In the future, big data will be a critical component of Industry 4.0 to reduce human tasks.
- It will be utilised to collect and disseminate information for Industry 4.0.
- It will present a fantastic chance for businesses to make goods more quickly and effectively with fewer personnel.
- It will provide the potential to turn the world into a luxury society for everyone.
- It will enhance the flexibility of automation and provide the driving force for an intelligent manufacturing system.
- It will simplify several tasks while creating new work opportunities.
- With the support of this data collection facility, Industry 4.0 will provide a better approach to improve manufacturing in the future.

6.10 Disadvantages of big data

The significant disadvantages of big data being perceived are as follows [1,4]:

- Employees may lose their jobs.
- The essential need to collect precise data.
- A cultural shift is required.
- Extensive requirements of hardware infrastructure.
- Information gathered does not help create learning activities.
- Cybersecurity dangers are a threat to the security and privacy of individuals.
- New legal requirements.

6.11 Conclusion

Big data is used to collect massive volumes of data through many feed sources. After receiving the data, it is filtered, normalised, and structured to identify significant process correlations for scientific research. Big data is utilised in Industry 4.0 to address appropriate safety, effectiveness, and performance concerns, guide clinical trial design, and identify promising leads in the research process.

Every stage of industrial processing generates massive volumes of data that can be analysed or stored using big data. It delivers real-time analytics for improved visualisation of outcomes, enhancing the production system's efficiency and predictability. Industry 4.0 will examine the role of big data analytics as a driving force in this new production environment. This technology is essential for increasing production efficiency. It exemplifies the Industry 4.0 solution that uses an in-process surveillance system.

Big data applications for Industry 4.0 include data collection, failure prediction, manufacturing automation, predictive analysis, storing relevant data from all operations, financial record maintenance, decision-making process, proper machine control, proper management during smart factory, manufacturing, fraud detection, digital manufacturing, and transportation system control. Industry 4.0 can be successfully implemented through big data applications, leading to significant advancements in manufacturing facilities.

References

[1] M. Javaid, A. Haleem, R.P. Singh, R. Suman, Significant applications of big data in industry 4.0, Journal of Industrial Integration and Management 6 (2021) 429–447.

[2] J. Yan, Y. Meng, L. Lu, L. Li, Industrial big data in an industry 4.0 environment: challenges, schemes, and applications for predictive maintenance, IEEE Access 5 (2017) 23484–23491.

[3] G. Li, J. Tan, S.S. Chaudhry, Industry 4.0 and big data innovations, Enterprise Information Systems 13 (2019) 145–147.

[4] S.K. Jagatheesaperumal, M. Rahouti, K. Ahmad, A. Al-Fuqaha, M. Guizani, The duo of artificial intelligence and big data for industry 4.0: applications, techniques, challenges, and future research directions, IEEE Internet of Things Journal 9 (2021) 12861–12885.

[5] M.S. Reis, G. Gins, Industrial process monitoring in the big data/industry 4.0 era: from detection, to diagnosis, to prognosis, Processes 5 (2017) 35.

[6] M.Y. Santos, J.O. e Sá, C. Andrade, F.V. Lima, E. Costa, C. Costa, et al., A big data system supporting bosch braga industry 4.0 strategy, International Journal of Information Management 37 (2017) 750–760.

[7] T. Nguyen, R.G. Gosine, P. Warrian, A systematic review of big data analytics for oil and gas industry 4.0, IEEE Access 8 (2020) 61183–61201.

[8] R. Rezazadegan, M. Sharifzadeh, Applications of artificial intelligence and big data in industry 4.0 technologies, Industry 4.0 Vision for Energy and Materials: Enabling Technologies and Case Studies, Wiley, (2022), pp. 121–158. Available from https://doi.org/10.1002/9781119695868.ch5.

[9] T. Papadopoulos, S.P. Singh, K. Spanaki, A. Gunasekaran, R. Dubey, Towards the next generation of manufacturing: implications of big data and digitalisation in the context of industry 4.0, Production Planning & Control 33 (2022) 101–104.

[10] M.L. Tseng, R.R. Tan, A.S. Chiu, C.F. Chien, T.C. Kuo, Circular economy meets industry 4.0: can big data drive industrial symbiosis? Resources, conservation and recycling 131 (2018) 146–147.

[11] M. Obitko, V. Jirkovský, Big data semantics in industry 4.0, International Conference on Industrial Applications of Holonic and Multi-Agent Systems, Springer International Publishing, Cham, (2015), pp. 217–229.

[12] A. Patil, A. Dwivedi, M.A. Moktadir, Big data-industry 4.0 readiness factors for sustainable supply chain management: towards circularity, Computers & Industrial Engineering 178 (2023) 109109.

[13] P. Dobos, P. Tamás, B. Illés, R. Balogh, Application possibilities of the big data concept in industry 4.0, IOP Conference Series: Materials Science and Engineering 448 (2018) 12011. Available from https://doi.org/10.1088/1757-899X/448/1/012011.

[14] K. Witkowski, Internet of Things, big data, industry 4.0–innovative solutions in logistics and supply chain management, Procedia Engineering 182 (2017) 763–769.

[15] L. Stojanovic, M. Dinic, N. Stojanovic, A. Stojadinovic, Big-data-driven anomaly detection in industry (4.0): an approach and a case study, 2016 IEEE International Conference on Big Data (Big Data), IEEE, (2016), pp. 1647–1652.

[16] N. Velásquez, E.C. Estevez, P.M. Pesado, Cloud computing, big data and the industry 4.0 reference architectures, Journal of Computer Science & Technology 18 (2018).

[17] A. Sharma, H. Pandey, Big data and analytics in industry 4.0, A Roadmap to Industry 4.0: Smart Production, Sharp Business and Sustainable Development, Springer International Publishing, Cham, (2019), pp. 57–72.

[18] J. Yan, Y. Meng, L. Lu, C. Guo, Big-data-driven based intelligent prognostics scheme in industry 4.0 environment, 2017 Prognostics and System Health Management Conference (PHM-Harbin), IEEE, (2017), pp. 1–5.

[19] Allen, R. What are the Types of Big Data?: https://www.selecthub.com/big-data-analytics/types-of-big-data-analytics/.

[20] J. Park, H. Bae, Big data and AI for process innovation in the industry 4.0 era, Applied Sciences 12 (2022) 6346.

[21] L. Wang, G. Wang, Big data in cyber-physical systems, digital manufacturing and industry 4.0, International Journal of Engineering and Manufacturing (IJEM) 6 (2016) 1–8.

[22] D. Jiang, Y. Wang, Z. Lv, S. Qi, S. Singh, Big data analysis-based network behavior insight of cellular networks for industry 4.0 applications, IEEE Transactions on Industrial Informatics 16 (2019) 1310–1320.

[23] V. Pilloni, How data will transform industrial processes: crowdsensing, crowdsourcing and big data as pillars of industry 4.0, Future Internet 10 (2018) 24.

[24] Pedamkar, P. Introduction to Big Data: https://www.educba.com/introduction-to-big-data/.

[25] R. Sharma,Characteristics of big data: types & 5V's, 2022. https://www.upgrad.com/blog/characteristics-of-big-data/.

[26] I. Berges, V.J. Ramírez-Durán, A. Illarramendi, A semantic approach for big data exploration in industry 4.0, Big Data Research 25 (2021) 100222.

[27] J. Lee, H.A. Kao, S. Yang, Service innovation and smart analytics for industry 4.0 and big data environment, Procedia CIRP 16 (2014) 3–8.

[28] R. Sahal, J.G. Breslin, M.I. Ali, Big data and stream processing platforms for industry 4.0 requirements mapping for a predictive maintenance use case, Journal of Manufacturing Systems 54 (2020) 138–151.

[29] H. Khayyam, A. Jamali, A. Bab-Hadiashar, T. Esch, S. Ramakrishna, M. Jalili, et al., A novel hybrid machine learning algorithm for limited and big data modeling with application in industry 4.0, IEEE Access 8 (2020) 111381–111393.

[30] T. Niesen, C. Houy, P. Fettke, P. Loos, Towards an integrative big data analysis framework for data-driven risk management in industry 4.0, 2016 49th Hawaii International Conference on System Sciences (HICSS), IEEE, (2016), pp. 5065–5074.

[31] M.A. Berawi, Utilising big data in industry 4.0: managing competitive advantages and business ethics, International Journal of Technology 3 (2018) 430–433.

[32] T. Latinovic, D. Preradović, C.R. Barz, A. Pop Vadean, M. Todić, Big data as the basis for the innovative development strategy of the industry 4.0, IOP Conference Series: Materials Science and Engineering 477 (2019) 12045. Available from https://doi.org/10.1088/1757-899X/477/1/012045.

[33] Q. Qi, F. Tao, Digital twin and big data towards smart manufacturing and industry 4.0: 360-degree comparison, IEEE Access 6 (2018) 3585–3593.

[34] G. Lampropoulos, Artificial intelligence, big data, and machine learning in industry 4.0, Encyclopedia of Data Science and Machine Learning, IGI Global Scientific Publishing, (2023), pp. 2101–2109.

[35] A.G. Frank, L.S. Dalenogare, N.F. Ayala, Industry 4.0 technologies: implementation patterns in manufacturing companies, International Journal of Production Economics 210 (2019) 15–26.

[36] H. Han, S. Trimi, Towards a data science platform for improving SME collaboration through industry 4.0 technologies, Technological Forecasting and Social Change 174 (2022) 121242.

[37] H. Singh, Big data, industry 4.0 and cyber-physical systems integration: a smart industry context, Materials Today: Proceedings 46 (2021) 157–162.

[38] R. Bonnard, M.D.S. Arantes, R. Lorbieski, K.M.M. Vieira, M.C. Nunes, Big data/analytics platform for industry 4.0 implementation in advanced manufacturing context, The International Journal of Advanced Manufacturing Technology 117 (2021) 1959–1973.

[39] R. Kumar, S.P. Singh, K. Lamba, Sustainable robust layout using big data approach: a key towards industry 4.0, Journal of Cleaner Production 204 (2018) 643–659.

[40] A.K. Rana, S. Sharma, Industry 4.0 manufacturing based on IoT, cloud computing, and big data: manufacturing purpose scenario, International Conference on Advanced Communication and Computational Technology, Springer Nature Singapore, Singapore, (2019), pp. 1109–1119.

[41] D. Li, H. Tang, S. Wang, C. Liu, A big data-enabled load-balancing control for smart manufacturing of industry 4.0, Cluster Computing 20 (2017) 1855–1864.

[42] F. Tao, J. Cheng, Q. Qi, M. Zhang, H. Zhang, F. Sui, Digital twin-driven product design, manufacturing and service with big data, The International Journal of Advanced Manufacturing Technology 94 (2018) 3563–3576.

[43] S. Ma, W. Ding, Y. Liu, S. Ren, H. Yang, Digital twin and big data-driven sustainable smart manufacturing based on information management systems for energy-intensive industries, Applied Energy 326 (2022) 119986.

[44] C.L. Stergiou, K.E. Psannis, Digital twin intelligent system for industrial internet of things-based big data management and analysis in cloud environments, Virtual Reality & Intelligent Hardware 4 (2022) 279–291.

[45] X. Li, H. Liu, W. Wang, Y. Zheng, H. Lv, Z. Lv, Big data analysis of the internet of things in the digital twins of smart city based on deep learning, Future Generation Computer Systems 128 (2022) 167–177.

[46] E. González-Sarmiento, J. Roa-Perez, L. Ortiz-Ospino, Big data and artificial intelligence in the development of industry 4.0: a bibliometric analysis, IOP Conference Series: Materials Science and Engineering 1154 (2021) 12008.

[47] M. Nagy, M. Figura, K. Valaskova, G. Lăzăroiu, Predictive maintenance algorithms, artificial intelligence digital twin technologies, and Internet of Robotic Things in big data-driven industry 4.0 manufacturing systems, Mathematics 13 (2025) 981.

[48] D. Cemernek, H. Gursch, R. Kern, Big data as a promoter of industry 4.0: lessons of the semiconductor industry, 2017 IEEE 15th International Conference on Industrial Informatics (INDIN), IEEE, (2017), pp. 239–244.

[49] A. Bousdekis, G. Mentzas, Enterprise integration and interoperability for big data-driven processes in the frame of industry 4.0, Frontiers in Big Data 4 (2021) 644651.

[50] J.H. Kim, A review of cyber-physical system research relevant to the emerging IT trends: industry 4.0, IoT, big data, and cloud computing, Journal of Industrial Integration and Management 2 (2017) 1750011.

[51] P. Figueiras, L. Lourenço, R. Costa, D. Graça, G. Garcia, R. Jardim-Gonçalves, Big data provision for digital twins in industry 4.0 logistics processes, 2021 IEEE International Workshop on Metrology for Industry 4.0 & IoT (MetroInd4.0&IoT), IEEE, (2021), pp. 516–521.

[52] K. Rinat, G. Thakur, M. Gupta, T.N.P. Madhuri, S. Bansal, Comparative analysis of big data computing in industry 4.0 and industry 5.0: an experimental study, BIO Web of Conferences 86 (2024) 1068. Available from https://doi.org/10.1051/bioconf/20248601068.

[53] L.D. Xu, L. Duan, Big data for cyber-physical systems in industry 4.0: a survey, Enterprise Information Systems 13 (2019) 148–169.

7

5G for Industry 4.0

7.1 5G: an introduction

5G is the fifth-generation mobile communications technology. This technology develops new services across various sectors to increase capacity and improve database performance, enabling users to get more out of their services and applications [1,2]. 5G technology is used in Industry 4.0 to deliver high-speed and flexible connections, which help to improve the overall production system. This allows for a variety of innovative applications in Industry 4.0. 5G benefits the whole value chain, including consumers and companies, in keeping with its long-term objective of digitalisation [3–5].

The creation of a mobile communications standard is centred on human digital connectedness and communications between devices, sensors, and machines in the Internet of Things (IoT). 5G is becoming a competitive aspect due to the increasing connectivity. This fifth generation of mobile communications has the potential to enhance and speed up manufacturing and logistical processes. This technology is used in every aspect of modern life to address the massive increase in the number of daily electronic devices. This technology improves quality control by enabling more inspections that rely on real-time analysis for quick, accurate detection [6–8].

5G wireless networks offer a new type of internet access that is not dependent on fixed lines and provides high speed, low latency, and large capacity. It enables a broader range of applications to be connected. 5G helps to update the present infrastructure of the production sites for the modern communication standard [9,10]. The launch of 5G improves some existing applications and creates new ones that current technology cannot handle. By testing and evaluating their roles, operators can ensure the benefits of 5G. The 5G revolutionises our understanding of today's fundamental industrial paradigm. This network can meet the demand for enough capacity and speed infrastructure [11–13].

7.2 Brief about first five Gs

The 'G' stands for 'Generation'. When we are connected to the Internet, the speed of our connection depends on the signal strength, which is shown on our home screen as 2G, 3G, 4G, etc. Every generation is a set of telephone network standards that outline the technical implementation of a certain mobile phone system. The technology needed to attain speed rises along with it. For instance, 1G provides 2.4 kbps, 2G is based on GSM and offers 64 kbps, 3G is based on long-term evolution (LTE) technology and offers 144 kbps–2 Mbps, and 4G is based on LTE technology and offers 100 Mbps–1 Gbps [14–16].

Each new generation of services represents a significant step toward wireless communication's ultimate goal, providing high-quality, reliable communication comparable to wired

Digital Twin and Industry 4.0. DOI: https://doi.org/10.1016/B978-0-443-54019-6.00006-6

communication. The voyage of progress began with 1G in 1979 and continues to 5G. Thus to use the G nomenclature officially, specific requirements for each generation must be completed. Organisations standardise each generation of mobile technology. Every generation has requirements, such as throughput, latency, and other elements, that must be met to be included in that generation [17–20]. Some significant developments in first five Gs are shown in Fig. 7.1.

7.2.1 1G – first generation

This was the first generation of mobile phone technology. Launched in the late 1970s and early 1980s, the first generation of commercial cellular networks established general standards. In 1987 Australia got its first cellular mobile phone network, which used 1G analogue technology. Because 1G is an analogue technology, phones frequently had poor sound quality, short battery lives, and occasionally dropped calls. Established in the 1980s, these analogue communication standards were in use until 2G digital communications replaced them. The highest speed of 1G is 2.4 Kbps.

7.2.2 2G – second generation

The transition from 1G to 2G in cell phones was the first notable advancement. Analogue radio signals are used by 1G networks, whilst digital signals are used by 2G networks. This is the main difference between the two mobile phone networks. This generation's main objective was to provide safe and reliable routes of communication. The concepts of CDMA and GSM were put into practice. Second generation, 2G cellular phone networks were commercially introduced in Finland by Radiolinja in 1991. 2G capabilities are made feasible by multiplexing, which allows many users over a single channel. During the 2G era, cellular phones were used for both voice and data. The maximum 2G speeds for enhanced data rates for GSM evolution and general packet radio service are 1 Mbps and 50 Kbps, respectively.

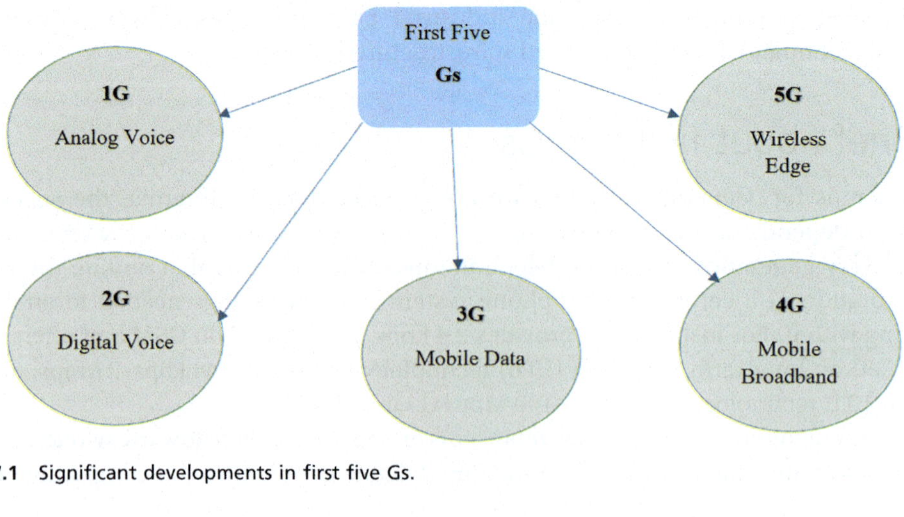

FIGURE 7.1 Significant developments in first five Gs.

7.2.3 3G – third generation

The great majority of the wireless technology we take for granted today was developed by this generation. Web browsing, email, downloading videos, sharing images, and other smartphone features were all part of the third generation of smartphones. With the stated goals of expanding voice and data capacity, supporting a wider range of applications, and facilitating less expensive data transfer, the third generation of mobile communication went on sale in 2001.

The 3G standard's main network architecture is the universal mobile telecommunication system. This network offers a significantly greater data rate by fusing new technology and protocols with aspects of the 2G network. Networks and services that adhere to the International Telecommunication Union's International Mobile Telecommunications-2000 (IMT-2000) standards are used by mobile devices and mobile telecommunications. IMT-2000 required a minimum speed of 200Kbps in order for a service to be considered 3G.

Multimedia service support and streaming are becoming more widespread over 3G. Universal access and mobility across different types of devices are possible with 3G. 3G boosted the frequency spectrum's efficiency by enhancing audio compression during conversations, allowing for more simultaneous calls in the same frequency range. The International Telecommunication Union IMT-2000 standard, which stipulates fixed speeds of 2 Mbps and roaming rates of 384 kbps, must be followed by a 'real' 3G network.

Similar to how 2G changed into 3G and 3.75G as new features were introduced to produce 4G, 3G changed into 3.5G and 3.75G. A 3G phone cannot connect to a 4G network, whereas a 4G phone can connect to a 3G or even a 2G network. Backward compatibility is typically a feature of newer phone generations.

7.2.4 4G – fourth generation

In contrast to 3G, 4G is a distinct technology that has only been made possible by advancements in technology over the past 10 years. Its objective is to provide customers with high-speed, high-quality, and high-capacity services while improving security and lowering the cost of IP-based internet access, multimedia, telephone, and data services. Potential and current applications include cloud computing, 3D television, video conferencing, gaming services, IP telephony, modified mobile web access, and high-definition mobile TV.

The key technologies that have made this possible are orthogonal frequency division multiplexing and multiple-input multiple-output (MIMO). Worldwide interoperability for microwave access and LTE, the two main 4G technologies, are also being replaced. When a device is moving, a 4G network's maximum speed is 100 Mbps; for low-mobility communication, like when standing or strolling, it is 1 Gbps. Along with a significant reduction in congestion, latency has dropped from approximately 300 ms to less than 100 ms.

4G was quicker than 3G when it was first made accessible. 4G LTE is not the same as 4G standards, even if it is very close to achieving them. A 4G phone can connect to a 3G or even 2G network because newer phone generations are frequently made backwards-compatible.

7.2.5 5G – fifth generation

5G will soon surpass 4G in performance. Significantly reduced latency, higher connection density, and notably faster data throughput are some advantages of 5G. The goals of 5G include increased cellular coverage, more energy efficiency, and device-to-device connectivity. More than 35 times faster than 4G, the highest 5G speed target is 35.46 Gbps. It's crucial to keep an eye on technologies like millimetre-wave mobile communications and massive MIMO. Recent developments like massive MIMO, millimetre wave, small cells, and Li-Fi could be used to enable connections for at least 100 billion devices and supply 10 Gbps to customers with previously unheard-of low latency. The date of the commercial rollout of 5G networks has been predicted in a number of ways.

7.3 Need for 5G in the current scenario

Future technology will enable the connection of an increasing number of devices, many of which will be used in industry. The new communications standard will provide a 5G connection for the IoT and individuals for the first time [21–25]. The primary needs of 5G in the current scenario are as follows:

7.3.1 Excellent for fast communications among different devices

With 5G, the existing mobile communication standards—2G, 3G, and 4G—are being improved. It is now possible to download data from the Internet considerably faster. The term '5G' refers to the creation of a mobile communications standard that emphasises human digital connectivity as well as communications between sensors, devices, and machines in the IoT. 5G is up to 20 times faster than 4G, with peak data transfer rates of 20 Gbps. For the first time, 'vital communications' in real time are now possible thanks to 5G networks. This suggests that wireless applications requiring the highest levels of security and dependability may be supported by 5G. This holds for both a factory and a remote-controlled crane. Whereas in the past, the performance of existing wireless technologies like Wi-Fi fell short, 5G offers new domains of application in business.

7.3.2 The competitive advantage of 5G

Nowadays, the number of IoT devices is several times more than the number of actual internet users. Over 70 billion gadgets will be connected by 2025, with many of them being used for industrial reasons. The new communications standard will be essential to linked and autonomous vehicles, and the smart factory of the future will be required. In turn, 5G is turning into a competitive factor as connectivity increases.

7.3.3 Production becomes safer, better, and more adaptable

In the hypothetical future factory, only the floor, ceiling, and walls will be immovable. Every other component is easily reconfigurable, portable, and adjustable. It is necessary to have a high-performance wireless infrastructure that enables efficient communication between facilities, equipment, and people. 5G will make it possible to deploy whole new industrial models. One example of 5G in production is mobile operating panels, which allow employees to connect flexibly to various systems and facilities and support even safety-critical applications like kill switches. These devices are now frequently permanently attached to all machines. With 5G, fewer people might utilise these devices, which would result in cheaper costs. Facilitating rapid and simple access to the equipment will help improve workers' working conditions. With the new mobile communications standard, augmented reality also offers significant benefits.

7.3.4 Data sovereignty is possible to ensure

Businesses can now create and manage small networks specifically designed for industrial applications thanks to the introduction of the new 5G mobile communications standard. Plant owners also have total authority over all pertinent security elements. Even in situations where the public network is overloaded, for example, local 5G networks will be able to operate and oversee the entire system. Campus networks will allow us to assume personal accountability for the toolkit, security, and configuration of the network. Data transit via open networks, like the nternet, is facilitated by dedicated interfaces that allow for constant monitoring.

7.3.5 The conversion made it easy for Industry 4.0

5G speeds up the IoT, accelerating the fourth industrial revolution. Control units and sensors installed in buildings, factories, and throughout logistical chains benefit from this fast and reliable real-time data flow, and a percentage of G-capable infrastructure is needed.

7.4 Advantages of 5G technology

The significant advantages of 5G are as follows [26–29]:

- Bi-directional massive bandwidth shaping with high resolution.
- 5G allows for faster data transfers and reduces the time between data transmission and reception.
- It is simple to manage with prior generations.
- The use of this technology makes the subscriber easier to use.
- In Industry 4.0, this 5G technology permits the real-time linking of IoT data, connected autos, and other smart products to the network.
- The 5G network's architecture significantly boosts resource provisioning capabilities.
- This technology brings all networks together on a single platform.
- 5G will play a critical role in enabling smart cities and smart homes, logistics monitoring, rethinking industrial production and maintenance operations, and transforming other IoT-ready enterprises
- This technology offers the infrastructure for the massive amounts of data required for AI to succeed.

- 5G technology eliminates risks for businesses and protects data sovereignty.
- This speeds up IoT and makes it more effective and efficient.
- Increases safety and greatly facilitates the integration of diverse technologies.
- Allow supply chains to connect to a website of interconnected devices from diverse locations in order to satisfy customer demand at the lowest cost.
- Provide a global connection that is uniform, uninterrupted, and constant.
- Wireless industrial monitoring system.
- Enhances the computer storage speed, allowing crucial independent gadgets to support all sorts of smart devices.
- With mobile 5G data, wearables can provide seamless service continuity while away from the phone range.
- 5G network slicing is an excellent method for supplying energy to intelligent grids.

7.5 Roles of 5G deployment for Industry 4.0 applications

Most industrial organisations are familiar with 5G applications. Several businesses have already adopted the 5G network. This technology increases factory network infrastructure in Industry 4.0, which manufacturing companies prefer [30–34]. The adoption and implementation of 5G networks will significantly impact the manufacturing industry. A smart plant uses a 5G network to gain numerous benefits and increased productivity. This allows manufacturers to tailor the network to their specific requirements [35–41]. The significant roles of 5G for Industry 4.0 are as follows:

7.5.1 Connect millions of devices

5G connects millions of new connected devices online, necessitating the safe transmission of more data. Thus it is necessary to improve present wireless networks to enable such connections. This technology can effectively address the need for firms and organisations across industries to adapt and achieve digital transformation. It connects devices and collects and uses critical data for the organisation. 5G has the potential to open the door to broader IoT applications by introducing new devices across industries.

7.5.2 Increased bandwidth

The advancements in 5G technology are also driving the proliferation of networks. 5G has much bandwidth, high speed, low latency, and many IoT capabilities. It enables network cutting and power-saving wide-area networking, resulting in a complete, cost-effective ecosystem critical for efficient industrial networks. 5G enables seamless connections across numerous technologies and speedy upgrades, access, and administration. This technology can improve customer satisfaction by reducing the time it takes to handle claims, owing to faster access to data points combined with AI and data analytics.

7.5.3 New standards are being developed in detail

In addition to actively supporting the creation and technical deployment of the new standard and long-term communication solutions, Industry 4.0 is presently investigating the topic of industrial 5G. The potential of 5G wireless connectivity for industrial applications is highly anticipated. Smart manufacturing, logistics, and extremely flexible manufacturing for self-driving cars are among the industrial applications that the 5G network assesses. Additionally, it makes it possible for artificial intelligence to be used in industrial settings. Digital service providers must leverage 5G networks to deliver new services to diverse customer segments.

7.5.4 Digitisation and automation

5G will help drive automation and digitalisation to the next level, resulting in significant benefits and advancements in manufacturing. 5G can detect, tackle, and prevent moving threats and humans from colliding. These can be used to transmit situational information to management. The specially designed software may be downloaded instantly during an assembly at the exact moment it is needed, thanks to reliable 5G connectivity across the shop floor. The software is constantly updated and installed with the most recent version of this technology.

7.5.5 Customised solutions

In Industry 4.0, a 5G network provides customised solutions. The manufacturing industry is undergoing a significant transformation in terms of generation. In the digitalisation and advancement of industrial processes and ecosystems, Industry 4.0 aims to deliver product lines that are considerably more adaptable, effective, and sustainable. 5G helps businesses and their supply chain partners accomplish this shift by allowing them to employ new technologies like AI and machine vision to make their organisations visible in all sectors.

7.5.6 Optimisation of performance

A 5G network's primary benefits are speed and stability, contributing to instant performance improvements. Another advantage of this platform is real-time monitoring of data, enabling machine status monitoring without requiring access. It allows operators to work more effectively while improving information flow on the status of the entire production line.

7.5.7 Offer high speed

5G aims to provide the ultimate miles of connection for successful IoT deployment by offering speed, reliability, power, and mobility. The goal of 5G networks is to provide high-speed connections while maintaining service requirements. 5G is widely available for digitally connected systems and provides extensive advancement due to its high-speed capability.

7.5.8 Provide the proper status of the plant

Plant managers will be kept informed of the status and progress of the 5G installation. This provides continuous equipment monitoring, helping save money. Instead of replacing a machine part, the manufacturer may monitor it and replace it as needed. 5G has significant benefits for many businesses through network connectivity.

7.5.9 Boost manufacturing output

5G technology offers new opportunities to expand manufacturing output through digital transformation. Production facilities can monitor thousands of field devices, and real-time data analytics can be conducted on the edge. Site managers may monitor site operations and improve the manufacturing output.

7.5.10 Digital monitoring

The 5G network enhances automation and digital monitoring systems. Various problems may be recognised by AI-enabled 5G cameras that use video analytics to optimise flow in real time. 5G will transform the industry by delivering better speed and connectivity in manufacturing. The 5G smart factory is the fastest and most productive, with significant increases in production quality and efficiency.

7.5.11 Supply chain optimisation

5G improve supply chains since it is necessary to optimise operations for the real-time flow of information on the item location. Supply chains that demand high synchronisation of incoming parts will become more efficient using a 5G tracking system. The automobile plant that synchronises parts delivery from outside can search for shipment status and incorporate that tracking data into its workflow to provide real-time arrival information for parts, assemblies, and subassemblies.

7.5.12 Confront several issues

The industry can benefit from 5G networks as it faces challenges. 5G networks, unlike wired networks, are operated locally and have the specialised infrastructure for excellent local coverage. The networks are specifically built and customised for industrial applications with high dependability requirements. The 5G network specification supports millions of devices through many machine-to-machine connections from Industry 4.0, resulting in increased capacity and speed.

7.5.13 Data protection

The 5G network is specialised and autonomous, which provides data privacy and security. Overall, 5G networks offer control and flexibility through network slicing, large capacity, low latency, and low prices. 5G network provides tracking, automated operations, remote robotic production, remote product assistance, predictive maintenance, supply chain integration, and inventory management.

7.5.14 Information about machines

5G enables machines to communicate and share data. This opens the door to more advanced real-time visualisations and simulations. This can quickly evaluate production efficiency by looking at raw material availability, robot performance, and availability. 5G networks have significantly improved capabilities for connecting machines and other essential devices compared to the previous generation.

7.6 Roles of digital twin for 5G

Digital twins are computer-generated images of real-world systems, processes, or objects that can be utilised for study, monitoring, and simulation. Digital twins are essential for optimising network architecture, boosting performance, and guaranteeing effective resource deployment in the context of 5G planning. The capacity of digital twins to offer a comprehensive and up-to-date picture of the network environment is one of its main advantages in 5G planning [42–45]. It is essential to 5G because they give operators an effective tool for network infrastructure optimisation, performance enhancement, and resource deployment. Operators can make better judgements, proactively manage the network, and provide users with high-quality service by utilising the capabilities of digital twins. This enables better decision-making and more proactive network management, enhancing performance and reliability. This technology can also be used to detect potential problems or irregularities and to track the 5G network in real time [46–49]. The significant roles of digital twin for 5G are as follows:

- It optimises tower sites and coverage by simulating the deployment of 5G cells.
- This allocates spectrum to reduce interference and increase bandwidth effectiveness.
- To improve signal strength and reduce latency, beam-steering patterns are simulated using a digital twin.
- Predicts and minimises ultra-low latency problems by analysing real-time data.
- This helps low-latency 5G applications by simulating the installation of edge nodes.
- This technique forecasts situations of peak demand using real-time traffic data.
- It uses real-time twin models to optimise base station power consumption.
- This finds abnormalities in the network and forecasts 5G infrastructure failures.
- Facilitates automated setup and optimisation.
- This technology replicates quality of service for critical applications such as self-driving cars.
- This simulates how signals move through crowded smart cities.
- It simulates potential cyberattacks on 5G virtual networks.
- This simulates service restoration and traffic rerouting in an emergency.

7.7 Upcoming goals of 5G

The major upcoming goals of 5G are as follows [50–52]:

- In the future, 5G will play a critical role in the development of efficient communication networks and data transfer.
- More devices will be linked to one another, with many of them being used in business.
- High-speed network transfer rates on 5G connection devices will allow super-fast work from faraway locations.
- Data may now be downloaded from the internet much faster.
- Future 5G applications will have diverse needs for speed, coverage, latency, and reliability.
- This technology will focus on digital connectivity to improve communication between sensors, equipment, and machines.

7.8 Disadvantages of 5G

The significant disadvantages of 5G are as follows [53–55]:

- 5G frequency interruption is caused by numerous impediments, such as towers, walls, trees, and other structures, which have an impact on the network's overall performance.
- The technology is being deployed, and a study into its long-term effect undergoing.
- Another constraint is the expense, which is due to the costly maintenance required to maintain high-speed connectivity.
- The claimed speed of this technology appears challenging to attain due to the lack of technological assistance in most regions of the world.
- Battery-draining concerns have arisen.
- Many older gadgets will be incompatible with 5G, necessitating their replacement, which is a pricey proposition.
- Infrastructure development is costly.
- When there is a requirement for intensive cloud data uploading duties, the differences in uploading and downloading speeds exacerbate the situation.
- The question of security and privacy has yet to be resolved.

7.9 Conclusion

In Industry 4.0, 5G is a breakthrough that takes the manufacturing industry to a much greater degree of digitisation, primarily by increasing mobility, flexibility, and production adaptability. This technology has promise in industrial production environments, including process automation, warehousing, logistics, human-machine interaction, supervision, and maintenance. The supply chain's production planning and logistics are aided by 5G-enabled infrastructure, which enhances direct product manufacturing.

In Industry 4.0, many devices are connected to the floor. 5G connections between these devices might provide even better communication. It will provide a solid, safe option for achieving outstanding results and will help develop diverse, reliable, and easily adjustable systems. 5G technology will aid in developing exceptional assets by improving manufacturing concepts. This technology has a significant impact on the industry supply chain. Manufacturing operations are becoming smarter every day, using 5G technology that enables improved communication between networked equipment.

5G connectivity boosts industry automation and enables next-generation digital transformations, including virtual reality, robots, real-time translation, training, and e-health applications. 5G networks claim to be more versatile due to network slicing possibilities. This may significantly improve the customer experience with a better network connection. 5G technology is used in Industry 4.0 and helps to support various technologies used in this revolution.

References

[1] S.K. Rao, R. Prasad, Impact of 5G technologies on Industry 4.0, Wireless Personal Communications 100 (1) (2018) 145–159.

[2] D. Peraković, M. Periša, P. Zorić, I. Cvitić, Development and implementation possibilities of 5G in Industry 4.0, Design, Simulation, Manufacturing: The Innovation Exchange, Springer International Publishing, Cham, (2020), pp. 166–175.

[3] D. Mourtzis, J. Angelopoulos, N. Panopoulos, Smart manufacturing and tactile Internet based on 5G in Industry 4.0: challenges, applications and new trends, Electronics 10 (24) (2021) 3175.

[4] M. Javaid, A. Haleem, S. Rab, R.P. Singh, R. Suman, S. Mohan, Progressive schema of 5G for Industry 4.0: features, enablers, and services, Industrial Robot: The International Journal of Robotics Research and Application 49 (3) (2022) 527–543.

[5] M. Gundall, M. Strufe, H.D. Schotten, P. Rost, C. Markwart, R. Blunk, ... D. Wübben, Introduction of a 5G-enabled architecture for the realisation of Industry 4.0 use cases, IEEE Access 9 (2021) 25508–25521.

[6] I. Rodriguez, R.S. Mogensen, A. Fink, T. Raunholt, S. Markussen, P.H. Christensen, ... O. Madsen, An experimental framework for 5G wireless system integration into Industry 4.0 applications, Energies 14 (15) (2021) 4444.

[7] J. Ordonez-Lucena, J.E. Chavarria, L.M. Contreras, A. Pastor, The use of 5G Non-Public Networks to support Industry 4.0 scenarios, in: 2019 IEEE Conference on Standards for Communications and Networking (CSCN), 2019, October,pp. 1-7. IEEE.

[8] C. Stefanović, Industry 4.0 from 5G perspective: Use-cases, requirements, challenges and approaches, in: 2018 11th CMI International Conference: Prospects and Challenges Towards Developing a Digital Economy within the EU, IEEE, 2018 November, pp 44-48.

[9] S. Vakaruk, J.E. Sierra-García, A. Mozo, A. Pastor, Forecasting automated guided vehicle malfunctioning with deep learning in a 5G-based Industry 4.0 scenario, IEEE Communications Magazine 59 (11) (2021) 102–108.

[10] C. Mannweiler, B. Gajic, P. Rost, R.S. Ganesan, C. Markwart, R. Halfmann, A. Wich, Reliable and deterministic mobile communications for Industry 4.0: key challenges and solutions for the integration of the 3GPP 5G system with IEEE, Mobile Communication-Technologies and Applications; 24. ITG-Symposium, VDE, (2019, May), pp. 1–6.

[11] C.C. Lessi, A. Gavrielides, V. Solina, R. Qiu, L. Nicoletti, D. Li, 5G and beyond-5 G technologies enabling Industry 5.0: network applications for robotics, Procedia Computer Science 232 (2024) 675–687.

[12] T.A. Kuzovkova, O.I. Sharavova, M.M. Sharavova, The role of 5G and 6G mobile communication technologies for Industry 4.0 and the transition to Industry 5.0, in: 2024 International Conference on Engineering Management of Communication and Technology (EMCTECH), IEEE, 2024, October, pp. 1-6.

[13] J. Sasiain, A. Sanz, J. Astorga, E. Jacob, Towards flexible integration of 5G and IIoT technologies in Industry 4.0: a practical use case, Applied Sciences 10 (21) (2020) 7670.

[14] 1G Vs. 2G Vs. 3G Vs. 4G Vs. 5G. http://net-informations.com/q/diff/generations.html.

[15] Z. Diao, F. Sun, Application of Internet of Things in Smart Factories under the background of Industry 4.0 and 5G communication technology, Mathematical Problems in Engineering 2022 (1) (2022) 4417620.

[16] M. Gundall, J. Schneider, H.D. Schotten, M. Aleksy, D. Schulz, N. Franchi, J. Grießbach, 5G as enabler for industry 4.0 use cases: challenges and concepts, in: 2018 IEEE 23rd International Conference on Emerging Technologies and Factory Automation (ETFA), IEEE, 2018, September, vol. 1, pp. 1401-1408.

[17] W. Nakimuli, J. Garcia-Reinoso, J.E. Sierra-Garcia, P. Serrano, I.Q. Fernández, Deployment and evaluation of an Industry 4.0 use case over 5G, IEEE Communications Magazine 59 (7) (2021) 14–20.

[18] J. Ansari, C. Andersson, P. de Bruin, J. Farkas, L. Grosjean, J. Sachs, ... R.H. Schmitt, Performance of 5G trials for industrial automation, Electronics 11 (3) (2022) 412.

[19] P. Varga, S. Bácsi, R. Sharma, A. Fayad, A.R. Mandeel, G. Soos, ... D. Ficzere, Converging telco-grade solutions 5 G and beyond to support production in Industry 4.0, Applied Sciences 12 (15) (2022) 7600.

[20] D. Kim, A 2020 perspective on "A dynamic model for the evolution of the next generation Internet–Implications for network policies": towards a balanced perspective on the Internet's role in the 5G and Industry 4.0 era, Electronic Commerce Research and Applications 41 (2020) 100966.

[21] 5 reasons for 5G. https://www.bosch.com/stories/5g-industry-4-0/.

[22] W. Tärneberg, O. Hamsis, J. Hedlund, K. Brunnström, E. Fitzgerald, A. Johnsson, C. Kilinc, Towards intelligent industry 4.0 5G networks: a first throughput and QoE measurement campaign, in: 2020 International Conference on Software, Telecommunications and Computer Networks (SoftCOM), IEEE, 2020, September, pp. 1-6.

[23] M. Attaran, The impact of 5G on the evolution of intelligent automation and industry digitisation, Journal of Ambient Intelligence and Humanised Computing 14 (5) (2023) 5977–5993.

[24] J. García-Morales, M.C. Lucas-Estañ, J. Gozalvez, Latency-sensitive 5G RAN slicing for Industry 4.0, IEEE Access 7 (2019) 143139–143159.

[25] A. Mahmood, L. Beltramelli, S.F. Abedin, S. Zeb, N.I. Mowla, S.A. Hassan, ... M. Gidlund, Industrial IoT in 5G-and-beyond networks: vision, architecture, and design trends, IEEE Transactions on Industrial Informatics 18 (6) (2021) 4122–4137.

[26] X. Zou, K. Li, J.T. Zhou, W. Wei, C. Chen, Robust edge AI for real-time Industry 4.0 applications in 5 G environment, IEEE Communications Standards Magazine 7 (2) (2023) 64–70.

[27] J. Navarro-Ortiz, J.J. Ramos-Munoz, F. Delgado-Ferro, F. Canellas, D. Camps-Mur, A. Emami, et al., Combining 5G new radio, Wi-Fi, and LiFi for Industry 4.0: performance evaluation, Sensors 24 (18) (2024) 6022.

[28] S. Michaelides, S. Lenz, T. Vogt, M. Henze, Secure integration of 5G in industrial networks: state of the art, challenges and opportunities, Future Generation Computer Systems 166 (2025) 107645.

[29] A. Larrañaga, M.C. Lucas-Estañ, S. Lagén, Z. Ali, I. Martinez, J. Gozalvez, An open-source implementation and validation of 5G NR configured grant for URLLC in ns-3 5G LENA: a scheduling case study in Industry 4.0 scenarios, Journal of Network and Computer Applications 215 (2023) 103638.

[30] F. Longo, A. Padovano, G. Aiello, C. Fusto, A. Certa, How 5G-based industrial IoT is transforming human-centered smart factories: a quality of experience model for operator 4.0 applications, IFAC-PapersOnLine 54 (1) (2021) 255–262.

[31] L. Chinchilla-Romero, J. Prados-Garzon, R. Vasist, M. Goordazi, Economic feasibility of 5G-based autonomous mobile robots solutions for Industry 4.0, IEEE Communications Magazine 62 (11) (2024) 52–59.

[32] P. Varga, J. Peto, A. Franko, D. Balla, D. Haja, F. Janky, ... L. Toka, 5G support for industrial IoT applications—challenges, solutions, and research gaps, Sensors 20 (3) (2020) 828.

[33] C. Lahari, M.Y. Nayudu, V. Veeraiah, A. Gupta, R. Anand, N. Sindhwani,, et al., Leveraging 5G for Industry 4.0 optimising business processes and driving automation, in: 2025 12th International Conference on Reliability, Infocom Technologies and Optimisation (Trends and Future Directions)(ICRITO), IEEE, 2025, September,pp. 1-6.

[34] A. Mahmood, S.F. Abedin, T. Sauter, M. Gidlund, K. Landernäs, Factory 5G: a review of industry-centric features and deployment options, IEEE Industrial Electronics Magazine 16 (2) (2022) 24–34.

[35] L. Chettri, R. Bera, Industry 4.0: communication technologies, challenges and research perspective towards 5G systems, in: International Conference on Communication, Devices and Networking, Singapore, Springer Singapore, 2019, December, pp. 67-77.

[36] R.O. Flores-Castañeda, S.M. Ledesma, S. Olaya-Cotera, 5G networks and their impact on the evolution of real-time monitoring and control systems for Industry 4.0: a systematic review, Quality & Quantity 60 (2026) 4953–4975. Available from https://doi.org/10.1007/s11135-025-02476-1.

[37] Y.R. Wei, A.S. Keshavamurthy, R. Wittmann, A.R. Zahonero, A standalone 5G industrial testbed design considerations for Industry 4.0, in: 2022 52nd European Microwave Conference (EuMC), IEEE, 2022, September, pp. 884-887.

[38] R. Bolla, R. Bruschi, K. Burow, F. Davoli, Z. Ghrairi, P. Gouvas, A. Zafeiropoulos, From cloud-native to 5G-ready vertical applications: an Industry 4.0 use case, in: 2021 IEEE 22nd International Conference on High Performance Switching and Routing (HPSR), IEEE, 2021, June, pp. 1-6.

[39] A. Larrañaga, M.C. Lucas-Estañ, I. Martinez, J. Gozalvez, 5G configured grant scheduling for 5G-TSN integration for the support of industry 4.0, in: 2023 18th Wireless On-Demand Network Systems and Services Conference (WONS), 2023 Januarypp. 72-79.

[40] D. Xu, K. Yu, J.A. Ritcey, Cross-layer device authentication with quantum encryption for 5G-enabled IIoT in Industry 4.0, IEEE Transactions on Industrial Informatics 18 (9) (2021) 6368–6378.

[41] M.K. Hawari, N.A. Apandi, Industry 4.0 with intelligent manufacturing 5G mobile robot based on genetic algorithm, Indonesian Journal of Electrical Engineering and Computer Science 23 (3) (2021) 1376–1384.

[42] H.X. Nguyen, R. Trestian, D. To, M. Tatipamula, Digital twin for 5G and beyond, IEEE Communications Magazine 59 (2) (2021) 10–15.

[43] R. Ramirez, C.Y. Huang, S.H. Liang, 5G digital twin: a study of enabling technologies, Applied Sciences 12 (15) (2022) 7794.

[44] M.S. Rodrigo, D. Rivera, J.I. Moreno, M. Alvarez-Campana, D.R. López, Digital twins for 5 G networks: a modelling and deployment methodology, IEEE Access 11 (2023) 38112–38126.

[45] R. Dong, C. She, W. Hardjawana, Y. Li, B. Vucetic, Deep learning for hybrid 5G services in mobile edge computing systems: learn from a digital twin, IEEE Transactions on Wireless Communications 18 (10) (2019) 4692–4707.

[46] U.D. Maiwada, K.U. Danyaro, A.B. Sarlan, Enhancing 5G energy efficiency through digital twin networks: a comprehensive review, Digital Twins and Applications 1 (1) (2024) 4–25.

[47] I. Vilà, O. Sallent, J. Pérez-Romero, On the design of a network digital twin for the radio access network in 5G and beyond, Sensors 23 (3) (2023) 1197.

[48] C. Hu, W. Fan, E. Zeng, Z. Hang, F. Wang, L. Qi, et al., Digital twin-assisted real-time traffic data prediction method for 5G-enabled Internet of vehicles, IEEE Transactions on Industrial Informatics 18 (4) (2021) 2811–2819.

[49]] H. Tran-Dang, D.S. Kim, Digital twin-empowered intelligent computation offloading for edge computing in the era of 5G and beyond: a state-of-the-art survey, ICT Express 11 (1) (2025) 167–180.

[50] V. Chandra Shekhar Rao, P. Kumarswamy, M.S.B. Phridviraj, S. Venkatramulu, V. Subba Rao, 5G enabled industrial Internet of things (IIoT) architecture for smart manufacturing, in: Data Engineering and Communication Technology: Proceedings of ICDECT 2020, Singapore, Springer Singapore, 2021, pp. 193-201.

[51.] K. Muthineni, A. Artemenko, J. Vidal, M. Nájar, A survey of 5G-based positioning for Industry 4.0: state of the art and enhanced techniques, in: 2023 Joint European Conference on Networks and Communications & 6G Summit (EuCNC/6G Summit), IEEE, 2023, June, pp. 120-125.

[52] J.M. de Freitas Vilela, J.R. Faria Filho, Industry 4.0 and 5G technology on firms network: a balanced competitive expansion conceptual model development, Brazilian Journal of Operations & Production Management 19 (3) (2022) 1–18.

[53] L. Chinchilla-Romero, J. Prados-Garzon, P. Ameigeiras, P. Muñoz, J.M. Lopez-Soler, 5G infrastructure network slicing: E2E mean delay model and effectiveness assessment to reduce downtimes in Industry 4.0, Sensors 22 (1) (2021) 229.

[54] P. Maiti, S. Prabhu, P. Ashok, Decentralised defences: a framework for 5G IoT security in Industry 4.0 using blockchain crowdsourcing, in: 2024 3rd International Conference on Sentiment Analysis and Deep Learning (ICSADL), IEEE, 2024, March, pp. 534-538.

[55] M. Noor-A-Rahim, J. John, F. Firyaguna, H.H.R. Sherazi, S. Kushch, A. Vijayan, ... E. Armstrong, Wireless communications for smart manufacturing and industrial IoT: existing technologies, 5G and beyond, Sensors 23 (1) (2022) 73.

8

Robotics for Industry 4.0

8.1 Introduction to robotics

Robotics is an area of engineering and computer science that is also useful in manufacturing, operations, and complex task performance. Robotics is the study of creating devices that can replace people and mimic their behaviours [1,2]. This technology can be employed across various scenarios and objectives. Robotics is an essential Industry 4.0 technology offering a wide range of capabilities in industry. Adopting this technology has improved automation systems, enabling them to perform repetitive tasks with greater accuracy and at lower cost [3–5].

Robotics is gradually paving the way for the production of high-quality goods while preserving the value of existing collaborative schemes. Robots rely on actuator systems such as DC motors, servo motors, and stepper motors to perform real-world motion and manipulation. The components used must meet the robots' operational and architectural requirements. The component size is progressively raised to meet the performance standards. The actuators are sized based on a single operating speed/operating torque point [6–8].

Robotics is employed in the automobile sector and is now used in various fields. This protects employees from tedious, worldly, and harmful repetitive tasks and replaces them with more appropriate professions like engineering, programming, administration, and machinery repair [9,10]. Robotics enables companies to leverage their employees' skills across other market areas. Manufacturing robots are more efficient, allowing companies to thrive on larger contracts. In a context where green production is becoming more critical, robots save resources by eliminating the need for climate control or lighting, resulting in cleaner environments [11,12].

8.2 Common types of robots

The engineering discipline of robotics is concerned with the development, design, manufacture, and application of robots. The goal of robotics is to create intelligent devices that can assist humans in many ways. A robot can be a machine that mimics a human or a robotic programme, like robotic process automation, that simulates how humans use software to perform repetitive, rule-based tasks. As robotics manufacturers continue to make advancements in capabilities, cost, and form factors, robotics solutions are being utilised in a variety of industries and applications [13–17]. The common types of robotics are as follows:

Digital Twin and Industry 4.0. DOI: https://doi.org/10.1016/B978-0-443-54019-6.00015-7

8.2.1 Autonomous mobile robots

Autonomous mobile robots (AMRs) make snap decisions while travelling the world. Sensors and cameras allow them to gather information about their surroundings. Onboard processing equipment allows them to evaluate the circumstances and make an intelligent decision, whether they are moving to avoid an approaching employee, picking the best cargo, or deciding which surface to disinfect. These are portable solutions that require minimal human involvement.

8.2.2 Automated guided vehicles

Automated guided vehicles (AGVs) frequently need human monitoring and depend on rails or preestablished paths, whereas AMRs can move independently. Distributing goods and moving items are common uses for regulated spaces like warehouses and production floors.

8.2.3 Articulated robots

The purpose of articulated robots is to replicate human arm movements. Usually, these have 2 to 10 rotary joints. Because each additional joint or axis offers a greater range of motion, these are ideal for arc welding, machine tending, material handling, and packing.

8.2.4 Humanoids robots

The term is widely used to describe robots that carry out human-centric jobs and often have human-like shapes, even though many mobile humanoid robots may be classified as AMRs. They use many of the same technical components as AMRs to perceive, plan, and act when carrying out tasks like delivering instructions or concierge services.

8.2.5 Cobots

Cobots are collaborative robots that assist or work directly for people. Unlike the majority of other robots, which carry out their tasks alone or in separate workplaces, cobots can share workspaces with workers to facilitate more productive work. They are often used to eliminate time-consuming, dangerous, or taxing tasks from routine procedures. Sometimes, cobots are able to detect and respond to human movement.

8.2.6 Hybrids

Different types of robots are sometimes combined to create hybrid systems that can carry out more difficult tasks. For example, an AMR and a robotic arm could be used to create a robot that handles goods inside a warehouse. Consolidation of functionality into single solutions is closely related to the consolidation of processing capabilities.

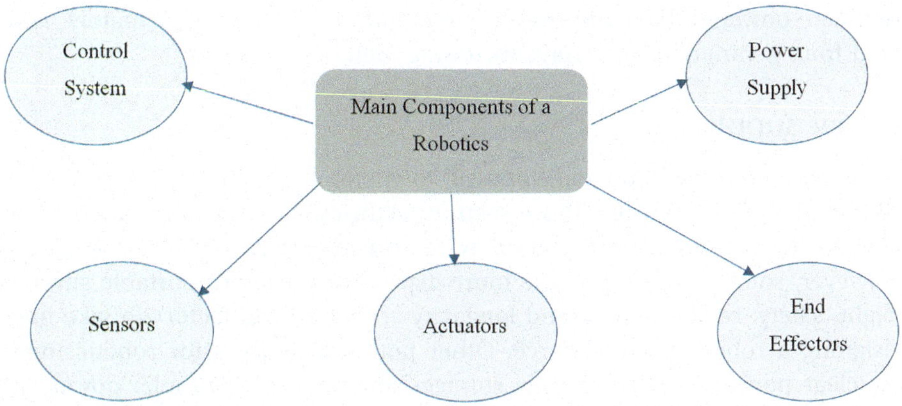

FIGURE 8.1 Main components of robotics.

8.3 Main components of a robotics

Robots require a wide range of specialised components to perform their jobs because they are made to meet several demands and serve multiple purposes. Nonetheless, specific parts, like a central processing unit or power supply, are necessary to build any robot [18–23]. As illustrated in Fig. 8.1, robotics components can be broadly categorised into five classes.

8.3.1 Control system

The components of a robot's control system, or central processing unit, are regarded as part of the computation. Control systems are designed to teach a robot how to use specific parts to complete a task, just like the human brain. These robotic occupations include assembly-line packaging and minimally invasive surgery.

8.3.2 Sensors

Through the use of sensors, which provide stimuli in the form of electrical impulses that the controller evaluates, robots can engage with their external surroundings. Photoresistors, which serve as eyes, video cameras, which serve as eyes, and microphones, which serve as ears, are common components of robots. With the help of these sensors, the robot can monitor its environment, determine the best course of action given the current situation, and use the controller to send commands to the other components.

8.3.3 Actuators

Only if a machine's body or frame can move can it be considered a robot. Actuators are the parts that make this movement. These components are made up of motors that cooperate to carry out the necessary movements after receiving instructions from the control system. Actuators are available in a variety of forms to effectively fulfil their specific functions.

They are often powered by compressed pneumatic or hydraulic actuators and can be constructed from a variety of materials, including metal or elastic.

8.3.4 Power supply

Similar to the human body, robots need power to operate. Usually, robots run on an internal battery. But stationary robots, like those seen in factories, can also run on AC power from a wall outlet. Due to their endurance and safety, lead-acid batteries are used by the majority of robots; however, some may employ the more expensive but more portable silver-cadmium kind. Weight, safety, replaceability, and longevity are all crucial factors to take into account when designing a robot's power source. Other potential sources for robotic improvement include nuclear power, flywheel energy storage, solar power, hydraulic power, pneumatic power from compressed gases, and organic waste through anaerobic digestion.

8.3.5 End effectors

The physical, external components that allow robots to carry out their tasks are called end effectors. These days, robots are widely employed in manufacturing, even for interchangeable tools like drills and paint sprayers. Scalpels are a common feature of surgical robots, whereas other robot types may have hands or grabbing claws for jobs like packing and distribution.

8.4 Different industries using robotics

There are several ways that organisations employ robots. All commonly used robots are employed to improve results and lighten staff workload, allowing staff to concentrate on the most critical and valuable tasks [14,24–28]. The major industries which use robotics are as follows:

8.4.1 Manufacturing

The manufacturing industry has long been at the forefront of using different types of robots to make money. Often employed in conjunction with programmable logic controllers, AMRs, AGVs, articulated robots, and cobots are utilised on production floors and warehouses to expedite processes, boost productivity, and improve safety. They are used for a number of tasks, including welding, assembling, transporting goods, and upholding warehouse security.

8.4.2 Farming and agriculture

By utilising their exceptional intelligence, AMRs for farming and agriculture assist farmers in harvesting crops more efficiently and rapidly. Agricultural robots are able to recognise when a crop is ready, remove leaves and branches, and delicately and accurately pick the produce without causing any damage.

8.4.3 Healthcare

In the healthcare industry, using different kinds of robots enhances patient satisfaction. AMRs are used for medicine delivery, surface disinfecting, and mobile telepresence. These can assist medical staff throughout rehabilitation or help nurses give better patient care.

8.4.4 Logistics and supply chain

Robots help transportation and logistics companies deliver goods and services quickly and efficiently. They use AMRs and AGVs to process items, which increases accuracy and speed. They also utilise AMRs to ensure the safe delivery of goods to customers by carrying them the last mile.

8.4.5 Retail and hospitality

There are several ways to use robotics to enhance the experience of customers or guests. Businesses in the retail and hospitality sectors are using robotics to automate inventory processes, provide wayfinding or concierge services, clean different areas, and help customers with their bags or valet parking.

8.4.6 Smart cities

Robotics helps create safer, more intelligent cities. Information and navigation services are offered by humanoid robots. AMRs are used for both delivery and routine security. Additionally, robotics can collect data for building models, expedite building development, and conduct site surveys.

8.5 Advantages of robotics

The significant advantages of Robotics are as follows [29–33]:

- Perform multiple functions at the same time.
- Major food firms have introduced robots in their production facility to limit contamination.
- In the automobile sector, robots are also utilised to screw, assemble, manipulate, mark, and control quality.
- They can measure, cut, pack, and palletise anything.
- It provides better safety and sustains high temperatures without affecting operations.
- This technology provides improved quality work.
- Reduce wastage as the work is precise.
- Perform work rapidly with improved process efficiency.
- Conduct complicated optical procedures with pinpoint accuracy.
- Ability to work in dangerous environments with lower risks.
- Can operate at all times, undertake cyclic jobs quite well, which increases productivity.

- Deliver food and other essential items at designated locations with lower human intervention.
- It became more versatile in manufacturing and carrying out creative activities.
- Picking, packing, and palletising can be done easily.
- Now have extensive heavy-duty operations and save time.

8.6 Roles of robotics for Industry 4.0

Robotics is a cutting-edge technology used in many new sectors. Due to their monotonous, repetitive work, many floor employees suffer from fatigue, weakness, and other physical discomforts. This technology reduces or eliminates physical stress, helps the floor, and appropriately supports staff in their duties [34–38]. The following are important functions of robotics in Industry 4.0:

8.6.1 Automation

Robotics provides automation, which helps to enhance the manufacturing system. It enables flexible automation by connecting several machine tools with a material-handling system controlled by a computer. A variety of automation tasks can be easily performed in Industry 4.0.

8.6.2 Work in a high-temperature environment

Employees may use robots to accomplish duties that are too dangerous or complex for them. High operating temperatures, hazardous pollutants, and prolonged exposure to loud sounds may wreak havoc. Dangerous employee operations can now be delegated to robots.

8.6.3 Performing a difficult task

This technology can efficiently perform complex tasks. It uses electronic equipment to perform physically demanding tasks, enhancing its personnel's health and well-being. Automation will automatically comprehend and adapt the material in the 'factory of the future' to deliver a specified, dependable operation. Finding painting professionals is hard, and the work is highly hazardous. This technology can efficiently perform painting tasks with high precision.

8.6.4 Streamlined production

Production may run more smoothly from planning to execution. Automated machines can offer finished items for sale in warehouses, bringing supply chains closer and more effectively. Most importantly, the number of interaction jobs for robotic machines has decreased, making them more versatile and easier to programme. Predictive maintenance was improved. As a result, they led to a smooth production system and better outcomes for Industry 4.0.

8.6.5 Boost productivity and dependability

In several industries, robotics is utilised to improve productivity and reliability while cutting production costs. Robots collaborate with employees to complete routine, single-task, or complex tasks under their supervision and guidance. Robotics development is also becoming more secure. This can use cameras, alerts, and automated shut-offs to detect and prevent unauthorised entry into the workplace. In the public safety industry, robotics is deployed in various applications to increase productivity.

8.6.6 Extraction of data

The data generated by robots for performing daily operations is enormous. Much of this data may be structured, supplied regularly, and backed up to ensure data consistency and availability. Companies are urged to make more efficient decisions based on precise performance statistics. This boosts innovation and allows unprecedented efficiency levels in Industry 4.0.

8.6.7 Self-management

Because of their capacity to observe, create, and interpret information, robots can not only complete a mission without human assistance, but they are also self-managed. This allows machines to communicate with one another. The growth of decentralised systems and robots' ability to communicate with people via integrated interfaces has made collaborative work more accessible.

8.6.8 Increased production effectiveness

Industrial robotics, including the Industrial Internet of Things and 3D printing robots, can have higher production efficiency and accuracy than other technologies. The continuous running of an industrial robot helps to increase production speed. It reduces cycle time and improves performance. Robots reduce the risk of worker injury, which is especially important in hostile settings where production is required.

8.6.9 Working nonstop and for lengthy periods

This technology can operate continuously for extended periods. To complete the entire workflow, the department's automation monitors application validation using this technology. Robots enable the rapid, flexible, and cost-effective automation of internal product flows.

8.6.10 Movement of materials

Internal transportation is typically handled manually by highly skilled people, which harms performance. Now, in many corporations, internal transportation is handled by robotics without interfering with manufacturing facility infrastructure. Autonomous robots may navigate complex environments using their sensing systems and protection algorithms. Robots are used to transport materials and workpieces from one location to another.

8.6.11 Efficient for assembly lines

Robotics provides efficient performance in assembly lines. This precisely performs the entire assembly-line operation at a lower cost and in less time. This technology can assemble smaller components at rapid rates. Robots accomplish other tasks, including screw driving, tyre assembly, and windshield installation.

8.6.12 Inspection

Inspection and quality control have advanced uses of robotic technology. They deployed cameras to observe their surroundings, resulting in some of the most profitable and efficient outcomes. Numerous uses are made possible by attaching a smart camera to the end of a robotic arm, which may move over the component being examined to track several attributes. For robot-based inspection and guidance, machine vision has advanced significantly, delivering practical outcomes through image capture and interpretation.

8.6.13 Management of the whole process

Automobile manufacturers are now experimenting with robotics in a variety of procedures. This technology creates a more powerful, precise, scalable, and dependable environment. The automobile industry has remained one of the most automated production chains in the world and one of the most significant users of robots.

8.6.14 Dark factories

Dark factories are fully automated manufacturing facilities that operate without any human beings; they can run 24/7 and function in complete darkness. It is said that they produce goods at lower cost and with greater efficiency. Dark factories are built with the support of specialised robotics and AI. They have stationary industrial robots to do high-precision tasks and mobile robotics: AMRs. Further, AGVs are widely used to transport materials and tools across the floor without human oversight. It also employs humanoid robots to handle complex tasks, such as automotive quality inspection. Here, in a dark factory, digital twin is designed to serve as the factory's central nervous system. Here, the digital twin serves as the primary interface for remote management between the physical manufacturing operations and off-site human supervisors, providing a real-time, bidirectional data link.

8.7 Significant role of digital twin for robotics

Digital twins are accurate digital replicas of the real world that mimic, assess, track, and enhance performance using dynamic data. Digital twins enable us to test every motion and procedure in production virtually before they are carried out in real life. They are used in a variety of industries, such as aerospace, where engineers and designers use digital models of aircraft engines to track performance, forecast maintenance, and improve safety. The digital twin is currently reshaping the adoption of robotics in these important industries. The digital twin transforms from a one-time simulation tool into a live, breathing replica of the real robot, replicating its actions in real time, evaluating performance, and making recommendations for enhancements. The twins' models identify anomalies, inefficiencies, or possible dangers. This covers anomaly detection, optimisation recommendations, and predictive maintenance warnings [39–43]. The major roles of digital twins for robotics are as follows:

- Before actual manufacturing, robotic systems are simulated through virtual prototyping.
- Through virtual testing, the robot's joints, structure, and actuators are improved.
- This virtual environment tests path and trajectory planning.
- These detect possible collisions in intricate robotic environments.
- Evaluates robot performance in real time by comparing it to digital models.
- Predicts when parts in robotic arms and manipulators may wear out and fail.
- This replicates a safe teamwork situation.
- It helps robots use power more efficiently.
- This technology makes it possible to monitor robots in real time using digital copies.
- This process finds the underlying reasons why robots malfunction.
- Enhances automated production lines' robotic workflow.
- This synchronises virtual models with real-time sensor data.
- Verifies system and firmware updates in a secure manner.
- Verifies that industrial safety regulations are being followed.
- This technology replicates how a robot would interact with a changing environment like a human being.

8.8 Upcoming goals of robotics

The upcoming goals of robotics are as follows [2,44]:

- Robotics will increase the manufacturing line's versatility while enabling the integration of numerous demanding operations.
- With multiple providers and a wider range of applications, robotics automation will continue to rise.
- Product improvements and customisation will become easier.
- Provides innovative concepts to keep up with the increasingly automated operations.
- Robotics will efficiently conduct raw material processing, assembly, welding, fabrication, distribution, and packaging of any items at a lower cost and time.
- Being a digital-based technology, it can integrate quite well with other digital technologies, supporting Industry 4.0.
- This technology will be deployed in the automobile sector and is integral to many manufacturing processes.

8.9 Disadvantages of robotics

The significant disadvantages of robotics are as follows [3,4]:

- Create unemployment.
- Lack of creativity in operations.
- Constant and regulated power requirements during operations.
- Not capable of thinking out of the box during an emergency.
- Only perform a specific or given task.
- No emotion and intelligence with the machine.
- Expensive to install.
- Cause a cybersecurity issue.
- Not having decision-making abilities.
- Extra cost for installation, high cost of maintenance, and even control.

8.10 Conclusion

Robotics is increasingly being integrated into industries to enhance precision, efficiency, and adaptability. This technology boosts the efficiency and performance of Industry 4.0. This technology enhances workplace safety and ergonomics by removing dangerous or repetitive duties for workers. Robotics is becoming increasingly popular in a variety of industries, improving security, quality, and long-term economic sustainability. With the advent of Industry 4.0, robotics integration is beneficial in production for reasons including performance, accuracy, dependability, and resistance to dangerous situations. It can be used in Industry 4.0 to enable reliable and effective manufacturing processes. Several of the company's production processes have been made simpler by intelligent robots that operate incredibly fast and accurately. Remote diagnostics, inspections, and even repairs using this technology strengthen customer relationships, increase value, and help suppliers keep up with demand. Robots are employed to finish items by applying specific materials for aesthetic or protective purposes, such as paint, enamel, and metal flakes. They are important to be controlled by digital factories and dark factories. They are integral to many manufacturing processes and used in products ranging from automobiles and electronics to other essential sectors.

References

[1] H. Parmar, T. Khan, F. Tucci, R. Umer, P. Carlone, Advanced robotics and additive manufacturing of composites: towards a new era in Industry 4.0, Materials and Manufacturing Processes 37 (2022) 483–517.

[2] M. Javaid, A. Haleem, R.P. Singh, R. Suman, Substantial capabilities of robotics in enhancing Industry 4.0 implementation, Cognitive Robotics 1 (2021) 58–75.

[3] K. Aggarwal, S.K. Singh, M. Chopra, S. Kumar, F. ColaceNedjah, N., Abd El-Latif, A.A., Gupta, B.B., Mourelle, L.M., Deep learning in robotics for strengthening Industry 4.0: opportunities, challenges and future directions, in: N. Nedjah, A.A. A1bd El-Latif, B.B. Gupta, L.M. Mourelle (Eds.), Robotics and AI for

Cybersecurity and Critical Infrastructure in Smart Cities. Studies in Computational Intelligence, vol 1030, Springer, Cham, (2022), pp. 1–19. Available from https://doi.org/10.1007/978-3-030-96737-6_1.

[4] B. Bayram, G. İnce, Advances in Robotics in the era of Industry 4.0, Industry 4.0: Managing the Digital Transformation, Springer International Publishing, Cham, (2017), pp. 187–200.

[5] Y. He, L. Liang, Application of robotics in higher education in the Industry 4.0 era, Universal Journal of Educational Research 7 (2019) 1612–1622.

[6] M. Soori, R. Dastres, B. Arezoo, F.K.G. Jough, Intelligent robotic systems in Industry 4.0: a review, Journal of Advanced Manufacturing Science and Technology 4 (2024) 2024007.

[7] M.R. Mia, J. Shuford, Exploring the synergy of artificial intelligence and robotics in Industry 4.0 applications, Journal of Artificial Intelligence General Science 1 (1) (2024) 31. Available from https://doi.org/10.60087/jaigs.v1i1.31.

[8] Z. Gao, T. Wanyama, I. Singh, A. Gadhrri, R. Schmidt, From industry 4.0 to robotics 4.0-a conceptual framework for collaborative and intelligent robotic systems, Procedia Manufacturing 46 (2020) 591–599.

[9] A. Weiss, A. Huber, J. Minichberger, M. Ikeda, First application of robot teaching in an existing Industry 4.0 environment: does it really work? Societies 6 (2016) 20.

[10] L. Sanneman, C. Fourie, J.A. Shah, The state of industrial robotics: emerging technologies, challenges, and key research directions, Foundations and Trends® in Robotics 8 (2021) 225–306.

[11] I. Karabegović, R. Turmanidze, P. Dašić, Robotics and automation as a foundation of the fourth industrial revolution (Industry 4.0), Grabchenko's International Conference on Advanced Manufacturing Processes, Springer International Publishing, Cham, (2019), pp. 128–136.

[12] H. Lee, J.L. Enriquez, G. Lee, Robotics 4.0: challenges and opportunities in the 4th industrial revolution, Journal of Internet Services and Information Security 12 (2022) 39–55.

[13] N. Velásquez Villagrán, P. Pesado, E. Estevez, Cloud robotics for Industry 4.0: a literature review, Conference on Cloud Computing, Big Data & Emerging Topics, Springer International Publishing, Cham, (2020), pp. 3–15.

[14] Types of Robots: How Robotics Technologies Are Shaping Today's World. Available from: https://www.intel.in/content/www/in/en/robotics/types-and-applications.html#:~:text=The%20six%20most%20common%20types,humanoids%2C%20cobots%2C%20and%20hybrids.

[15] R. Bader, A. Ali, N.M. Mirza, AI and robotics leading Industry 4.0, 2022 9th International Conference on Internet of Things: Systems, Management and Security (IOTSMS), IEEE, (2022), pp. 1–4.

[16] R. Sekhar, P. Shah, I. Iswanto, Robotics in industry 4.0: a bibliometric analysis (2011-2022), Journal of Robotics and Control (JRC) 3 (2022) 583–613.

[17] What is robotics? Robotics is a branch of engineering that involves the conception, design, manufacture. Available from: https://www.techtarget.com/whatis/definition/robotics#:~:text=Robotics%20is%20a%20branch%20of,on%20a%20number%20of%20forms.

[18] Daley S. (2024). Robotics: What Are Robots? Robotics Definition & Uses: https://builtin.com/robotics.

[19] A. Misra, A. Agrawal, V. Misra, Robotics in Industry 4.0, Handbook of Smart Materials, Technologies, and Devices: Applications of Industry 4.0, Springer International Publishing, Cham, (2021), pp. 1–35.

[20] J. Braun, A.O. Júnior, G. Berger, V.H. Pinto, I.N. Soares, A.I. Pereira, et al., A robot localisation proposal for the RobotAtFactory 4.0: a novel robotics competition within the Industry 4.0 concept, Frontiers in Robotics and AI 9 (2022) 1023590.

[21] A. Huber, A. Weiss, Developing human-robot interaction for an Industry 4.0 robot: how industry workers helped to improve remote-HRI to physical-HRI, In Proceedings of the Companion of the 2017 ACM/IEEE International Conference on Human-Robot Interaction (2017) 137–138.

[22] R.D. Campilho, F.J. Silva, Industrial process improvement by automation and robotics, Machines 11 (2023) 1011.

[23] A. Khang, K.C. Rath, S.K. Satapathy, A. Kumar, S.R. Das, M.R. Panda, Enabling the future of manufacturing: integration of robotics and IoT to smart factory infrastructure in Industry 4.0, Handbook of Research on AI-Based Technologies and Applications in the Era of the Metaverse, IGI Global Scientific Publishing, (2023), pp. 25–50.

[24] A. Baratta, A. Cimino, M.G. Gnoni, F. Longo, Human robot collaboration in industry 4.0: a literature review, Procedia Computer Science 217 (2023) 1887–1895.

[25] S. Bragança, E. Costa, I. Castellucci, P.M. Arezes, A brief overview of the use of collaborative robots in Industry 4.0: human role and safety, Occupational and environmental safety and health (2019) 641–650.

[26] M.A.K. Bahrin, M.F. Othman, N.H.N. Azli, M.F. Talib, Industry 4.0: a review of industrial automation and robotics, Jurnal Teknologi (Sciences & Engineering) 78 (6-13) (2016). Available from https://doi.org/10.11113/jt.v78.9285.

[27] A. Adel, The convergence of intelligent tutoring, robotics, and IoT in smart education for the transition from Industry 4.0 to 5.0, Smart Cities 7 (2024) 325–369.

[28] D. Mourtzis, S. Fotia, N. Boli, E. Vlachou, Modelling and quantification of industry 4.0 manufacturing complexity based on information theory: a robotics case study, International Journal of Production Research 57 (2019) 6908–6921.

[29] A. Kattepur, H.K. Rath, A. Simha, A. Mukherjee, Distributed optimisation in multi-agent robotics for Industry 4.0 warehouses, SAC '18: Proceedings of the 33rd Annual ACM Symposium on Applied Computing, Association for Computing Machinery, New York, NY, (2018), pp. 808–815. Available from https://doi.org/10.1145/3167132.3167221.

[30] C. Urrea, J. Kern, Recent advances and challenges in industrial robotics: a systematic review of technological trends and emerging applications, Processes 13 (2025) 832.

[31] J.M. Flores Gonzalez, E. Coronado, N. Yamanobe, ROS-compatible robotics simulators for Industry 4.0 and Industry 5.0: a systematic review of trends and technologies, Applied Sciences 15 (15) (2025) 8637. Available from https://doi.org/10.3390/app15158637.

[32] M. Gooroochurn, Robotics and automation in industry 4.0: emerging trends and research directions, Intelligent and Sustainable Engineering Systems for Industry 4.0 and Beyond, CRC Press, Boca Raton, (2025), pp. 71–91.

[33] R. Antunes, M.L. Aguiar, P.D. Gaspar, A dynamic STEM-Driven approach through mobile robotics to enhance critical thinking and interdisciplinary skills for empowering Industry 4.0 competencies, Technologies 11 (6) (2023) 170. Available from https://doi.org/10.3390/technologies11060170.

[34] Z. Buri, J.T. Kiss, Robotics and automation for energy efficiency and sustainability in the Industry 4.0 era: a review. energies, Energies 18 (20) (2025) 5399. Available from https://doi.org/10.3390/en18205399.

[35] A. Mazumder, M.F. Sahed, Z. Tasneem, P. Das, F.R. Badal, M.F. Ali, et al., Towards next-generation digital twin in robotics: trends, scopes, challenges, and future, Heliyon 9 (2) (2023) e13359. Available from https://doi.org/10.1016/j.heliyon.2023.e13359.

[36] L. Girletti, M. Groshev, C. Guimarães, C.J. Bernardos, A. de la Oliva, An intelligent edge-based digital twin for robotics, 2020, IEEE GlobeCom Workshops (GC Wkshps, IEEE, (2020), pp. 1–6.

[37] Y. Feddoul, N. Ragot, F. Duval, V. Havard, D. Baudry, A. Assila, Exploring human-machine collaboration in industry: a systematic literature review of digital twin and robotics interfaced with extended reality technologies, The International Journal of Advanced Manufacturing Technology 129 (2023) 1917–1932.

[38] V. Kuts, N. Cherezova, M. Sarkans, T. Otto, Digital twin: industrial robot kinematic model integration to the virtual reality environment, Journal of Machine Engineering 20 (2) (2020) 53–64. Available from 10.36897/jme/120182.

[39] H. Lee, S.D. Kim, M.A.U. Al Amin, Control framework for a collaborative robot using imitation-learning-based teleoperation from a human digital twin to a robot digital twin, Mechatronics 85 (2022) 102833.

[40] A.K. Inkulu, M.V.A. Raju Bahubalendruni, A. Dara, K. SankaranarayanaSamy, Challenges and opportunities in human robot collaboration context of Industry 4.0 -a state of the art review, Industrial Robot 49 (2) (2021) 226–239. Available from https://doi.org/10.1108/IR-04-2021-0077.

[41] R. Prasad, S.K. Upadhaya, A. Kumar, Integration of computer vision systems in robotics and industry 4.0, 2024 International Conference on IoT, Communication and Automation Technology (ICICAT, IEEE, (2024), pp. 1492–1497.

[42] Imoh, I.E., ANIEKAN, E., 2024. A technical survey on the role of robotics in conventional manufacturing process: an element of Industry 4.0. FUPRE Journal of Scientific & Industrial Research, 8, ("Tantawi, K.H., Sokolov, A., & Tantawi, O. (2019). Advances in industrial robotics: From industry 3.0 automation to industry 4.0 collaboration. In 2019, the 4th Technology Innovation Management and Engineering Science International Conference (TIMES-iCON) (pp. 1-4). IEEE.," no date).

[43] N. Ghodsian, K. Benfriha, A. Olabi, V. Gopinath, A. Arnou, Mobile manipulators in industry 4.0: a review of developments for industrial applications, Sensors 23 (2023) 8026.

[44] A. Weiss, A.K. Wortmeier, B. Kubicek, Cobots in Industry 4.0: a roadmap for future practice studies on human–robot collaboration, IEEE Transactions on Human-Machine Systems 51 (2021) 335–345.

9 Blockchain for Industry 4.0

9.1 Introduction to blockchain

Blockchain is a decentralised, distributed directory that powers smart contracts and allows for payment applications, supply chain automation, traceability, record management, and other commercial operations. Blockchain technology creates an immutable record of virtually real-time replication amongst a network of business partners [1,2]. This method considers the previously saved information in the company's enterprise resource planning. Several advantages of blockchain allow businesses to better understand their clients, especially on the demand side [3,4].

Innovations in cutting-edge digital technology, such as blockchain, are part of Industry 4.0. For both small and large businesses, blockchain may improve security, privacy, and data transparency. This well-known technology has the ability to enhance supply chains and industrial settings. These days, a number of fields provide fascinating perspectives on the advantages of blockchain. Industry 4.0 is evolving due to many breakthroughs and the prospect of enormous potential. This technology would work together to improve the process and deliver magnified results [5–7].

In the world of technology, the blockchain is a permanent distributed directory that keeps track of all value transactions. The organisation controls the entire blockchain, with the exception of the nodes that provide it an advantage over traditional data storage methods [8,9]. This technology provides a new database that stores transaction data in a decentralised and transparent manner. There is no single point of failure because a network of computers called nodes manages the database, and information can be accessed in real time. The industry is well aligned with the notion that blockchain technology is affecting the market [10–12].

To properly adopt Industry 4.0, it is vital to comprehend blockchain. This technology can give trust and potential benefits in some industries, such as financial transaction applications. In Industry 4.0, blockchain can be related to the product and its assembly's identifying portion. It serves as a reminder of how useful it is to recognise items with flaws. This technology secures a product's data, including its subassemblies, parts, and sales channels. It lowers the cost and inconvenience of retrieval at any point in the supply chain [13–15].

Blockchain is the ideal technology for tracking records and information, and it can overcome tremendous obstacles. This technology replenishment in the corporate world improves transaction efficiency. Blockchain-enabled sustainable production supports product lifecycle management in Industry 4.0. This will provide a greater level of process safety and

Digital Twin and Industry 4.0. DOI: https://doi.org/10.1016/B978-0-443-54019-6.00013-3

flourish in production environments with fewer risks [16–18]. The risk level may be reduced, and because technology is constantly improving, organisational executives must recognise the technology's positive influence on their business.

This technology offers a wide range of educational opportunities, healthcare services, transportation assistance, logistics, and organisations. The overarching blockchain idea aimed to improve these sectors from the ground up, reaching the practical heights of Industry 4.0. This technology is helpful in various sectors, including data privacy, fraud prevention, effective care, effective procurement, ease of record-keeping, cost and time savings, and improved transparency [19–21].

9.2 Blockchain and its types

A blockchain is a distributed database or ledger that is shared by nodes in a computer network. A blockchain functions as an electronic database for digital data storage. Maintaining a safe, decentralised record of transactions in cryptocurrency systems like Bitcoin is the most well-known application of blockchain technology. The novelty of blockchain technology is that it fosters confidence in data integrity and security without a reliable third party [22–24].

In a blockchain, data is organised into groups called blocks, which include data sets. Blocks have specific storage capabilities, and when filled, they are sealed and connected to the preceding block to create the data chain known as the blockchain. When the newly added block is finished, it is added to the chain along with any further information that follows [25,26].

Whereas a database normally groups its data into tables, a blockchain organises its data into linked blocks. By design, this data structure produces an irreversible data chronology when it is decentralised. A block is added to the timeline and sealed once it is finished. A block is given an exact timestamp upon its addition to the chain [27,28]. The goal of blockchain technology is to make it possible to capture and share digital data without alteration. Transaction records or immutable ledgers that cannot be altered, deleted, or destroyed are based on a blockchain [29–32]. The major types of blockchains are as follows:

9.2.1 Public blockchains

Decentralised computer networks known as public blockchains are accessible to anybody who wants to request or validate a transaction. Miners are rewarded for verifying transactions. Public blockchains employ consensus processes known as proof-of-work or proof-of-stake. Two popular public blockchains are Ethereum and Bitcoin.

9.2.2 Private blockchains

Private blockchains are not public and have access restrictions. Everyone who wants to join must have approval from the system administrator. They are often managed by a single organisation, making them centralised. Hyperledger is one example of a private, permissioned blockchain.

9.2.3 Hybrid blockchains or consortiums

Consortia integrate public and private blockchains with both centralised and decentralised components. Note that opinions on whether these phrases are different are not all in accord. Some perceive them as the same, while others make a distinction between the two.

9.2.4 Sidechains

A blockchain that coexists with the main chain is called a sidechain. It allows users to transfer digital assets across two different blockchains and improves scalability and efficiency. One illustration of the Liquid Network is a sidechain.

9.3 Major areas of blockchain

The significant areas of applications of blockchain technologies are shown in Fig. 9.1 [5,33–37].

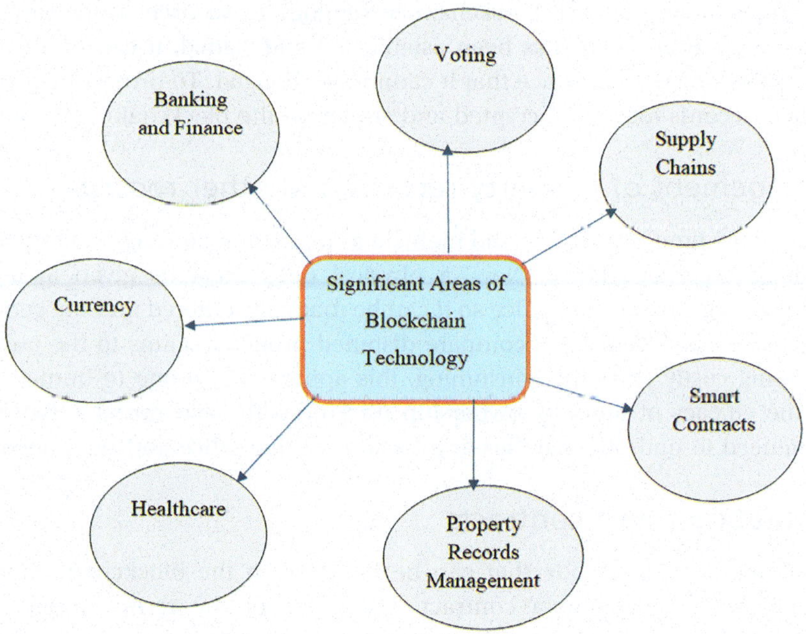

FIGURE 9.1 Major areas of blockchain technology.

9.3.1 Banking and finance

One industry that might benefit the most from integrating blockchain technology into its business operations is banking. Only during regular business hours, usually 5 days a week, are financial institutions open. We probably won't see the money in our account until the bank's business day if we try to deposit a cheque over the weekend. Even if we make our deposit during business hours, it may still take 1 to 3 days for the transaction to be validated due to the massive volume of transactions banks must settle. Consumers can expect banks to use blockchain technology to complete transactions in as little as 10 minutes.

9.3.2 Currency

Blockchain is the basis for cryptocurrencies such as Bitcoin. Theoretically, under this central authority structure, a user's information and funds are at the mercy of their bank. If a client's bank is hacked, their confidential information could be compromised. If the client lives in a country with an unstable government or their bank fails, the value of their money could be at risk. By spreading their operations over a network of computers, blockchain allows Bitcoin and other cryptocurrencies to operate in a decentralised manner. This eliminates high administrative and transaction costs and reduces risk.

9.3.3 Healthcare sector

Healthcare professionals can use blockchain technology to securely store patients' medical records. Once a medical record has been established and signed, it can be recorded on the blockchain, giving patients assurance that it cannot be changed. To protect their privacy, these private medical records may be encrypted and stored on the blockchain using a secret key.

9.3.4 Management of property records and other records

We are aware of the time-consuming and inefficient nature of recording property rights, having visited the local recorder's office. Now, a physical deed must be given to a government employee at the county recording office so it can be manually entered into the central database and public index. It is necessary to compare disputed property claims to the public index. In addition to being costly and time-consuming, this approach is prone to human error, which diminishes the efficacy of property ownership tracking with each mistake. Blockchain could eliminate the need to find physical files at a local recording office and scan documents.

9.3.5 Facilitating smart contracts

A smart contract, computer code that can be included in the blockchain, can be used to facilitate, validate, or negotiate a contract. Users accept the terms under which smart contracts operate. When certain conditions are met, such as when a potential tenant wishes to rent an apartment using a smart contract, the terms of the agreement will be automatically fulfilled. The landlord consents to give the tenant the apartment's door code if the renter pays the security deposit. On the lease start date, the smart contract would automatically exchange the door code for the security deposit after receiving and holding the renter's and landlord's payments.

9.3.6 Supply chains and logistics

The blockchain can be used by suppliers to trace the origins of the materials they have purchased. This will allow companies to verify the authenticity of their goods, as well as popular labels such as 'Organic', 'Local', and 'Fair Trade'. The food industry is increasingly using blockchain technology to monitor the safety and location of food as it moves from the farm to the customer. There are appropriate uses for logistics maintenance.

9.3.7 Voting

Blockchain has the potential to enable an existing voting system. Blockchain voting has the potential to boost voter turnout and eradicate election fraud. Voter rigging would be nearly impossible with blockchain. In addition to maintaining electoral transparency, blockchain technology will reduce the number of personnel needed to conduct an election and provide officials with near-instant access to the results. Recounts would be unnecessary as a result, and there would be no justification for worrying about election fraud.

9.4 Advantages of adopting blockchain in the new environment

The significant advantages of blockchain in the new environment are as follows [38–41]:

- Enhancing transparency in a system.
- Improve the process of record-keeping.
- Network distribution.
- Helpful for environmental sustainability.
- Improved level of security and cybersecurity.
- Improvement in management systems.
- Proper tracing of vendors, goods, and other subjects.
- Improve efficiency and speed of transactions and their secure storage.
- Property sale, maintenance mapping of physical assets.
- Finance management.

9.5 Roles of blockchain technologies for Industry 4.0

In Industry 4.0, blockchain enables decentralised transactions and information exchange in many processing steps, all inside a secure framework that tracks all transactions and timelines. Companies may use blockchain not only to work confidently but also to test their confidence. The present digital revolution will bring innovation with the advent of blockchain technology [42–44]. Blockchain may be used to solve a variety of problems in Industry 4.0 [45–50]. The significant roles of blockchain for Industry 4.0 are as follows:

9.5.1 Digital purchasing

This technology has opened up new possibilities for future purchasers in the global real estate sector. Blockchain provides several advantages in the market, including authenticating digital purchases and establishing trust in the solution for buying and selling both commercial and residential properties.

9.5.2 Product and assembly identifications

Blockchains store data that may be used to identify items and assemblies. This may aid in the quantification and resolution of high-specific quality concerns. All information about components, subassemblies, and the business delivery route is available on the blockchain. In the contemporary environment, this technology is utilised to gather data at each stage, significantly reducing reminder costs and interruptions. This is the ideal method for managing products and their components.

9.5.3 Finance

The financial industry is where blockchain shines brightest. The use of blockchains for financial transaction management is possible. Blockchains might address foreign-currency concerns, and a regulated spectrum in supply transactions may be achieved. The banking and payments industry is critical for blockchain adoption in many other businesses.

9.5.4 Data security in manufacturing

In Industry 4.0, blockchain-based encryption might be used to safeguard data. This can help prevent unauthorised access to some data transferred over a public network. Within the intellectual property area, blockchain may be used to provide evidence of copyright, preserve records, facilitate registrations and the clear assignment of rights, monitor and track distribution rights, establish IP contracts, and even manage the acquisition of rights in the context of Industry 4.0.

9.5.5 Keeping track of transactions

A distributed ledger of transactions is called a blockchain. The transactions are documented in a global blockchain notebook that is kept on multiple computers across the globe. Transactions are entered and arranged into blocks in the ledger. The blockchain stores the previous block hash, which makes it easy to identify and stop attempts to alter or fake transactions. The electronic exchange of information between individuals is the most well-known use of blockchain technology.

9.5.6 Automotive

Blockchain may be used by a number of partners and projects in the automotive sector. Industry 4.0's potential for manufacturing is encouraging as it continues to expand. Blockchain has a significant influence across a variety of industries worldwide. Purchases and other data can be recorded digitally in the automobile business blockchain. Parts and finished products can be tracked, and an information platform is developed.

9.5.7 Information

All the information provided by blockchain is digitally stored, including details on how commodities are manufactured, shipped, and maintained. The intrinsic security features of blockchain are based on the data structure. Rather than utilising traditional methods, blockchain utilises the most advanced cryptographic technologies to keep knowledge more secure.

9.5.8 Business

This technology has spread to many industries beyond financial services in recent years. It starts thinking about how it might contribute to the development of infrastructure. It begins by building a bar that is connected to a network of thousands of computers and verified online. In order to get reliable information in real time from any source, the network is configured to update itself on a frequent basis. It is also apparent, effective, adaptable, and highly stable for corporate purposes.

9.5.9 Storage of data

This technology sends data from the sensor to storage and, ultimately, to the analysis. When data is transported over a private network, the risk is low. This offers several advantages, including low-cost computer capacity and payment. With a robust data exchange and storage system in place, this technology may also reduce the risk.

9.5.10 Supply chain and logistics management

Industrial firms are managing their supply chains more effectively through a number of recent advancements. The supply chain industry will search for organic ways to boost accountability, transparency, and productivity as blockchain technology develops. Industry 4.0 companies have total control over their internal operations and are able to make data-driven decisions regarding the future using centralisation of ERP programs and real-time data alerts.

9.5.11 System integration

Blockchain technology includes an organisation's partners, clients, and supply chains, expanding the opportunities for integration beyond the business's ecosystem. This aids in public awareness of the activity. Data may be accessed evenly by Industry 4.0, reducing the possibility of withheld or manipulated data. With the successful deployment of this technology, manufacturers can get positive outcomes.

9.5.12 Store transaction record

A distributed digital ledger that keeps track of both public and private transactions is called a blockchain. Every transaction has a distinct digital signature that is recognised and validated by the whole network, protecting it from revocation, manipulation, and tampering. As a result, blockchain could eliminate the need for new data exchange and asset transfer mechanisms.

9.5.13 Enables effective management

Various areas are better suited to Industry 4.0 blockchain deployments and their benefits. This enables effective management and the provision of product information required by the market. Blockchain may be developed to quickly collect more information than a person can use, using the latest data from cameras and sensors. This technology may be used to manage financial transactions and to oversee the proper use of digital property.

9.6 Role of digital twin for blockchain

A digital twin with blockchain can enable bidirectional, synchronised interactions between the digital and physical worlds. By facilitating predictive maintenance, streamlining production procedures, and cutting waste, digital twins are transforming the manufacturing industry. Digital twins mimic production lines in smart factories to test new parameters, increasing productivity and reducing downtime. Blockchain systems' performance, transparency, and dependability are all improved by digital twins [51–54]. A digital twin allows for real-time monitoring, simulation, and validation of transactions before their permanent recording by generating a virtual copy of blockchain nodes, smart contracts, and distributed network operations. It offers a dynamic link between decentralised digital records and physical systems. This allows for secure automation, real-time audits, and enhanced governance. Therefore, in smart and linked ecosystems, the convergence of blockchain and digital twin technologies enhances decision-making, operational efficiency, and system resilience [55–58]. The significant role of the digital twin for blockchain is as follows:

- Digital twins are being used to make digital copies of blockchain nodes to analyse their performance.
- This method simulates execution scenarios to get rid of coding errors.
- It helps examine the block creation and validation procedures before network integration.
- This uses predictive modelling to analyse transaction-per-second capacity.
- It assesses how long it takes for distributed nodes to confirm transactions.
- It evaluates network stability by simulating both soft and hard fork scenarios.
- It helps forecast how much electricity blockchain processes will require.
- It is used to simulate incentive systems, token circulation, and token issuance.

- It is used for examining data sharing and cross-chain communication systems.
- It is used to simulate compliance situations in different legal contexts.
- This makes sure that ledger entries and tangible assets line up.
- Predicts bottlenecks during periods of high transaction volume using network congestion prediction.
- The continuous verification of distributed ledger operations, which is made possible by real-time audit modelling.
- It helps verify protocol modifications before going online.

9.7 Upcoming goals of blockchain

The primary upcoming goals of blockchain technologies being envisaged are as follows [59–61]:

- In the future, blockchain will transform business practices in a variety of industries, like commercial finance, supply chain activities, operations management, process-based systems, and other related fields.
- This technology will aid in the improvement of financial services.
- This will reduce the amount of manual labour.
- Users will be able to see a product's complete history on the blockchain, making maintenance easy and quality control better.
- This technology will store and consistently exchange data for any organisation or process.
- It will allow the supply chain to be linked to any vendors, production locations, and fulfilment centres.

9.8 Disadvantages of using blockchain

The significant disadvantages of using blockchain are as follows [62–65]:

- High implementation costs for companies.
- Inefficient with a large number of network users.
- Blockchain implementation implies a colossal waste of energy.
- Leads to higher complexity in the system.
- Need for better regulations for users and those implementing.
- Storage problems, as it generates a large amount of data.
- Fear of unemployment, with its implementation.

9.9 Conclusion

Blockchain ensures reliable, fast data sharing and the creation of an immutable database of all communications exchanged among linked smart devices. This technology allows users to create their own secure and trustworthy digital identity. People can now utilise their blockchain identities for everything from simple activities to apps, software, and digital signatures. For smaller companies and suppliers, blockchain offers a dependable source of superior transactional knowledge. Numerous industries recognise blockchain's potential, which might have a number of positive effects. This technology is being used more and more by banks to expedite and reduce the cost of transactions and payments.

Blockchain is utilised for parts suppliers and plant cyber-physical infrastructure, allowing machines to order replacement parts safely and autonomously. Furthermore, blockchain's potential to enable frictionless, open financial transactions across any number of smart devices makes it critical to the economic transformations of Industry 4.0. Alarms, scanners, and barcodes will be used in the workplace to verify that personnel and materials are properly registered and located, utilising blockchain. Blockchain enhances the security of internet-enabled devices while also ensuring openness throughout ecosystems. This technology provides information on shipment status and capabilities to manage the supply chain effectively.

References

[1] U. Bodkhe, S. Tanwar, K. Parekh, P. Khanpara, S. Tyagi, N. Kumar, et al., Blockchain for industry 4.0: a comprehensive review, IEEE Access 8 (2020) 79764–79800.

[2] T. Alladi, V. Chamola, R.M. Parizi, K.K.R. Choo, Blockchain applications for Industry 4.0 and Industrial IoT: a review, IEEE Access 7 (2019) 176935–176951.

[3] Y. Chen, Y. Lu, L. Bulysheva, M.Y. Kataev, Applications of blockchain in Industry 4.0: a review, Information Systems Frontiers 26 (2024) 1715–1729.

[4] M. Javaid, A. Haleem, R.P. Singh, S. Khan, R. Suman, Blockchain technology applications for Industry 4.0: a literature-based review, Blockchain: Research and Applications 2 (2021) 100027.

[5] Hayes, A., 2022. Blockchain facts: what is it, how it work, and how it can be used. Available from: https://www.investopedia.com/terms/b/blockchain.asp.

[6] A. Mushtaq, I.U. Haq, Implications of blockchain in Industry 4.0, 2019 International Conference on Engineering and Emerging Technologies (ICEET, IEEE, (2019), pp. 1–5 February.

[7] T.B.D. Silva, E.S.D. Morais, L.F.F.D. Almeida, R.D. Rosa Righi, A.M. Alberti, Blockchain and Industry 4.0: overview, convergence, and analysis, in: R. Rosa Righi, A. Alberti, M. Singh (Eds.), Blockchain technology for Industry 4.0: Secure, decentralized, distributed and trusted industry environment, Springer, Singapore, (2020), pp. 27–58. Available from https://doi.org/10.1007/978-981-15-1137-0_2.

[8] B. Esmaeilian, J. Sarkis, K. Lewis, S. Behdad, Blockchain for the future of sustainable supply chain management in Industry 4.0, Resources, Conservation and Recycling 163 (2020) 105064.

[9] D.V. Lypnytskyi, Opportunities and challenges of blockchain in Industry 4.0, Economy of Industry 1 (85) (2019) 82–100.

[10] X.L. Liu, W.M. Wang, H. Guo, A.V. Barenji, Z. Li, G.Q. Huang, Industrial blockchain based framework for product lifecycle management in Industry 4.0, Robotics and computer-integrated manufacturing 63 (2020) 101897.

[11] A.S. Bale, T.P. Purohit, M.F. Hashim, S. Navale, Blockchain and its applications in Industry 4.0, in: J.M. Chatterjee, H. Garg, R.N. Thakur (Eds.), A roadmap for enabling Industry 4.0 by artificial intelligence, Wiley, (2022), pp. 295–313. Available from https://doi.org/10.1002/9781119905141.ch16.

[12] J. Leng, S. Ye, M. Zhou, J.L. Zhao, Q. Liu, W. Guo, et al., Blockchain-secured smart manufacturing in Industry 4.0: a survey, IEEE Transactions on Systems, Man, and Cybernetics: Systems 51 (2020) 237–252.

[13] A.K. Tyagi, S. Dananjayan, D. Agarwal, H.F. Thariq Ahmed, Blockchain—internet of things applications: opportunities and challenges for industry 4.0 and society 5.0, Sensors (Basel, Switzerland) 23 (2023) 947.

[14] J. Lee, M. Azamfar, J. Singh, A blockchain-enabled cyber-physical system architecture for Industry 4.0 manufacturing systems, Manufacturing Letters 20 (2019) 34–39.

[15] S.B. ElMamy, H. Mrabet, H. Gharbi, A. Jemai, D. Trentesaux, A survey on the usage of blockchain technology for cyber-threats in the context of Industry 4.0, Sustainability 12 (2020) 9179.

[16] Z. Wang, L. Yu, L. Zhou, Navigating the blockchain-driven transformation in industry 4.0: opportunities and challenges for economic and management innovations, Journal of the Knowledge Economy 16 (2025) 3507–3549.

[17] A. Shrivastava, K.M. Krishna, M.L. Rinawa, M. Soni, G. Ramkumar, S. Jaiswal, Inclusion of IoT, ML, and blockchain technologies in next generation industry 4.0 environment, Materials Today: Proceedings 80 (2023) 3471–3475.

[18] A. Aoun, A. Ilinca, M. Ghandour, H. Ibrahim, A review of Industry 4.0 characteristics and challenges, with potential improvements using blockchain technology, Computers & Industrial Engineering 162 (2021) 107746.

[19] M. Soori, R. Dastres, B. Arezoo, AI-powered blockchain technology in industry 4.0: a review, Journal of Economy and Technology 1 (2023) 222–241.

[20] T. Syafira, S. Jackson, A. Tambunan, Fintech integration with crowdfunding and blockchain in the Industry 4.0 era, Startupreneur Business Digital (SABDA Journal) 3 (2024) 10–18.

[21] M.M. Nuttah, P. Roma, G.L. Nigro, G. Perrone, Understanding blockchain applications in Industry 4.0: From information technology to manufacturing and operations management, Journal of Industrial Information Integration 33 (2023) 100456.

[22] S. Joshi, A.A. Pise, M. Shrivastava, C. Revathy, H. Kumar, O. Alsetoohy, et al., Adoption of blockchain technology for privacy and security in the context of Industry 4.0, Wireless Communications and Mobile Computing 2022 (2022) 4079781.

[23] D. Su, L. Zhang, H. Peng, P. Saeidi, E.B. Tirkolaee, Technical challenges of blockchain technology for sustainable manufacturing paradigm in Industry 4.0 era using a fuzzy decision support system, Technological Forecasting and Social Change 188 (2023) 122275.

[24] A. Kapitonov, I. Berman, S. Lonshakov, A. Krupenkin, Blockchain based protocol for economical communication in industry 4.0, 2018 Crypto Valley Conference on Blockchain Technology (CVCBT), IEEE, (2018), pp. 41–44 June.

[25] L. Li, Blockchain technology in Industry 4.0, Enterprise Information Systems 16 (2022) 2095535.

[26] P. Gangwani, A. Perez-Pons, S. Joshi, H. Upadhyay, L. Lagos, Integration of data science and IoT with blockchain for Industry 4.0, Blockchain and Its Applications in Industry 4.0, Springer Nature Singapore, Singapore, (2023), pp. 139–177.

[27] G. Bovenzi, G. Aceto, V. Persico, A. Pescape, Blockchain performance in industry 4.0: drivers, use cases, and future directions, Journal of Industrial Information Integration 36 (2023) 100513.

[28] D. Mehta, S. Tanwar, U. Bodkhe, A. Shukla, N. Kumar, Blockchain-based royalty contract transactions scheme for Industry 4.0 supply-chain management, Information Processing & Management 58 (2021) 102586.

[29] Types of BlockChain: https://blockchain.gov.in/Home/BlockChain?blockchain=type.

[30] P.W. Khan, Y.C. Byun, N. Park, IoT-blockchain enabled optimized provenance system for food Industry 4.0 using advanced deep learning, Sensors (Basel, Switzerland) 20 (2020) 2990.

[31] C. Hennebert, F. Barrois, Is the blockchain a relevant technology for Industry 4.0? 2020 2nd Conference on Blockchain Research & Applications for Innovative Networks and Services (BRAINS), IEEE, (2020), pp. 212–216 September.

[32] G. Rathee, M. Balasaraswathi, K.P. Chandran, S.D. Gupta, C.S. Boopathi, A secure IoT sensor communication in Industry 4.0 using blockchain technology, Journal of Ambient Intelligence and Humanised Computing 12 (2021) 533–545.

[33] C. Lin, D. He, X. Huang, K.K.R. Choo, A.V. Vasilakos, BSeIn: a blockchain-based secure mutual authentication with fine-grained access control system for Industry 4.0, Journal of Network and Computer Applications 116 (2018) 42–52.

[34] D. Muzylyov, N. Shramenko, Blockchain technology in transportation as a part of the efficiency in industry 4.0 strategy, Grabchenko's International Conference on Advanced Manufacturing Processes, Springer International Publishing, Cham, (2019), pp. 216–225, September.

[35] J. Leng, G. Ruan, P. Jiang, K. Xu, Q. Liu, X. Zhou, et al., Blockchain-empowered sustainable manufacturing and product lifecycle management in industry 4.0: a survey, Renewable and Sustainable Energy Reviews 132 (2020) 110112.

[36] M. Singh, Blockchain technology for data management in Industry 4.0, Blockchain Technology for Industry 4.0: Secure, Decentralised, Distributed and Trusted Industry Environment, Springer Singapore, Singapore, (2020), pp. 59–72.

[37] I. Al Ridhawi, M. Aloqaily, A. Abbas, F. Karray, An intelligent blockchain-assisted cooperative framework for Industry 4.0 service management, IEEE Transactions on Network and Service Management 19 (2022) 3858–3871.

[38] S. Qahtan, K.Y. Sharif, A.A. Zaidan, H.A. AlSattar, O.S. Albahri, B.B. Zaidan, et al., Novel multi-security and privacy benchmarking framework for blockchain-based IoT healthcare systems in Industry 4.0, IEEE Transactions on Industrial Informatics 18 (2022) 6415–6423.

[39] A.K. Sharma, M.S. Peelam, B.K. Chauasia, V. Chamola, QIoTChain: quantum IoT-blockchain fusion for advanced data protection in Industry 4.0, IET Blockchain 4 (2024) 252–262.

[40] R. Gupta, S. Tanwar, N. Kumar, S. Tyagi, Blockchain-based security attack resilience schemes for autonomous vehicles in Industry 4.0: a systematic review, Computers & Electrical Engineering 86 (2020) 106717.

[41] S. Benzidia, N. Makaoui, N. Subramanian, Impact of ambidexterity of blockchain technology and social factors on new product development: a supply chain and Industry 4.0 perspective, Technological Forecasting and Social Change 169 (2021) 120819.

[42] Y. Wu, H.N. Dai, H. Wang, Convergence of blockchain and edge computing for secure and scalable IIoT critical infrastructures in Industry 4.0, IEEE Internet of Things Journal 8 (2020) 2300–2317.

[43] N.V. Vafiadis, T.T. Taefi, Differentiating blockchain technology to optimize the processes quality in industry 4.0, 2019 IEEE 5th World Forum on Internet of Things (WF-IoT), IEEE, (2019), pp. 864–869. April.

[44] Z. Rahman, X. Yi, I. Khalil, Blockchain-based AI-enabled industry 4.0 CPS protection against advanced persistent threat, IEEE Internet of Things Journal (2022) 10F.B.D. Silva, E.S.D. Morais, L.F.F.D. Almeida, R.D. Rosa Righi, A.M. Alberti, Blockchain and industry 4.0: overview, convergence, and analysis, Blockchain Technology for Industry 4.0. Blockchain Technologies, Springer, Singapore, (2020), pp. 27–58. Available from https://doi.org/10.1007/978-981-15-1137-0_2.

[45] G.B. Raja, Impact of internet of things, artificial intelligence, and blockchain technology in Industry 4.0, Internet of Things, Artificial Intelligence and Blockchain Technology, Springer International Publishing, Cham, (2021), pp. 157–178.

[46] K. Sabri-Laghaie, S. Jafarzadeh Ghoushchi, F. Elhambakhsh, A. Mardani, Monitoring blockchain cryptocurrency transactions to improve the trustworthiness of the fourth industrial revolution (Industry 4.0), Algorithms 13 (2020) 312.

[47] S.B. Rane, Y.A.M. Narvel, Re-designing the business organization using disruptive innovations based on blockchain-IoT integrated architecture for improving agility in future Industry 4.0, Benchmarking: An International Journal 28 (2021) 1883–1908.

[48] R.E. Masa'deh, M. Jaber, A.A.A. Sharabati, A.Y. Nasereddin, A. Marei, The Blockchain Effect on Courier Supply Chains Digitalization and Its Contribution to Industry 4.0 within the Circular Economy, Sustainability 16 (16) (2024) 7218. Available from https://doi.org/10.3390/su16167218.

[49] I. Yaqoob, K. Salah, M. Uddin, R. Jayaraman, M. Omar, M. Imran, Blockchain for digital twins: recent advances and future research challenges, IEEE Network 34 (2020) 290–298.

[50] S. Suhail, R. Hussain, R. Jurdak, A. Oracevic, K. Salah, C.S. Hong, et al., Blockchain-based digital twins: research trends, issues, and future challenges, ACM Computing Surveys (CSUR) 54 (11s) (2022) 1–34.

[51] D. Lee, S.H. Lee, N. Masoud, M.S. Krishnan, V.C. Li, Integrated digital twin and blockchain framework to support accountable information sharing in construction projects, Automation in Construction 127 (2021) 103688.

[52] S. Huang, G. Wang, Y. Yan, X. Fang, Blockchain-based data management for digital twin of product, Journal of Manufacturing Systems 54 (2020) 361–371.

[53] P. Kumar, R. Kumar, A. Kumar, A.A. Franklin, S. Garg, S. Singh, Blockchain and deep learning for secure communication in digital twin empowered industrial IoT network, IEEE Transactions on Network Science and Engineering 10 (2022) 2802–2813.

[54] F. Tao, Y. Zhang, Y. Cheng, J. Ren, D. Wang, Q. Qi, et al., Digital twin and blockchain enhanced smart manufacturing service collaboration and management, Journal of Manufacturing Systems 62 (2022) 903–914.

[55] E.E.D. Hemdan, W. El-Shafai, A. Sayed, Integrating digital twins with IoT-based blockchain: concept, architecture, challenges, and future scope, Wireless Personal Communications 131 (2023) 2193–2216.

[56] G. Thakur, P. Kumar, S. Jangirala, A.K. Das, Y. Park, An effective privacy-preserving blockchain-assisted security protocol for cloud-based digital twin environment, IEEE Access 11 (2023) 26877–26892.

[57] X. Xu, J. Gu, H. Yan, W. Liu, L. Qi, X. Zhou, Reputation-aware supplier assessment for blockchain-enabled supply chain in Industry 4.0, IEEE Transactions on Industrial Informatics 19 (2022) 5485–5494.

[58] R. Rajmohan, T.A. Kumar, M. Pavithra, S.G. Sandhya, Blockchain: next-generation technology for Industry 4.0, Blockchain Technology, CRC Press, (2020), pp. 177–198.

[59] S. Otoum, I. Al Ridhawi, H. Mouftah, A federated learning and blockchain-enabled sustainable energy trade at the edge: a framework for Industry 4.0, IEEE Internet of Things Journal 10 (2022) 3018–3026.

[60] C. Zhang, Y. Chen, A review of research relevant to the emerging industry trends: Industry 4.0, IoT, blockchain, and business analytics, Journal of Industrial Integration and Management 5 (2020) 165–180.

[61] S.F. Wamba, M.M. Queiroz, Industry 4.0 and the supply chain digitalisation: a blockchain diffusion perspective, Production Planning & Control 33 (2022) 193–210.

[62] S.B. Rane, Y.A.M. Narvel, Data-driven decision making with blockchain-IoT integrated architecture: a project resource management agility perspective of Industry 4.0, International Journal of System Assurance Engineering and Management 13 (2022) 1005–1023.

[63] P.B. Sangode, Fusion of blockchain and Industry 4.0: a bibliometric analysis and detailed review, Operations Research Forum, Springer International Publishing, Cham, (2025). February.

[64] K. Kaushik, B. Kaushik, Role of blockchain in Industry 4.0, Intelligent Analytics for Industry 4.0 Applications, CRC Press, (2023), pp. 233–240.

[65] H. Yetis, M. Karakose, N. Baygin, Blockchain-based mass customisation framework using optimised production management for Industry 4.0 applications, Engineering Science and Technology, an International Journal 36 (2022) 101151.

10

Sensors for Industry 4.0

10.1 Introduction to sensor

A sensor responds with a measured digital signal after detecting an input stimulus from the physical environment, which can be any quantity, quality, or condition. Pressure, force, motion, flow, light, heat, wetness, or other environmental factors might serve as input stimuli. The output is generally an electrical signal, such as voltage, current, capacitance, resistance, frequency, and so on, which is displayed or communicated across a network for reading or other electronic dispensation. The most frequent applications are the level, temperature, electrical, pressure, flow, fluid speed, heat, and other levels [1–3].

Sensors are rapidly being used to develop automated manufacturing lines and to enable networking, Internet of Things (IoT) process control, data interchange, and logistical convergence. The long-range sensor, in various forms, plays an increasingly important role in the production process as Industry 4.0 transitions to big and small regions. The sensor can usually communicate and has built-in diagnostics. Intelligent sensors can offer a variety of alternatives and carry out a broad range of tasks. Because they give the system intelligence, sensors are essential to factory automation [4,5].

Industry 4.0 aims to increase productivity through process automation. To detect, analyse, measure, and process a variety of transitions in industrial production facilities, including changes in position, length, height, external circumstances, and dislocations, sensors are essential components of Industry 4.0. Sensors will see greater adoption in Industry 4.0 through data sharing, end-to-end linking, and logistical integration. In future years, sensor installations may become more common in automated manufacturing lines, process management, and digital supply chains [6–8].

Sensors are employed across various sectors for both routine and commercial applications. Sensor systems have recently become prevalent in industrial demonstrations, demonstrating their impressive capabilities [9,10]. Sensors link several systems and devices, enabling machinery to communicate and monitor equipment and systems across the facility. Sensors have become much smaller and more portable as their capabilities have grown, allowing them to be linked to hard-to-reach and sometimes dangerous objects, transforming them into high-tech geniuses [11,12].

Industry 4.0 production models leverage new sensor technology. These technologies enhance intelligent processes. Sensors may help enhance product quality, lower production costs, and open new markets. Sensors are now widely employed in daily life [13,14]. These can detect atmospheric conditions and modify operations accordingly, since they can also process onboard. This technology can help collect massive amounts of data, which can be analysed faster and more precisely than anyone could have predicted. It eliminates all hazards associated with human error, improving output and quality. Quantitative analyses, which can be completed quickly with this technology, provide significant benefits for maintenance [15,16].

Digital Twin and Industry 4.0. DOI: https://doi.org/10.1016/B978-0-443-54019-6.00009-1

Sensors provide valuable insights, increase production, enable competition, promote, reduce costs, and improve company performance and overall efficiency across many industrial and commercial contexts. With sensors, established operations are possible. Flow calculation and temperature sensors are critical for robot measurements in form and space in-process control applications. Cost-effective, reliable sensors that provide qualitative and quantitative material information would greatly assist future smart factories [17,18].

Sensors are also crucial for goods and solutions, as they can monitor critical processes that affect production, the economy, and safety. Wireless sensor networking is already a reality. The potential for growth across various disciplines will continue to expand. It can readily monitor, calibrate, and regulate automatic control functions, allowing proactive planning and maintenance [19,20].

Sensors measure the force and calculate the force distribution to properly determine the target's location and monitor the pickup position and end effector strength. Some touch sensors can detect temperature changes as well [21,22]. A wide range of businesses use chemical sensors. Its goal is to display fluid changes or detect air. By accurately and thoroughly knowing the reality of their assets and processes, the manufacturer may enhance the efficiency, health, and safety of their operations. Smart sensors to monitor operating characteristics are integrated into new industrial equipment. Among other things, it may regulate vibration and flow. Important data can be measured and computed by sensors' integrated CPUs and memory [23,24].

10.2 Different types of sensors

A 'sensor' is a part or apparatus that helps identify changes in electrical quantities like current or other kinds of energy, or in physical quantities like pressure or force. A sensor sends the observed input to a microcontroller or microprocessor after seeing the changes. An optical, electrical, or other readable signal that represents a change in the input signal is produced by a sensor. Every measurement system must have sensors. Several sensors are utilised to monitor physical attributes such as heart rate, pulse rate, speed, heat transfer, and temperature [25,26].

There are many sensors around the globe. Every activity engaged in daily is automated. Automation includes mobile phone-based activation of lights, fans, and TVs, as well as mobile app-based temperature management and smoke detector activation. Sensors are used to do all of this. These days, sensors are incorporated into every embedded system-based device. Applications include mobile-controlled CCTV cameras and weather-tracking and forecasting apps. In healthcare monitoring and detection, sensors are critical [27–29]. Different types of sensors are discussed as follows [23]:

10.2.1 Analogue sensors

Analogue sensors provide a constant signal about time and an analogue output. The amount of analogue output generated depends on the measured value or input supplied to the system. The output is usually either a current or an analogue voltage between 0 and 5 V. Many physical properties, including temperature, pressure, stress, displacement, and so forth, are represented by continuous signals. Accelerometers, speed, light, pressure, and temperature sensors are a few examples.

10.2.2 Infrared sensors

The near-infrared, middle-infrared, and far-infrared are the three regions of the infrared portion of the electromagnetic spectrum. The infrared spectrum spans a wider range of frequencies than visible light but a narrower range than microwaves. An infrared sensor generates and detects infrared radiation. This concept enables the use of an infrared sensor as an obstruction detector. There are two types of infrared sensors: active and passive.

10.2.3 Active infrared sensors

Active sensors consist of two components: a detector and an infrared source. LEDs and laser diodes are examples of infrared sources. Photodiodes or phototransistors are used in infrared detectors.

10.2.4 Passive infrared sensors

When a sensor does not require an infrared source to detect energy emitted by obstacles, it operates as a passive infrared sensor. Bolometers, thermocouples, and pyroelectric detectors are examples of passive sensors.

10.2.5 Temperature and thermocouple sensors

Over time, analogue sensor signals change continuously. At the microvolt or millivolt level, the sensor's output will be inadequate. Thus signal conditioning circuits are required for amplification. An analogue-to-digital converter converts the obtained analogue signal into a digital value. Temperature sensors measure and monitor temperature variations. Other types of temperature sensors include thermocouples, thermosiphons, resistive temperature devices, temperature sensor integrated circuits, and more.

10.2.6 Proximity sensors

One type of noncontact sensor used for object detection is the proximity sensor. It does not make direct contact with the object. The object whose distance needs to be ascertained is the target. Infrared light, or electromagnetic radiation, is used by proximity sensors. Proximity sensors are used for item detection, rotation identification, material detection, reverse parking sensors, item counting, and velocity measurement. There are several types of proximity sensors, such as inductive, capacitive, and ultrasonic.

10.2.7 Ultrasonic sensors

Ultrasonic sensors use ultrasonic waves to calculate trip time or distance. A source will emit ultrasonic waves. When the waves hit the target, they are reflected, and the detector collects the signal. The distance that both the sent and reflected waves travel is measured by an ultrasonic sensor. There are two distinct components in optical sensors: the transmitter and the receiver. An ultrasonic sensor uses a single element for both transmission and reception.

10.2.8 Accelerometer sensors

An accelerometer sensor detects motion to determine location, velocity, and vibration variations. Both digital and analogue versions are available. The amount of acceleration applied to an analogue accelerometer determines the continuous analogue voltage signal that is produced.

10.2.9 Gyroscope sensors

Gyroscope sensors measure angular velocity by using Earth's gravity to determine direction. The main difference between accelerometers and gyroscopes is that accelerometers can detect rotation, whereas gyroscopes cannot. In other words, an accelerometer cannot work near the centre of rotation and cannot distinguish between acceleration and rotation. A gyroscope, on the other hand, is independent of acceleration and can measure any rotation.

10.2.10 Digital sensors

When data is converted and sent digitally, digital sensors are utilised. Digital sensors produce only discrete output signals. Discrete signals can be transmitted serially or concurrently in 'bits' or 'bytes', but they eventually become noncontinuous. There will be a digital display of the measuring quantity. Digital output can be either Logic 1 or Logic 0 (ON or OFF). A transmitter, a wire, and a sensor make up a digital sensor. The sensor internally converts the measured signal to a digital signal, eliminating the need for an additional component. The cable is used for long-distance communication.

10.2.11 Pressure sensors

A device that gauges the pressure of liquids or gases is called a pressure sensor. The force needed to stop a fluid from expanding is called pressure, and it is commonly expressed as force per unit area. The transducer function is usually carried out by a pressure sensor, which generates a signal in response to the applied pressure. Pressure sensors are used for monitoring and control in many common applications. Pressure sensors can also be used to indirectly monitor other variables, including height, water level, speed, and fluid/gas flow. Apart from these names, other names for pressure sensors include piezometers, manometers, pressure indicators, pressure transducers, pressure transmitters, and senders.

10.2.12 Hall effect sensors

The Hall effect sensor operates on the principle of the magnetic effect. The input is a magnetic field, and the output is an electrical signal. An external magnetic field activates the Hall effect

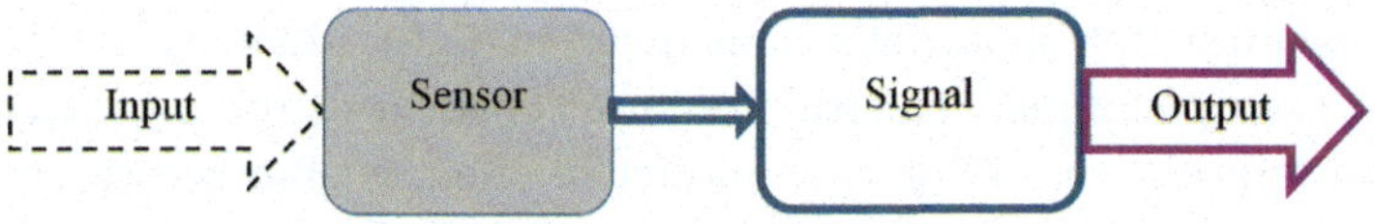

FIGURE 10.1 General working principle of sensors.

sensor. Flux density and polarity are the two key features shared by all magnets. Around an item, the magnetic flux density is constantly there. As a result, the Hall effect sensor's output will depend on the magnetic flux density.

10.3 Working principle of sensors

A sensor is a detection device that, in compliance with certain regulations, transforms measured data into electrical signals or other essential information outputs to satisfy the requirements of data transmission, processing, storage, display, recording, and control. Sensors are commonly classified by their operating principle, input data, and application range [30,31]. Fig. 10.1 depicts the general working principle of a sensor.

Sensors have various characteristics, such as systematisation, miniaturisation, digitalisation, intelligence, multifunctionality, and networking. Objects can experience touch, taste, and smell through sensors, allowing them to evolve into living things. The significant elements are thermal, gas-sensitive, photosensitive, force-sensitive, humidity-sensitive, magnetic-sensitive, sound-sensitive, colour-sensitive, radiation-sensitive, and taste-sensitive.

10.4 Classifications of sensors as per applications

Sensors are commonly classified by their operating principle, input data, and application range. Based on their various operating principles, they may be broadly divided into three types [32,33]: physical, chemical, and biological [28].

10.4.1 Physical sensors

Physical sensors are made with specialised functional materials and the distinct physical properties of particular changing components. Resistive sensors use changes in resistance in metals and semiconductors under the influence of the measured force; piezoelectric crystal sensors use the measured force to change resistance; and inductance and differential transformer sensors use magnetoresistance to change with the measured force. For example, piezoelectric sensors exploit the piezoelectric effect. A variety of sensors, such as pressure-, photosensitive-, and magnetic-sensitive sensors based on the piezoresistive effect, photoelectric effect, and Hall effect, have been developed to capitalise on the special properties of semiconductor materials. There are two types of physical sensors: structural and physical.

A physical sensor uses the properties and effects of specific functional materials to convert the observed value directly into electrical energy. Another example is a photoresistor, which changes the semiconductor material's sensitivity to light intensity to detect changes in conductivity. For example, a piezoelectric pressure sensor uses the piezoelectric effect of the material to measure pressure. The photoconductive effect is changed to create the sensor.

Structure-based structural sensors generate power from observed data using physical principles. An air-gap inductive sensor requires an armature with a specific gap with the iron core, a coil with a specific number of turns, and an iron core of a given size. If it meets certain structural and dimensional specifications, its safety can be guaranteed. An electric sensor, consisting of a magnetic shoe with a coil that travels within its gap, is another example. The right electrical signal is produced by the magnetoresistance changing in tandem with the variation in the distance between the armature and the iron core. Thus to induce an electromotive force in the coil that corresponds to the measured speed, this structure ensures that the coil travels while cutting the magnetic field line.

10.4.2 Chemical sensors

Chemical sensors convert the concentrations and compositions of organic and inorganic compounds into electrical signals through electrochemical reactions. Ion-selective electrodes, which measure a solution's pH or the activity of specific ions such as K^+, Na^+, and Ca^{2+}, are the most common type. Similar measurement concepts apply, even though the electrodes use different measuring instruments. They mostly depend on electrochemical interactions between the electrode interface (solid phase) and the measured solution (liquid phase), or on the electrode's selective reaction with ions in the solution. We can change something. The observed ion activity is measured because the work potential difference of the reaction process component, or the current value affected by it, is directly related to the logarithm of the measured ion activity.

Ion-selective membranes, which come in liquid and solid forms, are the main component. Solid films include glass, single-crystal films, and polycrystalline films. Liquid films include neutral carrier films and positively and negatively charged carrier films.

The ion-selective electrode, combining membrane and field-effect transistor technologies, has advanced at the fastest rate in recent years. It appears on the gate of the field-effect transistor. In addition to inorganic ions, this ion-selective electrode can detect glucose, gases, and certain serum components. In the chemical industry, chemical sensors are frequently used for chemical analysis, online detection, and environmental monitoring.

10.4.3 Biological sensors

The biosensor is another type of sensor that has become increasingly popular. It is a sensor that detects and measures biologically active chemicals. The functional conversion-recognition capability of a physiologically active agent is its specific affinity for a given molecule. After identifying whether a material exists and at what concentration, this single recognition capacity is used to alter electrical impulses via electrochemical processes.

The active recognition compound, one of the biosensor's two primary components, is used to specifically identify the medication being tested. Cells, enzymes, antigens, antibodies, bacteria, and viruses are a few examples of active recognition compounds. Curing these identifiers on a particular organic film using a specific process yields a functional film that can detect a range of low- to high-molecular-weight molecules. The purpose of this device is to convert the chemical reaction that occurs when the measured object is identified on the functional film into a straightforward electrical or optical signal. Hydrogen peroxide and oxygen electrodes are the most commonly used. By placing the functional layer on the field-effect transistor rather than on the gate-drain, biosensors with a small total volume have recently been developed.

10.5 Advantages of using sensors

The significant advantages of deploying sensors in different applications are as follows [34,35]:

- Speeds up the required process.
- It senses different types of materials.
- Collect real-time data on processes and assets.
- Improves the life of the system.
- Improves response.
- Accurately, consistently, and continually monitoring the processes and assets.
- Make the system predictable.
- Boost productivity while lowering the overall cost of ownership.
- Easy to install.
- Reduced energy wastage.
- Enhances capability.
- Now available at competitive pricing.
- Deployed to check the fluid level and measurements.
- Requires less power.
- Exhibit excellent repeatability and accuracy.

10.6 Roles of sensors for Industry 4.0

Sensors are used to acquire data from a physical environment. Sensor technology is deployed in Industry 4.0 to perform a range of tasks. Sensors are helpful for process-level information systems; workflows capture data on system and machine status. This is deployed throughout the system, making data collection easier without the need for computers or tablets [36–38]. The significant roles of sensors for Industry 4.0 are as follows:

10.6.1 Controls entire processes

Sensors are crucial in continually tracking and regulating operations in Industry 4.0. Various sensors are also placed in IoT-enabled sectors, and their remote sensing and tracking capabilities play a crucial role in industrial control. This detects material movement as part of customised automation and manufacturing lines. Sensors expand the possibilities for uninterrupted production automation.

10.6.2 Transmit information

Sensors gather data and transmit it to a local network. Intelligent sensors can collect vast amounts of data over time from ongoing production systems. Depending on their architecture, they may detect temperature, wetness, stress, and more. Sensors will restart manufacturing without slowing assembly lines handling items of varying dimensions or shapes.

10.6.3 Improve manufacturing efficiency

Sensors improve manufacturing efficiency and boost operations by allowing manufacturers to revolutionise their plants. Sensors generate data by connecting different devices and systems, allowing machines to communicate with one another. Manufacturers can lower their substitute asset value by reducing excessive scheduled servicing, replacement expenses, and market-disruption capacity. This technology helps companies transition from scheduled to automated maintenance.

10.6.4 Keep track of the entire process

This manufacturing technology is used to track the entire process. These are used to collect and analyse Industry 4.0 data by sending it to central cloud computing platforms. Sensors are also useful across sectors such as architecture, engineering, and the automobile industry. They monitor heat leaks in structures and industrial plant buildings.

10.6.5 Helpful in performing real-time operations

Sensors enable manufacturers to use agile approaches to increase production efficiency and support real-time operations. In Industry 4.0, sensor data make the plant more visible by providing visual depictions of peaks and flows. Intelligent sensors and manufacturing digitisation enable organisations to retain open, dependable, high-quality production. By using this technology, manufacturers can improve production performance more efficiently through greater precision in the plant.

10.6.6 Management of quality

Sensors may be used in manufacturing environments to ensure proper quality control and equipment tracking. Advances in sensor technology have attracted and improved the efficiency of potential manufacturers. Intelligent sensors connect multiple operating systems, enabling machines to interact and make informed decisions.

10.6.7 Calculate the machine's speed

Machine speed is calculated using sensors. It is used to estimate angular and rotational speed. Sensors continually generate data for each component of the manufacturing process, which may be gathered and analysed in real time. The manufacturing operations manager will then use this information and make real-time decisions.

10.6.8 Identify fundamental causes of manufacturing errors

Many operational efficiencies, including workforce, transportation, and quality-control expenses, are improved by deploying sensors. It allows engineers to spot manufacturing issues as they are being constructed, before they become failure points. Sensors can assist in generating reports in data-installed equipment, which can be used to demonstrate industry regulatory compliance. Furthermore, sensors are frequently used to improve manufacturing processes by providing product quality data and notifying of issues in real time.

10.6.9 To improve automation, sense, and convey data

Sensors are devices that may passively collect data throughout any stage of the production or logistics process. Sensors allow machines to collect and intelligently convey all of this data, and to identify and analyse it to improve automation. The sensors' data allows them to control equipment and processes, track specific production systems, and detect probable failures. It provides data so that precise information may be retrieved without calculation.

10.6.10 Determining the pressure

A pressure sensor transforms pressure into an electrical signal. For many systems to work, liquid or other pressure sources are necessary. Sensors make it possible to track pressurised systems and equipment in Industry 4.0. These sensors are beneficial to the entire water and heating systems and aid in their maintenance since they allow for the immediate observation of pressure fluctuations or dips. Ion-detection and water-quality sensors are the main tools used by water distributors to monitor water quality.

10.6.11 Optimise the production process

To optimise manufacturing activities and sensor-enabled job monitoring, sensors are employed to monitor quality in assembly lines and reduce unnecessary working time, enabling the physical-digital loop to be closed in minutes. Critical infrastructure is tied and tracked by sensors to realistically address any outages. Sensors are used in industrial manufacturing locations to detect changes in direction, weight, height, external, and dislocation.

10.7 Roles of digital twin for applications with sensors

By simulating physical sensing systems for real-time monitoring and analysis, digital twins play a critical role in improving the performance, dependability, and intelligence of sensors. Comparing real-time data with simulated models to identify deviations and abnormalities enables continuous sensor calibration, problem identification, and performance optimisation. By anticipating sensor failures before they happen, digital twins enable predictive maintenance, reducing maintenance costs and downtime. Digital twins combine sensor data with AI and IoT platforms in smart manufacturing and Industry 4.0 settings to facilitate intelligent decision-making, adaptive control systems, and enhanced system responsiveness [39–42]. The major roles of digital twins for sensors are as follows:

- Real-time virtual simulation of physical sensor behaviour.
- Constant accuracy check using actual and simulated data comparison.
- Forecasting predictive failure based on past and present tendencies.
- Simulate remote calibration without interfering with activities.
- Modelling for anomaly detection in anomalous signal patterns.
- Benchmarking performance versus optimal sensor models.
- Simulated environmental effects on sensor data.
- Modelling for compensation and data correction to increase accuracy.
- Sensing parameter optimisation for optimal performance.
- Analysis of fault root causes through virtual testing scenarios.
- Analysis of sensor response times under different loads.
- Optimisation of wireless sensor energy consumption.
- Modelling of multisensor synchronisation for coordinated operation.
- Extreme condition risk-free scenario testing.
- Using predictive insights to generate alerts in real time.
- Cyber-physical security surveillance of intelligent sensor systems.
- Intelligent interpretation through the integration of AI systems with integration modelling.

10.8 Upcoming goals of sensors

The upcoming goals of sensors are as follows [1,43]:

- In the future, new intelligent sensors and the required wireless communication links are probably going to hit the market.
- Sensors will become more prevalent in manufacturing and daily usage.
- The sensor will primarily be utilised for predictive maintenance.
- By collecting data regularly, operators can detect irregularities early.
- Sensors will boost productivity by providing additional insights into operations and enabling data collection and analysis.

10.9 Challenges with sensors

The significant challenges with sensors are as follows [9,44]:

- Sometimes requires physical contact, which is not feasible many times.
- Sometimes a prolonged response makes it ineffective.
- Sensing range affected by colour and other factors.
- There are distance limitations, and connectivity is required, while its cybersecurity is important.
- Very sensitive to extreme environmental changes.

- Some sensors have a limited shelf life.
- Sensors for very high pressure and extreme ranges are difficult.

10.10 Conclusion

Sensors detect and convert physical motion in a specific region into an electrical signal. Sensors can be adapted to various applications and have a wealth of knowledge for optimising them. Sensors are often used to provide data measurement and control devices. This technology is critical across various sectors because it provides data on pressure, temperature, velocity, gas composition, and location, significantly impacting processes and systems. Sensors play a crucial role in advancing technology in Industry 4.0. The information captured by this technology is utilised to improve the Industry 4.0 ecosystem. The sensors keep an eye on environmental variables, turbulence, and ground temperatures, giving the pilot vital information to handle both routine operations and emergencies. These technologies are used in manufacturing to identify, quantify, evaluate, and handle a variety of changes at production facilities, including position, displacement, weight, height, and presence.

References

[1] M. Javaid, A. Haleem, R.P. Singh, S. Rab, R. Suman, Significance of sensors for Industry 4.0: roles, capabilities, and applications, Sensors International 2 (2021) 100110.

[2] A. Schütze, N. Helwig, T. Schneider, Sensors 4.0–smart sensors and measurement technology enable Industry 4.0, Journal of Sensors and Sensor Systems 7 (2018) 359–371.

[3] T. Kalsoom, N. Ramzan, S. Ahmed, M. Ur-Rehman, Advances in sensor technologies in the era of smart factory and Industry 4.0, Sensors 20 (2020) 6783.

[4] G. Rathee, M. Balasaraswathi, K.P. Chandran, S.D. Gupta, C.S. Boopathi, A secure IoT sensor communication in Industry 4.0 using blockchain technology, Journal of Ambient Intelligence and Humanised Computing 12 (2021) 533–545.

[5] A. Varshney, N. Garg, K.S. Nagla, T.S. Nair, S.K. Jaiswal, S. Yadav, et al., Challenges in sensors technology for Industry 4.0 for futuristic metrological applications, Mapan 36 (2021) 215–226.

[6] V.P. Gupta, Smart sensors and industrial IoT (IIoT): a driver of the growth of Industry 4.0, Smart Sensors for Industrial Internet of Things: Challenges, Solutions and Applications, Springer International Publishing, Cham, (2021), pp. 37–49.

[7] K. Bhatt, C. Agrawal, A.M. Bisen, A review on emerging applications of IoT and sensor technology for industry 4.0, Wireless Personal Communications 134 (2024) 2371–2389.

[8] M. Borghetti, E. Cantù, E. Sardini, M. Serpelloni, Future sensors for smart objects by printing technologies in the Industry 4.0 scenario, Energies 13 (2020) 5916.

[9] O. Kanoun, S. Khriji, S. Naifar, S. Bradai, G. Bouattour, A. Bouhamed, et al., Prospects of wireless energy-aware sensors for smart factories in the Industry 4.0 era, Electronics 10 (2021) 2929.

[10] S. Rouf, A. Raina, M.I. Ul Haq, N. Naveed, Sensors and tribological systems: applications for Industry 4.0, Industrial Robot: The international journal of robotics research and application 49 (2022) 442–460.

[11] J.E. Naranjo, C.A. Mora, D.F. Bustamante Villagómez, M.G. Mancheno Falconi, M.V. Garcia, Wearable sensors in industrial ergonomics: enhancing safety and productivity in Industry 4.0, Sensors 25 (2025) 1526.

[12] M.J. Alenjareghi, F. Sekkay, C. Dadouchi, S. Keivanpour, Wearable sensors in Industry 4.0: preventing work-related musculoskeletal disorders, Sensors International 7 (2026) 100343.

[13] T. Ageyeva, S. Horváth, J.G. Kovács, In-mould sensors for injection moulding: on the way to Industry 4.0, Sensors 19 (2019) 3551.

[14] M. Majid, S. Habib, A.R. Javed, M. Rizwan, G. Srivastava, T.R. Gadekallu, et al., Applications of wireless sensor networks and internet of things frameworks in the industry revolution 4.0: A systematic literature review, Sensors 22 (2022) 2087.

[15] A.T. Rosário, J.C. Dias, How Industry 4.0 and sensors can leverage product design: opportunities and challenges, Sensors 23 (2023) 1165.

[16] A.S. Ullah, Modelling and simulation of complex manufacturing phenomena using sensor signals from the perspective of Industry 4.0, Advanced Engineering Informatics 39 (2019) 1–13.

[17] E. García, N. Montés, J. Llopis, A. Lacasa, Miniterm, a novel virtual sensor for predictive maintenance for the Industry 4.0 era, Sensors 22 (2022) 6222.

[18] J. Villalba-Diez, X. Zheng, D. Schmidt, M. Molina, Characterisation of industry 4.0 lean management problem-solving behavioral patterns using EEG sensors and deep learning, Sensors 19 (2019) 2841.

[19] D. Schmidt, J. Villalba Diez, J. Ordieres-Meré, R. Gevers, J. Schwiep, M. Molina, Industry 4.0 lean shopfloor management characterisation using EEG sensors and deep learning, Sensors 20 (2020) 2860.

[20] D. Gupta, V.H.C. de Albuquerque, A. Khanna, P.L. Mehta, Smart sensors for industrial Internet of Things: Challenges, Solutions and Applications, Springer International Publishing, Cham, (2021). Available from https://doi.org/10.1007/978-3-030-52624-5.

[21] S. Ağseren, S. Şimşek, Touch sensors used in Industry 4.0 to machines in the manufacturing industry on occupational health and safety, Sensor Review 44 (2024) 122–131.

[22] C.C. Lin, D.J. Deng, Z.Y. Chen, K.C. Chen, Key design of driving industry 4.0: joint energy-efficient deployment and scheduling in group-based industrial wireless sensor networks, IEEE Communications Magazine 54 (2016) 46–52.

[23] What is a Sensor? Different Types of Sensors with Applications, n.d. Avialble from: https://www.electricaltechnology.org/2018/11/types-sensors-applications.html.

[24] S. Farahani, N. Brown, J. Loftis, C. Krick, F. Pichl, R. Vaculik, et al., Evaluation of in-mold sensors and machine data towards enhancing product quality and process monitoring via Industry 4.0, The International Journal of Advanced Manufacturing Technology 105 (2019) 1371–1389.

[25] D. Jaurker, P. Gupta, A. Sahu, S.S. Joshi, I.A. Palani, Investigation of laser micro-textured triboelectric nanogenerator based self-powered vibration sensor for industry 4.0 application, Sensors and Actuators A: Physical 377 (2024) 115679.

[26] B. Bordel, R. Alcarria, T. Robles, Prediction-correction techniques to support sensor interoperability in Industry 4.0 systems, Sensors 21 (2021) 7301.

[27] M. Sujatha, N. Priya, A. Beno, T. Blesslin Sheeba, M. Manikandan, I.M. Tresa, et al., IoT and machine learning-based smart automation system for Industry 4.0 using robotics and sensors, Journal of Nanomaterials 2022 (2022) 6807585.

[28] Sensor Basics: Classification, Working Principle and Application. Available from: https://www.utmel.com/blog/categories/sensors/sensor-basics-classification-working-principle-and-application.

[29] J. Yan, Y. Meng, L. Lu, L. Li, Industrial big data in an Industry 4.0 environment: challenges, schemes, and applications for predictive maintenance, IEEE Access 5 (2017) 23484–23491.

[30] M. Faheem, G. Fizza, M.W. Ashraf, R.A. Butt, M.A. Ngadi, V.C. Gungor, Big data acquired by internet of things-enabled industrial multichannel wireless sensors networks for active monitoring and control in the smart grid Industry 4.0, Data in Brief 35 (2021) 106854.

[31] M. Shrivastava, P. Singhal, J. Bhuvana, Integrating sensor data and machine learning for predictive maintenance in Industry 4.0, Proceedings on Engineering Sciences 5 (S1) (2023) 55–62.

[32] R. Heynicke, D. Krush, G. Scholl, B. Kaercher, J. Ritter, P. Gaggero, et al., IO-Link wireless-enabled sensors and actuators for Industry 4.0 networks, AMA Conferences, (2017), pp. 134–138. Available from 10.5162/sensor2017/A8.1.

[33] A. Novak, D. Bennett, T. Kliestik, Product decision-making information systems, real-time sensor networks, and artificial intelligence-driven big data analytics in sustainable Industry 4.0, Economics, Management and Financial Markets 16 (2021) 62–72.

[34] A.J. Calderón Godoy, I. González Pérez, Integration of sensor and actuator networks and the scada system to promote the migration of the legacy flexible manufacturing system towards the Industry 4.0 concept, Journal of Sensor and Actuator Networks 7 (2018) 23.

[35] A. Ghosh, D.J. Edwards, M.R. Hosseini, R. Al-Ameri, J. Abawajy, W.D. Thwala, Real-time structural health monitoring for concrete beams: a cost-effective 'Industry 4.0'solution using piezo sensors, International Journal of Building Pathology and Adaptation 39 (2021) 283–311.

[36] L.W. Qin, M. Ahmad, I. Ali, R. Mumtaz, S.M.H. Zaidi, S.S. Alshamrani, et al., Precision measurement for Industry 4.0 standards towards solid waste classification through enhanced imaging sensors and a deep learning model, Wireless Communications and Mobile Computing 2021 (2021) 9963999.

[37] J. Koo, S. Yoon, Simultaneous in-situ calibration for physical and virtual sensors towards digital twin-enabled building operations, Advanced Engineering Informatics 59 (2024) 102239.

[38] Q. Qi, F. Tao, T. Hu, N. Anwer, A. Liu, Y. Wei, et al., Enabling technologies and tools for digital twin, Journal of Manufacturing Systems 58 (2021) 3–21.

[39] A.K. Ghosh, A.S. Ullah, R. Teti, A. Kubo, Developing sensor signal-based digital twins for intelligent machine tools, Journal of Industrial Information Integration 24 (2021) 100242.

[40] T. Wang, V.J. Gan, D. Hu, H. Liu, Digital twin-enabled built environment sensing and monitoring through semantic enrichment of BIM with SensorML, Automation in Construction 144 (2022) 104625.

[41] G. Valente, V. Muttillo, M. Muttillo, G. Barile, A. Leoni, W. Tiberti, et al., SPOF—Slave Powerlink on FPGA for smart sensors and actuators interfacing for Industry 4.0 applications, Energies 12 (2019) 1633.

[42] A. Koenig, Integrated sensor electronics with self-X capabilities for advanced sensory systems as a baseline for Industry 4.0, Sensors and Measuring Systems; 19th ITG/GMA-Symposium, VDE, (2018), pp. 1–4.

[43] A.Y. Alqahtani, S.M. Gupta, K. Nakashima, Warranty and maintenance analysis of sensor embedded products using internet of things in Industry 4.0, International Journal of Production Economics 208 (2019) 483–499.

[44] S. Raza, M. Faheem, M. Guenes, Industrial wireless sensor and actuator networks in Industry 4.0: exploring requirements, protocols, and challenges—a MAC survey, International Journal of Communication Systems 32 (2019) e4074.

11 Cloud computing for Industry 4.0

11.1 Introduction to cloud computing

Cloud computing uses hardware and software to provide services over the Internet. Many cloud solutions use state-of-the-art 'computing hardware', such as smartphones. Cloud apps can make some management and operational tasks easier. Cloud applications enable real-time information exchange, fostering collaboration. The cloud hosts data storage servers and helps Industry 4.0 complete required tasks. It delivers various security benefits, including automatic upgrades that improve infrastructure protection [1–3].

Companies that use collaborative supply chains in Industry 4.0 profit from various cloud-related factors. Management offers a constructive approach to real-time visibility of central knowledge for supply chain operations. Cloud computing is crucial for the supply chain and helpful for many other processes and businesses [4,5]. This technology is used to manage cloud infrastructure and is becoming increasingly important in the manufacturing and automotive industries, particularly for assembly lines and autonomous vehicles. The Internet of Things (IoT) enables basic cloud infrastructure applications to transform distribution and logistics, eliminate person-to-person operations, and replace them with computer-to-computer communication [6,7].

Cloud computing is now helping companies adapt to rapidly changing technology. Cloud computing allows businesses to react quickly to changing circumstances without losing data, and the sector is increasingly incorporating artificial intelligence (AI) and automation. Innovative new technologies may be tailored to meet demands on optimised cloud network services. It enables suppliers and merchants to track items throughout the supply chain while maintaining real-time management of the entire supply chain [8–10].

Cloud computing helps large businesses compete through innovation. The integration of capital and advanced technologies propels Industry 4.0. The cloud is a powerful, interactive platform that is revolutionising the data processing industry. Instead of fencing off information from competitors, companies now have a better chance of sharing and trading it using this technology. This endeavour to establish contact networks will aid the sector in achieving faster and more decisive outcomes [11–13].

11.2 Contemporary cloud computing technologies

Service providers utilise several primary methods and varieties of cloud computing in light of the most recent technological advancements [14–17]. The most well-known and widely used cloud computing technologies are as follows:

Digital Twin and Industry 4.0. DOI: https://doi.org/10.1016/B978-0-443-54019-6.00003-0

11.2.1 Grid computing

Grid computing is a cloud computing technology that enables the aggregation and interconnection of multiple computer resources to achieve a single objective. Grid computing is particularly useful for large-scale computing tasks that are too demanding for a single computer resource to complete. This kind of work is divided into smaller parts and dispersed throughout the grid. To solve the issue, the grid's computer resources collaborate and interact.

11.2.2 Virtualisation

Virtualisation technology in cloud computing uses specialised programming techniques to create a virtual version of a computer resource, making it cost-effective and practical for cloud service providers to distribute an application across many users within a single organisation or across multiple organisations. In addition to offering hardware flexibility, this significantly reduces the cost of IT infrastructure and upkeep.

11.2.3 Service-oriented architecture

The newest technology in cloud computing, known as service-oriented architecture, is an online architecture that enables the delivery of multiple services. To provide specific services to users, components of various apps or services communicate via a predefined protocol over a network.

11.2.4 Utility computing

Here, software service providers enable organisations to access cloud computing on a pay-per-use basis. Users can avoid a significant upfront expense by subscribing to their preferred services and paying only for what they use. This enables users to change how much cloud service they use or to cease using it completely.

11.3 Significant services of cloud computing

Cloud computing refers to any service that is hosted online. Servers, databases, software, networks, analytics, and other cloud-based computer tools are often included in these services. The service eliminates the need to be near physical hardware, allowing users to access data and apps stored in the cloud from any location. Spreadsheets and documents created by users, for example, used to need to be stored on a physical hard disc, USB device, or disc. Nowadays, few people worry about corrupted, lost, or broken USB drives or hard drives. Using the cloud, data is stored on a network of host computers that transmit information via the Internet. Papers can be accessed from any location thanks to cloud computing [18–20]. Fig. 11.1 shows the significant services of cloud computing.

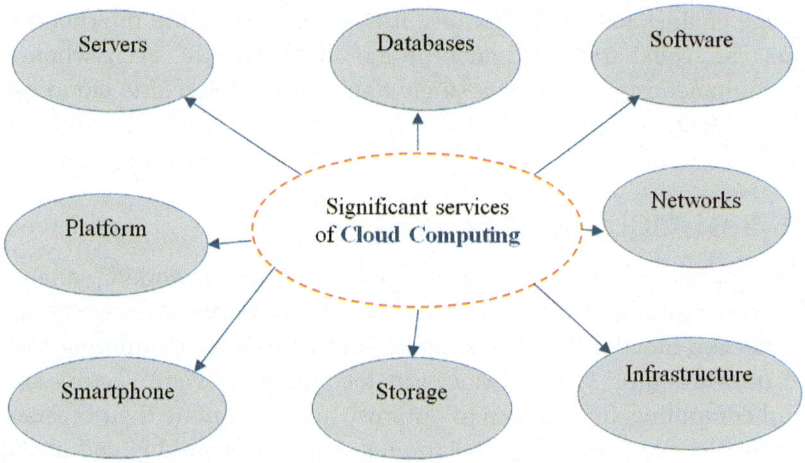

FIGURE 11.1 Major services of cloud computing.

Cloud computing is an application-based software architecture in which data is stored on remote servers and accessed over the Internet. To better understand cloud computing, the front-end and backend can be separated. Users can access cloud-based data through a web browser or cloud computing software. However, the backend is the primary component of cloud computing and is responsible for securely storing data. Included are central servers, PCs, servers, and databases [21,22].

The central server streamlines processes by adhering to a set of protocols. It facilitates faultless communication between computers and equipment connected through cloud computing by using middleware software. Cloud computing service providers usually keep multiple copies of data to reduce the risk of security threats, data loss, and data breaches. E-commerce helps small and medium-sized firms expand their market reach. Even medium-sized businesses can now become global brands through the virtual market. Businesses now benefit from technology.

Cloud computing is the most current and appropriate method for an internet experience [23,24]. They want to comprehend cloud computing, its nature, and how it functions. Sharing knowledge, resources, and software is its primary purpose. The unseen connection that ties everything together is the Internet. This includes real and virtual servers worldwide. Thus, a cloud may be used to illustrate how cloud computing works. It can travel anywhere in the world [25–27].

Cloud storage is a central repository for all Internet data, programmes, and computer resources. This enables access from any Internet-connected device. It may be a tablet, smartphone, or laptop. The information uploaded to the cloud computing system will be accessible for as long as there is Internet connectivity. This data is editable, transferable, and downloadable. The accessibility and portability that cloud storage offers can help us better understand cloud computing, what it is, and how it functions [28–30].

Cloud storage enables teamwork because it is accessible via the Internet from anywhere. Multiple users can collaborate and incorporate ideas quickly, even when separated by distance, by simultaneously accessing a single data source. Therefore, cloud storage enables remote working [31,32].

11.4 Areas of cloud computing

Cloud computing offers a wide range of services, including networking, analytics, and data storage, which run automatically with minimal user intervention. It is widely used to ensure the smooth, effective functioning of businesses and to support computing systems through software and infrastructure. With advances in technology, several sectors are increasingly adopting cloud computing applications to optimise and streamline their business operations. Numerous industries, including administration, social networking, education, and data storage, have found use for cloud computing [11,14,33–35]. The following are the key domains of cloud computing:

11.4.1 E-commerce

Another popular application is e-commerce software. The cloud computing application's user-friendly interface and efficient business processes are advantageous to e-commerce companies' users. The business benefits from tracking every detail, from the order received to the goods delivered, as well as expenses, refund rates, and damage rates. Since perfect solutions take less time and effort to deploy, most well-known companies use this software. The primary benefit is that there are no direct costs to administer these cloud computing application initiatives. One well-known company that uses cloud computing services is Amazon.

11.4.2 Healthcare

The operational, functional, and financial benefits of the cloud are advantageous to patients, physicians, and healthcare organisations. Healthcare facilities nowadays can view their unified medical records online. Cloud-based healthcare applications are safer and more efficient. Patients can schedule appointments with doctors using the cloud platform, and auto-reminders can be sent to both the patient and the doctor via email or SMS. Because healthcare cloud computing solutions are scalable, expanding or contracting the infrastructure in response to shifting business requirements is simple and rapid. A healthcare business may save money by switching from on-premise computer networking to cloud infrastructure.

11.4.3 Data storage

Computer programmes for data storage are another option for cloud computing applications. Additionally, it is one of many cloud apps that let you save files, data, photos, and other content to the cloud. It facilitates data access through the cloud application. These cloud computing applications are designed to be safe and secure, guaranteeing the safe backup of data. Data from Word, PDF, Excel, and other file types can all be retrieved and transformed.

11.4.4 Art

By providing a range of art application services, cloud computing enables the creation of aesthetically pleasing designs for books, cards, and other materials. Examples of cloud computing applications for the arts include Vistaprint, Moo, and Adobe Creative Cloud. These applications facilitate rapid card creation, printing, and design. Additionally, Adobe Creative Cloud is available, a cloud-based with outstanding professional editing capabilities.

11.4.5 File storage platform

Online file hosting services like Mediafire, Hotfile, Rapidshare, and others are examples of cloud-based applications that make it easier to host documents, photos, and videos. Users can utilise, upload, read, and share documents from these websites using a range of commercial and free cloud computing services.

11.4.6 Image editing

There are several apps available today that offer free image editing. These cloud computing services offer a wide range of functionality, such as graphical user interfaces and image resizing, editing, cropping, and special effects. Additionally, these programmes have programmable brightness and contrast settings. They also include sophisticated yet easy-to-use features. Two well-known examples are Adobe Creative Cloud and Fotor.

11.4.7 Antivirus

There are many antivirus applications available to help. These cloud application services guarantee the system's efficient operation. They offer clients a range of benefits, including system cleanup, malware detection, and virus removal. This is the best free antivirus software for a home computer. This application's primary purpose is to identify and patch malware by sending data to a cloud data centre. Among the most widely used cloud antivirus applications are Kaspersky Endpoint Security and Sophos Endpoint Protection.

11.4.8 Entertainment

Several cloud services are used by other entertainment applications to communicate with specific audiences. Cloud computing apps offer online leisure and gaming services. A quick and seamless connection is made possible by many online games.

11.4.9 URL conversion

There are several social media sites, one of which is connected to Twitter and helps condense lengthy URLs. A tool like Bitly is used to shorten long URLs so consumers can return to the original website. In addition to protecting the software from malware and hackers, it facilitates microblogging.

11.4.10 Meeting

Cloud computing apps offer convenient meeting options, including video conferencing and online meeting tools. We can schedule a meeting using these cloud application services for both work-related and private purposes. There is a seamless connection between meetings and other services, such as screen and presentation sharing. Some programmes that provide all the functionality required for basic video conferencing are 'GoToMeeting' and 'Zoom'.

11.4.11 Presentation

The programme enables the importation of PowerPoint presentations for presenting services by generating slides. One such programme that helps users produce polished presentations is 'SlideRocket'. You can use these cloud computing apps from any location in the world. There are premium and free versions of the programme.

11.4.12 Social media

Numerous people might communicate with one another every second using a range of social media platforms. Apps like Facebook, Twitter, Yammer, LinkedIn, and others enable real-time connections among users. These programmes facilitate the sharing of photos, videos, stories, and experiences.

11.4.13 GPS

GPS is another innovation in cloud computing that clients can utilise. These apps use a map as a reference to help users locate destinations online. Websites such as Google Maps, Yahoo Maps, and others are examples of cloud computing applications. Millions of people use these free programmes.

11.4.14 Accounting

Accounting software is a real-time cloud computing application that helps with business accounting management. Outright is another tool that larger businesses utilise to provide real-time daily accounting services. We were able to monitor our costs, earnings, and losses in real time. Cloud accounting software includes programmes such as KashFlow and Zoho Books.

11.4.15 Entertainment

To communicate with its target audience, the entertainment sector employs a multicloud strategy. Online gaming and video conferencing are two examples of cloud computing's recreational applications. Cloud gaming is one of the most significant forms of entertainment currently available. It provides a selection of cloud-hosted online games that can be controlled remotely. These days, the leading cloud gaming providers include Shadow, GeForce Now, Vortex, Project xCloud, and PlayStation. Video conferencing apps offer a quick and easy way to communicate. There are several advantages to using video conferencing, including reduced expenses, increased productivity, and interoperability.

11.4.16 Management

Cloud computing offers a range of management tools to support various cloud operations, including resource deployment, data integration, and disaster recovery. Additionally, these administration solutions provide control over apps, infrastructure, and platforms. With GoToMeeting's online meeting and video conferencing services, one can initiate a conference with business colleagues from any location at any time using a smartphone or tablet.

11.5 Advantages of cloud computing

The significant advantages of cloud computing are as follows [36–38]:

- Large amounts of data are kept centrally and are available through the Internet, etc.
- Imparts flexibility in business.
- Helpful in fulfilling an organisation's computing needs.
- Production quality control.
- Support collaboration in business.
- Automatically updates software.
- Cost savings.
- Helpful for processing resources.
- Reduce infrastructure expenses.
- It makes cloud storage more agile.
- Environmentally friendly.

11.6 Roles of cloud computing for Industry 4.0

Cloud computing has several exciting applications. In conjunction with a new generation of smart IoT edge devices, cloud storage appliances have the potential to completely transform the development chain, increasing performance and quality while keeping costs in check. Significant usage of cloud technology would be critical in encouraging enterprises to become more interested in digitalisation. This technology enables real-time visibility, efficiency, and quality adjustments [39–44]. The significant roles of cloud computing for Industry 4.0 are as follows:

11.6.1 Automation in the workplace

Various sectors can benefit from cloud storage for production and automation. This allows the industry to use real-time computation and analytics, opening many possibilities.

Manufacturing operations can be easily performed, and the number of items can grow as this technology is implemented.

11.6.2 Digital network

The cloud enables real-time data exchange, the creation and support of digital networks, and ecosystem integration. A shared cloud-enabled supply chain may provide real-time knowledge, motivate organisations to manage their supply chains proactively, increase efficiency, and improve risk management. It enables management to take the appropriate steps to meet customer demands.

11.6.3 Allows innovative and disruptive applications

With the power of computational resources, cloud systems allow innovative and disruptive applications. Cloud computing has unquestionably changed the way businesses conduct their day-to-day operations. To exploit the strength of a minimal code framework, organisations must adopt a modular, flexible, and agile paradigm. Cloud deployment has matched those expectations by introducing new ways to create and improve cost-effective applications.

11.6.4 Technological advancements

Cloud computing is increasingly important to technological advancement. Beyond storage and scalability, the latest cloud architecture lays the groundwork for specialised capabilities such as real-time learning, ongoing learning, data harvesting, prediction, and AI. Cloud infrastructure is critical to the growth of technologies across every area because it enables innovation.

11.6.5 Gain a competitive advantage

Cloud provides the essential creativity, agility, speed, and personalisation to gain a competitive advantage. Small and medium-sized businesses, in particular, will benefit from cloud technology. This would make it much easier to incorporate and use upcoming breakthroughs.

11.6.6 Analysis of a vast data collection

A cloud is a platform for aggregating and analysing massive datasets. Manufacturers will utilise the information to create remote service plans or data-as-a-service packages to help end-users maximise the performance of their equipment. This technology makes physical storage centres on-site significantly cheaper to maintain.

11.6.7 Confidentiality

For many firms, cloud suppliers offer to guarantee a certain level of anonymity, stay on top of emerging security issues, and prevent data breaches. The risk of data leakage is relatively low compared to the physical integrity of cloud services. Thus, they choose to keep more redundancies. Cloud-based hardware sharing will provide organisations with additional processing capacity.

11.6.8 Computerised service

Cloud storage provides access to computer services without requiring local resources. Cloud computing includes computer capacity and data storage. It is well-maintained and accessible from practically anywhere, providing security and uptime comparable to local PCs without hiring extra IT staff, purchasing equipment, or making large-scale infrastructure changes.

11.6.9 Better organisation outcomes

This technology provides better organisational outcomes and identifies opportunities for change. The cost of data storage and retrieval is high; therefore, cloud capacity might be computed at both extremes, financially and in terms of delayed reactions. This technology enhances outcomes in Industry 4.0.

11.6.10 Reduce technological expenditure

Cloud computing is ideal for businesses seeking to reduce technology expenditures and programming complexity. The cloud network focuses entirely on programmes and data rather than the infrastructure that supports them. A sophisticated cloud infrastructure will personalise creative apps to a company's demands. Cloud computing has the potential to be a game-changing strategic solution for mid-sized to large businesses.

11.6.11 Strategic planning and decision-making

The technology enables data-driven as well as strategic planning and decision-making. With cloud computing, manufacturing concepts and behaviour components may be transferred to more efficient hardware at a remote location, commonly over the Internet via satellite or other transmission modes. Cloud computing can collect far more data for long-term storage.

11.6.12 Solution for backup and recovery

When mistakes or events occur, they result in data loss. Cloud computing provides a backup and recovery option. With the right cloud platform, system data will be safe and secure in the event of a catastrophic data failure. This will aid in the efficient execution of the Industry 4.0 strategy.

11.6.13 Learning process

Educational institutions provide cloud-based solutions for both staff and students. IT departments will keep student information secure and confidential using cloud technologies, ensuring compliance with various regulations. The cloud also encourages students to share their experiences by enabling them to connect and communicate with other students using the same programme or service. Many collaboration platforms are adaptable and can be used to build a comprehensive school infrastructure.

11.6.14 Increase digital transformation

Digital transformation is crucial to implementing cloud computing technologies within Industry 4.0. Data will be a valuable asset for businesses, and its future role will be critical. As a result, the prospects in Industry 4.0 may be realised, accelerating digital transformation.

11.6.15 Effectively interact with the customer

This allows more effective customer interaction and helps supply what they require. Because all manufacturers and producers share data in a cloud environment, all parties are aware of which components are accessible. Cloud-based data processing solutions can help identify the customer's essential needs.

11.7 Roles of digital twin for cloud computing

By facilitating real-time data integration, advanced analytics, and scalable simulation environments, digital twins significantly enhance the use of cloud computing. Digital twins generate continuous streams of data by creating virtual copies of physical assets, systems, or processes, which are stored and processed on cloud platforms. To facilitate intelligent decision-making, scalability, and collaborative access, digital twins [45–47] are used with cloud computing. Through cloud-based interfaces, digital twin dashboards are made accessible to multiple stakeholders, thereby facilitating coordinated planning and control across geographically dispersed locations. Cloud makes it easier for organisations to integrate big data analytics, machine learning, and AI to enhance scenario analysis, predictive maintenance, and performance predictions. Cloud computing and digital twins work together to form a potent cyber-physical ecosystem that improves operational transparency, adaptability, and creativity in smart businesses [48–51]. The major roles of digital twins for cloud computing in different areas are as follows:

- Virtual modelling of the performance of cloud infrastructure.
- Simulating workload before cloud deployment.
- Monitoring the use of cloud resources in real time.
- Forecasting capacity bottlenecks and server overloads.
- Analysis of latency in dispersed cloud networks.
- Data centre energy usage modelling.
- Predicting faults in storage systems and cloud servers.
- Decision support that automatically scales according to consumption trends.
- Simulating network traffic flow in cloud architecture.
- Modelling security risks and identifying vulnerabilities.

- Testing disaster recovery scenarios without actual interruption.
- Cloud service performance benchmarking.
- Modelling cost optimisation for cloud resource utilisation.
- Predicting service availability and uptime.
- Virtual integration testing of cloud-based apps.
- Tracking the performance of multicloud interoperability.
- Cloud data centre thermal behaviour modelling.
- Optimisation modelling for redundancy and storage performance.
- Provides a feedback loop for ongoing enhancement of cloud system performance.

11.8 Upcoming goals of cloud computing

The upcoming goals of cloud computing are as follows [52–54]:

- This technology is being developed to bring changes in the way people think and function.
- Businesses will implement cloud-based technologies, which are supposed to increase the efficiency of production or service.
- Cloud computing is enabling businesses to act more effectively and helps Industry 4.0 to provide all possibilities.
- Cloud-based solutions will be upgraded and adjusted as market needs and infrastructure change.
- The development of cloud systems may enable businesses to access the most up-to-date data, procedures, and equipment safely without investing in costly physical infrastructure.

11.9 Disadvantages of cloud computing

The significant disadvantages of cloud computing are as follows [3,10]:

- Risk of data confidentiality.
- Cloud attack.
- Minimal control of customer requirements.
- Configuration complexities.
- Required extra cost for setup and maintenance.
- High dependence on internet connectivity.
- Risk of data leakage and privacy.

11.10 Conclusion

An Industry 4.0-enabled factory is an intelligent manufacturing facility that uses data processing, machine learning, cloud computing, and other advanced technologies to produce customised products. Today, cloud-based solutions are often offered as infrastructure options. It enables devices to communicate and work for the masses across the IoT and the Internet of Services. Cloud computing has enabled a wide range of Industry 4.0 applications.

Cloud computing entails storing data and software and providing smooth access over the Internet rather than the hard disk. Cloud storage allows data and applications to be downloaded, transferred, and viewed using any digital device through the Internet rather than on a computer's hard drive. This contemporary manufacturing era is transforming businesses worldwide with an increasingly digital future in which the cloud delivers unique processing, storage, and networking capabilities. This technology is essential to Industry 4.0, as it recognises breakthroughs and investigates how to adapt them to meet current customer needs.

References

[1] N. Velásquez, E.C. Estevez, P.M. Pesado, Cloud computing, big data and the Industry 4.0 reference architectures, Journal of Computer Science & Technology 18 (2018) .

[2] M. Sharifzadeha, H. Malekpoura, E. Shojab, Cloud computing and its impact on industry 4.0: an overview, Industry 4.0 Vision for the Supply of Energy and Materials: Enabling Technologies and Emerging Applications, Wiley, (2022), pp. 99–120. Available from 10.1002/9781119695868.ch4.

[3] J.H. Kim, A review of cyber-physical system research relevant to the emerging IT trends: Industry 4.0, IoT, big data, and cloud computing, 2 (2017) 1750011 Journal of Industrial Integration and Management.

[4] M. Azadi, Z. Moghaddas, T.C.E. Cheng, R. Farzipoor Saen, Assessing the sustainability of cloud computing service providers for Industry 4.0: a state-of-the-art analytical approach, International Journal of Production Research 61 (2023).

[5] N.N. Grigoriou, A. Fink, Cloud computing: key to enabling smart production and Industry 4.0, The Future of Smart Production for SMEs: A Methodological and Practical Approach Towards Digitalisation in SMEs, Springer International Publishing, Cham, (2022), pp. 315–322.

[6] G. Aceto, V. Persico, A. Pescapé, Industry 4.0 and health: Internet of Things, big data, and cloud computing for healthcare 4.0, Journal of Industrial Information Integration 18 (2020) 100129.

[7] C. Matt, Fog computing: complementing cloud computing to facilitate Industry 4.0. Business & Information Systems Engineering, 60(Technology Q&A), 2018, pp. 351-355. https://www.hcltech.com/technology-qa/how-does-cloud-computing-work#:~:text=It%20uses%20a%20software%2C%20middleware,loss%2C%20data%20breach%2C%20etc.

[8] S.S.S. Mokhtar, A.S.B. Mahomed, Y.A. Aziz, S.A. Rahman, Industry 4.0: the importance of innovation in adopting cloud computing among SMEs in Malaysia, Polish Journal of Management Studies 22 (2020) .

[9] P. O'Donovan, C. Gallagher, K. Leahy, D.T. O'Sullivan, A comparison of fog and cloud computing cyber-physical interfaces for Industry 4.0 real-time embedded machine learning engineering applications, Computers in Industry 110 (2019) 12–35.

[10] G. Gil, D. Corujo, P. Pedreiras, Cloud native computing for Industry 4.0: challenges and opportunities, 2021, the 26th IEEE International Conference on Emerging Technologies and Factory Automation (ETFA), IEEE, (2021), pp. 01–04.

[11] A. Sarangam, 15+ important applications of cloud computing, 2022 https://www.jigsawacademy.com/blogs/cloud-computing/applications-of-cloud-computing/.

[12] C.T. Yen, Y.C. Liu, C.C. Lin, C.C. Kao, W.B. Wang, Y.R. Hsu, An advanced manufacturing solution to the Industry 4.0 trend through sensing network and cloud computing technologies, 2014 IEEE International Conference on Automation Science and Engineering (CASE), IEEE, (2014), pp. 1150–1152.

[13] M. Singh, S. Tuli, R.J. Butcher, R. Kaur, S.S. Gill, Dynamic shift from cloud computing to Industry 4.0: eco-friendly choice or climate change threat? IoT-based Intelligent Modelling for Environmental and Ecological Engineering: IoT Next Generation EcoAgro Systems, Springer International Publishing, Cham, (2021), pp. 275–293.

[14] D. Sharma, W. Salehi, B. Bhardwaj, M. Chand, H. Salihy, Dovetailing the human resource management with the cloud computing in the era of Industry 4.0: a review, Financial Metrics in Business 5 (2025) 340–351.

[15] Y. Liu, X. Xu, Industry 4.0 and cloud manufacturing: a comparative analysis, Journal of Manufacturing Science and Engineering 139 (2017) 34701.

[16] S.S. Aravinth, A.S.R. Krishnan, R. Ranganathan, M. Sasikala, M.S. Kumar, R. Thiyagarajan, Cloud computing—everything as a cloud service in Industry 4.0, Digital Transformation: Industry 4.0 to Society 5.0, Springer Nature Singapore, Singapore, (2024), pp. 103–121.

[17] I. Castelo-Branco, F. Cruz-Jesus, T. Oliveira, Assessing Industry 4.0 readiness in manufacturing: evidence for the European Union, Computers in Industry 107 (2019) 22–32.

[18] T. Bayer, L. Moedel, C. Reich, A Fog-Cloud Computing Infrastructure for Condition Monitoring and Distributing Industry 4.0 Services, CLOSER, (2019), pp. 233–240.

[19] R. Petrasch, R. Hentschke, Process modeling for Industry 4.0 applications: towards an Industry 4.0 process modeling language and method, 2016 13th International Joint Conference on Computer Science and Software Engineering (JCSSE, IEEE, (2016), pp. 1–5.

[20] G. Pedone, I. Mezgár, Model similarity evidence and interoperability affinity in cloud-ready Industry 4.0 technologies, Computers in Industry 100 (2018) 278–286.

[21] W. Caesarendra, T. Wijaya, B.K. Pappachan, T. Tjahjowidodo, Adaptation to Industry 4.0 using machine learning and cloud computing to improve the conventional method of deburring in the aerospace manufacturing industry, 2019 12th International Conference on Information & Communication Technology and System (ICTS), IEEE, (2019), pp. 120–124.

[22] K.G. Monsang, M. Kirola, K. Joshi, R. Kumar, H.C. Anandaram, R. Gupta, Industry 4.0: IoT and cloud computing enabled smart waste management system, 2023 International Conference on Electrical, Electronics, Communication and Computers (ELEXCOM), IEEE, (2023), pp. 1–6.

[23] D. Mourtzis, Introduction to cloud technology and Industry 4.0, Design and Operation of Production Networks for Mass Personalization in the Era of Cloud Technology, Elsevier, (2022), pp. 1–12.

[24] S.K. Routray, Cloud and Edge Computing for Industry 4.0. IT Professional, 27(Technology Q&A),2025, pp. 48-53. https://www.hcltech.com/technology qa/how-does-cloud-computing-work#:~:text=It%20uses%20a%20software%2C%20middleware,loss%2C%20data%20breach%2C%20etc.

[25] A. Haleem, M. Javaid, Ahatsham, R. Suman, S. Rab, R.P. Singh, Cloud computing applications for Industry 4.0: a literature-based study, Journal of Industrial Integration and Management (2025). Available from https://doi.org/10.1142/S2424862225300030.

[26] L. Thames, D. Schaefer, Software-defined cloud manufacturing for Industry 4.0, Procedia CIRP 52 (2016) 12–17.

[27] Y.H. Hung, Investigating how cloud computing is transforming the development of industries, IEEE Access 7 (2019) 181505–181517.

[28] K.W. Chun, H. Kim, K. Lee, A study on research trends of technologies for Industry 4.0: 3D printing, artificial intelligence, big data, cloud computing, and Internet of Things, International Conference on Multimedia and Ubiquitous Engineering, Springer Singapore, Singapore, (2018), pp. 397–403.

[29] M. Abaee, M. Saeedi, H. Taherdoost, IoT and cloud computing for sustainable development goals in Industry 4.0: a bibliometric analysis, Reshaping Environmental Science Through Machine Learning and IoT (2024) 98–118.

[30] Cloud Computing Applications. https://www.javatpoint.com/cloud-computing-applications.

[31] T.M. Fernández-Caramés, P. Fraga-Lamas, M. Suárez-Albela, M. Vilar-Montesinos, A fog computing and cloudlet-based augmented reality system for the Industry 4.0 shipyard, Sensors 18 (2018) 1798.

[32] O. Givehchi, H. Trsek, J. Jasperneite, Cloud computing for industrial automation systems—a comprehensive overview, 2013 IEEE 18th Conference on Emerging Technologies & Factory Automation (ETFA), IEEE, (2013), pp. 1-4.

[33] O.A. Shonubi, The role of digital B2B platforms within Industry 4.0 technological ecosystems (the integration of cloud computing, artificial intelligence, and the Internet of Things) as a growth lever, Sustainable Futures 10 (2025) 101041.

[34] M. Azadi, Z. Moghaddas, R. Farzipoor Saen, Evaluating sustainable cloud computing service providers in Industry 4.0: an enhanced Russell directional distance method, Journal of Business Analytics (2025) 1–16.

[35] Z. Jan, F. Ahamed, W. Mayer, N. Patel, G. Grossmann, M. Stumptner, et al., Artificial intelligence for Industry 4.0: systematic review of applications, challenges, and opportunities, Expert Systems with Applications 216 (2023) 119456.

[36] Y.W. Ma, Y.C. Chen, J.L. Chen, SDN-enabled network virtualisation for Industry 4.0 based on IoTs and cloud computing, 2017, 19th International Conference on Advanced Communication Technology (ICACT), IEEE, (2017), pp. 199–202.

[37] M. Bortolini, E. Ferrari, M. Gamberi, F. Pilati, M. Faccio, Assembly system design in the Industry 4.0 era: a general frame work, IFAC-Papersonline 50 (2017) 5700–5705.

[38] G. Subramanian, B.T. Patil, B.B. Gardas, Evaluation of enablers of cloud technology to boost Industry 4.0 adoption in manufacturing micro, small, and medium-sized enterprises, Journal of Modelling in Management 16 (2021) 944–962.

[39] J. Wan, M. Xia, Cloud-assisted cyber-physical systems for the implementation of Industry 4.0, Mobile Networks and Applications 22 (2017) 1157–1158.

[40] R. Reznikov, The economic impact of cloud technologies on the Industry 4.0 development. Economic Herald of the Donbas, (Technology Q&A), 2023, pp. 74. https://www.hcltech.com/technology-qa/how-does-cloud-computing-work#:~:text=It%20uses%20a%20software%2C%20middleware,loss%2C%20data%20breach%2C%20etc.

[41] M.R.A. Bakar, N.A.M. Razali, K.K. Ishak, M.N. Ismail, T.M.T. Sembok, Adoption of Industry 4.0 with cloud computing as a mediator: evaluation using TOE framework for SMEs, JOIV: International Journal on Informatics Visualization 8 (2024) 554–563.

[42] H. Zhou, C. Yang, Y. Sun, Intelligent ironmaking optimisation service on a cloud computing platform by digital twin, Engineering 7 (2021).

[43] Q. Qi, D. Zhao, T.W. Liao, F. Tao, Modeling of cyber-physical systems and digital twin based on edge computing, fog computing and cloud computing towards smart manufacturing, Proceedings of the ASME 2018 13th International Manufacturing Science and Engineering Conference. Volume 1: Additive Manufacturing; Bio and Sustainable Manufacturing. College Station, Texas, USA. June 18–22, 2018. V001T05A018, American Society of Mechanical Engineers, (2018). Available from https://doi.org/10.1115/MSEC2018-6435.

[44] Y.C. Wang, N. Zhang, H. Li, J. Cao, Research on digital twin framework of military large-scale UAV based on cloud computing, Journal of Physics: Conference Series 1738 (1) (2021) 012052 IOP Publishing.

[45] H.V. Dang, M. Tatipamula, H.X. Nguyen, Cloud-based digital twinning for structural health monitoring using deep learning, IEEE Transactions on Industrial Informatics 18 (2021) 3820–3830.

[46] Y. Wang, J. Fang, Y. Cheng, H. She, Y. Guo, G. Zheng, Cooperative end-edge-cloud computing and resource allocation for digital twin-enabled 6G industrial IoT, IEEE Journal of Selected Topics in Signal Processing 18 (2023) 124–137.

[47] N. Susila, A. Sruthi, S. Usha, Impact of cloud security in digital twin, Advances in Computers 117 (1) (2020) 247–263 Elsevier.

[48] H. Feng, L. Qiao, Z. Lv, Innovative soft computing-enabled cloud optimisation for next-generation IoT in digital twins, Applied Soft Computing 136 (2023) 110082.

[49] D. Soni, N. Kumar, Machine learning techniques in emerging cloud computing integrated paradigms: a survey and taxonomy, Journal of Network and Computer Applications 205 (2022) 103419.

[50] J. He, G. Jia, G. Han, H. Wang, X. Yang, Locality-aware replacement algorithm in flash memory to optimise cloud computing for the smart factory of Industry 4.0, IEEE Access 5 (2017) 16252–16262.

[51] B.W. Shaheen, I. Németh, Integration of maintenance management system functions with Industry 4.0 technologies and features—a review, Processes 10 (2022) 2173.

[52] B. Sundarakani, R. Kamran, P. Maheshwari, V. Jain, Designing a hybrid cloud for a supply chain network of Industry 4.0: a theoretical framework, Benchmarking: An International Journal 28 (2021) 1524–1542.

[53] B. Syairudin, A.A.R. Nabila, Development model of cloud computing adoption for Industrial 4.0 implementation strategy for improve MSMEs performance, E3S Web of Conferences 483 (2024) 3005 EDP Sciences.

[54] N. Krishnamoorthy, S. Umarani, Implementation and management of cloud security for Industry 4. O-data using hybrid elliptical curve cryptography, The Journal of High Technology Management Research 34 (2023) 100474.

12

3D scanning for Industry 4.0

12.1 3D scanning

Three-dimensional (3D) scanning is becoming a popular method for physically imaging objects as 3D digital models. This is done using 3D scanners. Nowadays, a 3D scanner is a noncontact, nondestructive digital instrument that converts the geometry of a physical object into computer-aided design (CAD) data. It creates a point cloud, or a collection of data points, in a coordinate system that precisely portrays the size and shape of an actual item [1,2]. 3D scanners enable the prompt collection and transmission of reliable data and help Industry 4.0 perform various required tasks. This technology can gather information about an item, such as its height, width, and depth, using various approaches. 3D scanners range from low-cost to high-end [3–5].

The resolution of 3D scanning systems varies with the distance between points captured. There are various types of 3D scanners available on the market. Among these, the handheld scanner is more convenient and efficient, effectively performing the required task. Handheld 3D scanners may be manually moved and have fewer size constraints. These scanners can capture 3D photos of everything from a room to small industrial products. Handheld high-end scanning devices cover many applications and bridge the gap for everything that requires precise measurement [6–8].

Handheld 3D scanners can also capture product data in real time, making them ideal for human assessments in ergonomic and industrial applications. Because of its ability to scan and analyse structures in great detail, 3D scanning is increasingly making architectural services more attractive. The user may fully utilise, inspect, and edit CAD data and provide precise measurements using 3D scanning. 3D scanning allows for in-depth analysis of any object. It can swiftly investigate an infinite number of materials at any given moment [9,10].

3D scanning is a great way to reverse-engineer and survey for various purposes, whether for exterior construction or interior furnishings. Engineering is an important sector where 3D scanning is used to its full capacity. The ability to scan and modify anything using CAD simplifies reverse engineering and prototyping [11,12]. In an Industry 4.0 context, detailed measurements and mechanical parameters may be explored more readily. Introducing 3D scanning opens up many possibilities, ranging from better topographies to optical measurements and archiving [13,14].

When combined with 3D printing, 3D scanning improves part design and production efficiency. Some scanners generate massive mesh files that prevent further operations. Scanner software fills in small gaps and speeds up scanning, making CAD data easier to understand. Product designers and researchers utilise 3D scanners to swiftly and efficiently build digital models, whether by reverse-engineering existing designs or by capturing the exact geometry of the human body. 3D scanning can help with quality control after manufacturing and with evaluating the accuracy of a 3D-printed component [15–17].

Digital Twin and Industry 4.0. DOI: https://doi.org/10.1016/B978-0-443-54019-6.00016-9

12.2 Working steps of a 3D scanner

The exact size and shape of a natural object can be captured by 3D scanning and displayed digitally as a 3D model. This technique captures the complex geometry, captivating shape, colour, texture, and other characteristics of the scanned 3D object. A 3D scanner collects information about both the surrounding environment and the object being scanned. In essence, 3D scanners produce a digital replica of an actual thing. The user may then modify this digital copy of the 3D file and have it 3D-printed as needed. A 3D file can also be converted to other 3D modelling procedures [18–20]. The working steps of 3D scanning are shown in Fig. 12.1.

A complete representation of the scanned object cannot be recreated from a single scan. Typically, it takes many scans to gather data from different angles and perspectives. Integrating these scans requires using an alignment/registration system as a common reference point. The final model is built by combining several images. 3D scanning refers to the complete process of mixing and merging various images [21–23].

12.3 Types of 3D scanning technologies

The following are types of commercially available 3D scanning technologies [24].

12.3.1 Laser-based 3D scanners

Trigonometric triangulation is a technique used by laser-based 3D scanners to record 3D geometry with precision, as millions of points. Laser scanners send one or more laser lines onto the target surface to capture an object's reflection. One or more sensors then pick up the reflection. The sensors' locations relative to the laser's source. The laser light's reflection angle may then be determined to get precise point measurements. There are many different types and models of laser scanners. They include single-point, long-range, arm-based, portable, handheld, and portable devices.

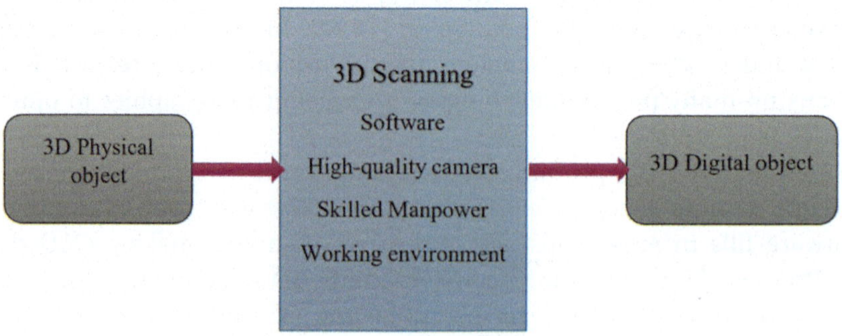

FIGURE 12.1 Working steps of 3D scanning.

12.3.2 Advantages of 3D laser scanners

The significant advantages of 3D laser scanners are as follows:

- Examine hard surfaces, such as those with bright or black finishes.
- Less sensitive to ambient light and varying lighting conditions.
- Transportable.
- Low cost, simple design, and ease of usage.

12.3.3 Structured light 3D scanners

Traditionally referred to as 'white light', 3D scanners, such as structured light 3D scanners, usually use a blue or white LED light that is projected. These 3D scanners project a pattern of light onto a block, bar, or other shape. The 3D scanner has one or more sensors that detect the edges of those patterns or structural elements to determine the object's 3D shape. The same trigonometric triangulation method used in laser scanners is used to calculate the distance between the sensors and the light source. Structured light scanners can be tripod-mounted or portable.

12.3.4 Advantages of structured light 3D scanners

The significant advantages of structured light 3D scanners are as follows:

- Scanning takes only a few seconds, up to two seconds for every scan.
- Generous scanning area, up to 48 inches in one scan.
- High resolution with a point spacing of 16 microns every scan, and as many as 16 million points.
- Extremely high precision, that is up to 10 microns.
- Use one system with numerous lenses to scan various small and oversized items.
- Handheld systems are highly portable.
- Safe for both human and animal 3D scanning.
- Price ranges from affordable to pricey, depending on resolution and accuracy.

12.3.5 Photogrammetry

The technology is simple. It involves combining pictures of an object taken from different angles. The images are captured with a camera, even a smartphone, using specific camera settings, and sophisticated software is used to stitch them together. The software combines photos by identifying pixels that represent the same physical position. To create an accurate model, the user needs to enter information into the application, such as the lens's focal length and distortion.

12.3.6 Advantages of photogrammetry

The significant advantages of photogrammetry are as follows:

- Quick and economical data collection.
- Exact, trustworthy, quantitative outcomes.

- Increased efficiency.
- It gives a genuine and long-lasting photographic record.

12.3.7 Contact scanning

With this scanning method, a probe and the object's surface come into direct physical touch. To stop it from moving, the object is first firmly fixed in position. The object's details and all the 3D data needed to create a digital file are then collected by dragging the touching probe over it. The surface must have enough points sampled to create a precise model. It is occasionally possible to precisely manipulate the contact probe and take measurements from multiple angles and configurations using an articulated arm.

Even clear, shiny surfaces can be successfully scanned using the contact scanning technique, which involves physical contact with the item's surface. This is its main advantage over other scanning technologies. Contact scanning can inspect freshly produced parts for any deformations or defects.

12.3.8 Advantages of contact scanning

The major advantages of contact scanning are as follows:

- Reduce expenses, boost productivity, and enhance access to essential information.
- Faster compliance, discoveries, and audits.
- Improved data preservation.
- Emergency recovery.
- Enhanced security.
- Support openness and transparency.
- Better records management and document retention.

12.4 Application areas of 3D scanning

The technology of 3D scanners has advanced significantly over time, becoming more accessible, smaller, and simpler to operate [25–28]. The major areas of 3D scanning are as follows:

12.4.1 Designing

The design of most products relies heavily on reverse engineering and rapid prototyping. When working with demanding clients, strategies that help achieve better results and a quicker turnaround are often used. Without a doubt, the development of 3D scanning has made this possible. As we develop more valuable materials for various applications, such as plant structures and even fossils, we often try to imitate natural materials for their superior properties.

12.4.2 Research

With 3D scanning technology, everything can be thoroughly examined. At any given time, research fellows are investigating an infinite number of topics. The introduction of 3D scanning opens up a vast array of possibilities, ranging from optical measurements and archiving to complex topography. The process of creating tiny robotic replicas of dinosaur fossils through 3D scanning, printing, and assembly has just begun.

12.4.3 Architecture

Because 3D scanning can scan and evaluate structures in great detail, it is a preferred solution for architectural surveys. Users may take extremely precise measurements and fully utilise, display, and change data with CAD thanks to 3D scanning. Quick and accurate information collection saves time and increases productivity, which is important when delivering results to a client. The ideal technique for reverse engineering is 3D scanning, which can be used for surveying as well as for exterior and some interior furnishings of a building.

12.4.4 Education

Classrooms are increasingly using 3D scanning. Students with minimal 3D CAD modelling skills can build and 3D print unique objects that would be impossible to make with software alone by using sculpting clay or a comparable medium. Students can create goods that are the appropriate size and shape by designing, scanning, and printing them. Many instructors agree that 3D scanning and 3D printing work together perfectly to teach key curriculum elements and enable the creation of unique 3D models.

12.4.5 Art

Artists and art historians increasingly use 3D scanning. One of the primary uses in this field is the scanning of artefacts for preservation and curatorial purposes, or to create replicas that can be 3D-printed. With 3D scanning, every kind of cultural asset may be categorised, quantified, examined, and even disseminated across the research community. Artists can incorporate real-world images and objects into their works by using 3D scanning. One home-goods designer told us he wanted to scan natural objects like leaves and feathers and use them in his designs, such as on a lovely bowl or vase.

12.4.6 Healthcare

3D technology has emerged as a crucial tool in the healthcare industry in recent years. One of the most popular applications of 3D scanning technology in this industry is prosthetics. Prosthetic limbs can be precisely designed and manufactured using 3D scanning, ensuring optimal comfort, mobility, and pain management. Compared with more traditional 3D printing methods, this technology requires minimal time and money. Medical professionals are using 3D scanning to inspect body parts, such as organs, before surgery. For physicians and surgeons to make better decisions and improve the chances of successful surgeries, the site needs to be well designed and prepped.

12.4.7 Engineering

Another sector with significant potential for 3D scanning is engineering. The ability to scan an object and then alter it in CAD makes rapid prototyping and reverse engineering extremely easy. It is simpler to take accurate measurements and assess mechanical properties.

12.4.8 Virtual reality and animation

3D scanning is a key component of the rapidly growing gaming industry. Therefore, people and locations are scanned to use gamification to generate realistic simulations. The advent of affordable virtual reality headsets has spurred subsequent developments and increased the use of 3D scanning. While major game firms have long used 3D scanning to generate virtual avatars, smaller and medium-sized game and app developers will be able to employ it to improve user experiences as the technology becomes more widely available and affordable.

12.4.9 Quality control

One of the most popular applications of 3D scanning is quality control. Since every company wants to produce the right components as quickly as possible, quality control is essential in today's highly competitive markets. Every production strategy must include precision, accuracy, and repeatability since superior manufacturing practices can differentiate a business from its competitors. 3D scanning software can help companies improve the quality and effectiveness of their inspections by creating precise 3D models of manufactured components and offering procedures customised for particular industries. Each part can be scanned, and its digital representation examined to ensure that every position on the component is within the required tolerances.

12.4.10 Prototyping

Prototyping and 3D printing are two more common uses for 3D scanning software. A 3D scanner can be used to capture a physical product as a digital representation, such as an STL file. Afterwards, we can use the digital model to create prototypes, which are then produced on 3D printers. The high degree of accuracy achievable with high-quality 3D scanners greatly accelerates this process, as increasing the granularity of scan data drastically reduces the manual time required to produce a solid model.

12.5 Advantages of 3D scanning

The significant advantages of 3D scanning are as follows [29–31]:

- Rapid data capturing capability.
- Greater flexibility.

- High speed.
- Lesser risk.
- Environmentally friendly.
- Improve the safety of industries.
- Provide high production output.
- Noncontact measuring capability.
- Ability to capture detailed records of the product.
- Save the cost of designing.
- More secondary involvement in the workforce.
- Enhance innovation.
- Allows integration.
- Inspect the overall size of the product.
- Ensure parts will fit together.

12.6 Roles of 3D scanning for Industry 4.0

3D scanners have significantly advanced, making them more accessible and affordable. Some real-world 3D scanning applications can be examined in the context of Industry 4.0. This technology is becoming widely used in classrooms worldwide. In many situations, companies use 3D data from commodities to help create products. The most beneficial uses of 3D scanning for Industry 4.0 include planning future operations and prototypes. This technology produces 3D models used to capture multiple viewpoints [32–35]. The significant roles of 3D scanning for Industry 4.0 are as follows:

12.6.1 Increased visibility

3D scanners have less precision but provide enough product detail. Thus, to increase visibility, several 3D scanners use turntables. Scanners may spin an item along various axes, allowing them to imagine it from several perspectives. 3D scanners are an effective way to produce small product designs with better visibility.

12.6.2 Maintaining a high level of quality

It is critical to ensure that the product meets the established quality requirements throughout the manufacturing process. 3D scanning may rapidly evaluate each component before sending it to a lab for quality testing. This technology provides detailed data that can be processed and analysed using various software options, part modifications, or historical item papers for future reference.

12.6.3 Industrial design

3D scanners are widely used in industrial design, engineering, and production because they capture data quickly and precisely. It would be necessary to obtain measurements using outdated manual techniques, which might be extremely costly and time-consuming. Using this technology, measurements can be taken quickly. 3D scanning solutions can significantly speed up workflows across each stage of product design, testing, development, launch, and aftercare, help avoid costly mistakes, and boost productivity.

12.6.4 Enhancing employee abilities

Employee abilities are improved through 3D scanning, including design, scanning, and printing. A range of instructional programmes combining 3D scanning and 3D printing will be developed in the future. Along with 3D printing and the development of unique 3D models, many teachers believe that 3D scanning is excellent for forming crucial curricular concepts.

12.6.5 Measurement of a complex-shaped object

3D scanners quickly and accurately measure complex objects in real-world environments, greatly enhancing design productivity. Collecting and altering basic shapes helps create printed components that precisely match current products of all kinds. This technology converts physical objects into digital formats that can be viewed from multiple angles.

12.6.6 Full-colour models

By displaying the data in full 3D colour, 3D scanning enables the manufactured product to be compared with its original CAD design, and variations to be identified promptly. Large museums worldwide successfully use 3D scanners to preserve some of the most famous objects from different civilisations. The scanner records the object's volume and material composition. The size, stability, and design of an object can all be measured in millimetres.

12.6.7 Product development

Any stage of the manufacturing process may benefit from 3D scanning technology. It is straightforward to construct and manufacture new items or modify existing ones using scanned data, which is helpful for the product design and development process. In the conceptualisation phase, digitising and using renderings saves time, energy, and money.

12.6.8 Information on the manufacturing process

This technology assists in collecting data from existing objects and in creating new objects based on the acquired data. Users can utilise 3D scanners to scan geometries and digitise physical objects, obtaining data for individual models, packs, etc. They may be used to capture appropriate information regarding the manufacturing process.

12.6.9 Reverse engineering

In combination with 3D scanning and 3D printing, reverse engineering is a powerful method for creating digital designs from physical parts and can be a helpful tool for prototyping.

3D scanners can scan existing broken objects, and further design of that part is easily possible using various software.

12.6.10 Helpful to create a prototype

This technology is used to create an accurate prototype using a 3D printer. This technique can reduce the number of prototype cycles necessary during design and production. It collects all scanning data and measures cross-sections more quickly and at a lower cost.

12.6.11 Assisting architects and construction sites

The laser scanner is also deployed on the construction site to scan the buildings. Scanning software can digitally process the creation of the desired image of any plant. These applications provide architects with a better idea for designing new buildings. 3D scanning technology has enabled more factories to collect measurement data on machined components and finished products, enabling faster, more accurate inspections and improved quality control.

12.6.12 Tool wear prediction

3D scanning may be used to track and measure tool wear throughout manufacturing, and to predict and prevent tool failure. Data acquired during tool wear can be used to rebuild optimised components. 3D scanning is a cost-effective and precise method for inspection and quality control.

12.7 Roles of digital twin for 3D scanning

In industries, digital twins are essential for improving the efficiency and precision of 3D scanning. The digital twin enables real-time comparison of digital and physical objects by integrating scanned data into a dynamic virtual model. This aids in the highly accurate identification of structural irregularities, surface errors, and dimensional aberrations. By facilitating simulation, analysis, and prediction insights, digital twins expand the potential of 3D scanning [36–39]. Without physically changing the component, the scanned model can be utilised for performance testing, what-if analysis, and simulating structural behaviour after it is integrated into the digital twin environment. This is especially helpful for design optimisation, predictive maintenance, and reverse engineering. Digital twins enable collaborative decision-making, lifecycle management, and continuous improvement by integrating 3D scanning equipment with cloud platforms and analytics tools. This enhances productivity and creativity in Industry 4.0 ecosystems [40–43]. The significant roles of digital twin for 3D scanning are as follows:

- Instantaneous comparison of the source CAD models and scanned geometry.
- Finding dimensional variations in manufactured parts.
- Identifying surface errors by virtually analysing scan data.

- Automated verification of tolerance in relation to design specifications.
- Creating models for reverse engineering from scanned artefacts.
- Multiple 3D scan alignment and registration optimisation.
- Using scanned geometry data, structural behaviour is simulated.
- Analysis of component deformation under operating loads.
- Evaluation of wear and tear for planning predictive maintenance.
- Use deviation trend analysis to identify process errors.
- Component lifecycle tracking with updated scan inputs.
- Automation of quality checks in intelligent production lines.
- Using scanned part integration, assembly accuracy is validated.
- Benchmarking produced the parts' performance over time.
- Cloud platform integration for remote scan analysis.
- Using precise 3D scan models to improve virtual prototyping.
- Digital preservation of tangible assets for later duplication.
- Redesign procedures are optimised using insights from scans.
- Adaptive geometry modelling supports mass customisation.
- A feedback loop for continuous improvement between the digital model and practical production.

12.8 Upcoming goals of 3D scanning

The upcoming goals of 3D scanning are as under [1,44,45]:

- 3D scanning will be employed with precision material amounts in the future.
- This will sustain the working environment of industrial sectors in the approaching days.
- This technology will allow for the creation of any complex shapes in less time and at a lower cost.
- During an integrated architectural construction project, it will be helpful for innovative design.

12.9 Disadvantages of 3D scanning

The significant disadvantages of 3D scanning are as follows [46–48]:

- The initial cost is high.
- Specialised maintenance and calibration.
- Any surface outside the scanner's field of view cannot be measured.
- Scanning accuracy is affected by temperature change.
- It cannot capture hidden or internal geometry.
- It may take up many roles as done by a human, like automating inspection and measurements, and many people may lose their jobs.
- It cannot determine the material type and structure.

- It cannot capture shiny surfaces accurately.
- Some scanner sizes are big, which occupies space.
- This technology does not scan transparent materials.
- An expensive scanner is required for better accuracy.
- Requirements of scanning software, other support software, and hardware.
- Being temperature-sensitive, atmospheric control and care are required.

12.10 Conclusion

In many sectors, 3D scanning technology has become a standard tool. Measuring systems are designed to the highest quality standards in today's product development and manufacturing processes and have proven invaluable for quality control. 3D scanning is a technique for creating accurate 3D representations of real-world objects. This technology dramatically reduces the time required to develop a working 3D model and lowers its cost. Various uses of 3D scanning may be found in manufacturing. Scanning provides the data needed to create a point cloud of the item's surface. 3D scanning for Industry 4.0 involves collecting and converting real-world data into a 3D representation. A 3D scanner expands the possibilities of 3D printing, enabling the manufacture of complex products from a wide range of nonconventional materials. Both technologies provide comprehensive digital workflows that simplify and advance processes across industries.

References

[1] A.J. Muminović, J. Smajić, I. Šarić, N. Pervan, 3D scanning in Industry 4.0, Basic Technology Model Implementation Index 4 (2023) 231–240.

[2] T.M. Onyia, I.A. Olarinoye, S.A. Jimoh, Advancements and challenges in 3D scanning: a comprehensive review of engineering applications, African Journal of Advances in Science and Technology Research 18 (2025) 191–206.

[3] A. Haleem, M. Javaid, R.P. Singh, S. Rab, R. Suman, L. Kumar, et al., Exploring the potential of 3D scanning in Industry 4.0: an overview, International Journal of Cognitive Computing in Engineering 3 (2022) 161–171.

[4] K.R. da Silva Santos, W.R. de Oliveira, E. Villani, A. Dttmann, 3D scanning method for robotised inspection of industrial sealed parts, Computers in Industry 147 (2023) 103850.

[5] Y. Zong, J. Liang, W. Pai, M. Ye, M. Ren, J. Zhao, et al., A high-efficiency and high-precision automatic 3D scanning system for industrial parts based on a scanning path planning algorithm, Optics and Lasers in Engineering 158 (2022) 107176.

[6] O. Vozisova, S. Eroshenko, E. Koksharova, A. Khalyasmaa, S. Dmitriev, Application of 3D scanning and printing technologies in the electric power industry, International Conference on Industrial Technology (ICIT), IEEE, (2016), pp. 892–897.

[7] K. Kubieniec, The role of 3D scanning technology in advanced quality control systems, Quality 64 (2025) 18-18.

[8] C.P. Nielsen, A.A. Malik, D.G. Hansen, A. Bilberg, Low-cost 3D scanning in a smart learning factory, Procedia Manufacturing 38 (2019) 824–831.

[9] T.N. Thulasy, P.N. Nohuddin, I.F. Nusyirwan, M.H. Ahmad Hijazi, M.M. Abdul Zahra, Application of 3D scanning technology in Royal Malaysian Air Force Industrial Revolution 4.0-based aircraft maintenance, IET Networks (2022).

[10] J. Hegedűs-Kuti, J. Szőlősi, D. Varga, J. Abonyi, M. Andó, T. Ruppert, 3d scanner-based identification of welding defects—clustering the results of point cloud alignment, Sensors 23 (2023) 2503.

[11] K. Bartol, D. Bojanić, T. Petković, T. Pribanić, A review of body measurement using 3D scanning, IEEE Access 9 (2021) 67281–67301.

[12] P. Trebuňa, M. Mizerák, J. Trojan, L. Rosocha, 3D scanning as a modern technology for creating 3D models, Acta Tecnologia 6 (2020) 21–24.

[13] C. Afteni, V. Paunoiu, M. Afteni, V. Teodor, Using 3D scanning in assessing the dimensional accuracy of mechanically machined parts, IOP Conference Series: Materials Science and Engineering 2351 (2022) 12071.

[14] D. Klimecka-Tatar, M. Krynke, Reverse engineering tools-3D scanning-as support for precise quality control in automated special processes, Procedia Computer Science 253 (2025) 1933–1942.

[15] S. Mönchinger, M.M. Schmidt, S. Dreßen, P. Wissmann, R. Stark, Automatised 3D-scanning application for the virtualisation of large components, Proceedings of the ASME 2019 Gas Turbine India Conference. Volume 2: Combustion, Fuels, and Emissions; Renewable Energy: Solar and Wind; Inlets and Exhausts; Emerging Technologies: Hybrid Electric Propulsion and Alternate Power Generation; GT Operation and Maintenance; Materials and Manufacturing (Including Coatings, Composites, CMCs, Additive Manufacturing); Analytics and Digital Solutions for Gas Turbines/Rotating Machinery. Chennai, Tamil Nadu, India. December 5–6, 2019. V002T11A002, American Society of Mechanical Engineers, (2019). Available from https://doi.org/10.1115/GTINDIA2019-2388.

[16] M. Paneva, P. Panev, N. Stoimenov, S. Gyoshev, Methodology for 3D scanning of objects, WSEAS Transactions on Applied and Theoretical Mechanics 18 (2023) 216–220.

[17] C.S. Tang, L.P. Veelenturf, The strategic role of logistics in the Industry 4.0 era, Transportation Research Part E: Logistics and Transportation Review 129 (2019) 1–11.

[18] A. Luximon, Y. Luximon, New technologies—3D scanning, 3D design, and 3D printing, Handbook of Footwear Design and Manufacture, Woodhead Publishing, (2021), pp. 477–503.

[19] A. Kantaros, T. Ganetsos, F.I.T. Petrescu, Three-dimensional printing and 3D scanning: emerging technologies exhibiting high potential in the field of cultural heritage, Applied Sciences 13 (2023) 4777.

[20] A. Jandyal, I. Chaturvedi, I. Wazir, A. Raina, M.I.U. Haq, 3D printing–a review of processes, materials and applications in Industry 4.0, Sustainable Operations and Computers 3 (2022) 33–42.

[21] M. Abdelmomen, F.O. Dengiz, M. Tamre, Survey on 3D technologies: case study on 3D scanning, processing and printing with a model, 2020 21st International Conference on Research and Education in Mechatronics (REM), IEEE, (2020), pp. 1–6.

[22] A. Singh, G. Madaan, S. Hr, A. Kumar, Smart manufacturing systems: a futuristics roadmap towards application of Industry 4.0 technologies, International Journal of Computer Integrated Manufacturing 36 (2023) 411–428.

[23] M. Javaid, A. Haleem, R.P. Singh, S. Rab, R. Suman, R. Kumar, Studies on the metrological need and capabilities of 3D scanning technologies, Journal of Industrial Integration and Management 8 (2023) 321–339.

[24] K. Kamani, (2020). Types of 3D Scanning Technologies and 3D Scanners: https://www.linkedin.com/pulse/types-3d-scanning-technologies-scanners-karan-kamani/

[25] Printlab. The Real Value of 3D Scanning and its Applications: https://weareprintlab.com/blog/the-real-value-of-3d-scanning-and-its-applications

[26] M. Kowalski, D. Rybarczyk, A. Milecki, Stationary 3D scanning system for IoT applications, Applied Sciences 14 (2024) 11587.

[27] B.D. Lee, T.H. Kim, Utilisation of 3D scanner and BIM for uprising 4th industrial revolution, Journal of Korean Association for Spatial Structures 18 (2018) 19–26.

[28] P. Zheng, H. Wang, Z. Sang, R.Y. Zhong, Y. Liu, C. Liu, et al., Smart manufacturing systems for Industry 4.0: conceptual framework, scenarios, and future perspectives, Frontiers of Mechanical Engineering 13 (2018) 137–150.

[29] J. Yan, Y. Meng, L. Lu, L. Li, Industrial big data in an Industry 4.0 environment: challenges, schemes, and applications for predictive maintenance, IEEE Access 5 (2017) 23484–23491.

[30] S. Verykokou, C. Ioannidis, An overview on image-based and scanner-based 3D modeling technologies, Sensors 23 (2023) 596.

[31] G. Prashar, H. Vasudev, D. Bhuddhi, Additive manufacturing: expanding 3D printing horizon in Industry 4.0, International Journal on Interactive Design and Manufacturing (IJIDeM) 17 (2023) 2221–2235.

[32] R. Hughes, M.A. A Rahman, C. Lin, P.G. Maropoulos, E. Mohamad, M.R. Salleh, et al., Implementation of a 3D scanning system for quality inspection in manufacturing, Implementation of a 3D scanning system for quality inspection in manufacturing. Proceedings of International Conference on Design and Concurrent Engineering & Manufacturing Systems Conference, The Japan Society of Mechanical Engineers, (2023), p. 10.

[33] C.A. Hu, L. Kong, F. Lv, Application of 3D laser scanning technology in engineering field, in: E3S Web of Conferences, EDP Sciences, 2021, vol. 233, p. 4014.

[34] T.A. Bin Abdul Razak, K.S. Bin Shafee, K.A. Bin Shamsuddin, M.R. Bin Ibrahim, B.H.T. Bin Baharuddin, Application of 3D scanning onto automotive door panel for quality, 2016 International Conference on Robotics and Automation Engineering (ICRAE, IEEE, (2016), pp. 31–34.

[35] M. Rozmus, J. Tokarczyk, D. Michalak, M. Dudek, K. Szewerda, M. Rotkegel, et al., Application of 3D scanning, computer simulations and virtual reality in the redesigning process of selected areas of underground transportation routes in coal mining industry, Energies 14 (2021) 2589.

[36] J.S. Fan, M.N. Xu, Y.C. Li, Y.F. Ding, Y.F. Liu, B. Geng, et al., Mechanics-based digital twin model for structural construction using 3D scanning and FE model updating, Automation in Construction 173 (2025) 106109.

[37] S. Sepahi-Boroujeni, F. Khameneifar, Digital twin-enabled error and uncertainty mapping for 3D scanning, Precision Engineering 88 (2024) 527–539.

[38] H.H. Hosamo, M.H. Hosamo, Digital twin technology for bridge maintenance using 3D laser scanning: a review, Advances in Civil Engineering 2022 (2022) 2194949.

[39] A. Javanmardi, K. Ghaedi, M. Rashidi, F. Huang, H. Khatibi, A brief overview of 3D scanning techniques for structures within the context of digital twin, International Conference on Concrete Engineering and Technology, Springer Nature Singapore, Singapore, (2024), pp. 541–553.

[40] C. Rausch, R. Lu, S. Talebi, C. Haas, Deploying 3D scanning based geometric digital twins during fabrication and assembly in offsite manufacturing, International Journal of Construction Management 23 (2023) 565–578.

[41] M. Grau, W. Korol, J. Lützenberger, J. Stjepandić, Automated generation of a digital twin of a process plant by using 3D scan and artificial intelligence, Transdisciplinary Engineering for Resilience: Responding to System Disruptions, IOS Press, (2021), pp. 93–102.

[42] J. Zhang, S. Li, Z. Zhao, Y. Gao, D. Liu, J. Wang, et al., Highly sensitive three-dimensional scanning triboelectric sensor for digital twin applications, Nano Energy 97 (2022) 107198.

[43] M. Minos-Stensrud, O.H. Haakstad, O. Sakseid, B. Westby, A. Alcocer, Towards automated 3D reconstruction in SME factories and digital twin model generation, 2018 18th International Conference on Control, Automation and Systems (ICCAS), IEEE, (2018), pp. 1777–1781.

[44] G. Arvanitis, A.S. Lalos, K. Moustakas, Robust and fast 3-D saliency mapping for industrial modeling applications, IEEE Transactions on Industrial Informatics 17 (2020) 1307–1317.

[45] A. Vodilka, M. Kočiško, J. Kaščak, Design of a prototype of an innovative 3D scanning technology for use in the digitization of hard-to-reach places, Applied Sciences 15 (2025) 2817.

[46] M. Javaid, A. Haleem, S. Khan, S. Luthra, Different flexibilities of 3D scanners and their impact on distinctive applications: an analysis, International Journal of Business Analytics (IJBAN) 7 (2020) 37–53.

[47] Z. Nagy, A. Kelemen, A. Sánduly, 3D scanning applications in structural design, IABSE Symposium: Challenges for Existing and Oncoming Structures, Prague, Czech Republic, 25-27 May 2022, IABSE Symposium Prague, (2022), pp. 1079–1086. Available from 10.2749/prague.2022.1079.

[48] A.R. Ali, M.W. Ramadan, M. Helal, Optimising product design workflows through 3D scanning and reverse engineering technologies, 2025 7th Novel Intelligent and Leading Emerging Sciences Conference (NILES), IEEE, (2025), pp. 149–152.

13

Virtual reality for Industry 4.0

13.1 Introduction to virtual reality

Virtual reality (VR) is an innovative digital technology that uses glasses to replace the real world with a three-dimensional (3D) virtual environment that involves other users. This technology operates in a fully replicated digital world, making it a perfect platform for gaming, modelling, 3D movies, and many industrial applications. Today, it is being utilised to reshape various sectors worldwide, including manufacturing and healthcare. In Industry 4.0, it is helpful to speed up the design process and perform multiple tasks [1–3].

VR enables researchers and designers to replicate real-world virtual settings accurately. This technology is utilised for research and predictive analysis to modify and enhance current parts, and for an efficient learning process. It is used to evaluate manufacturers, helping them streamline the process and reduce costs [4,5]. Product creation, refinement, and manufacturing processes are the most significant advantages of immersive VR in Industry 4.0. VR enables designers to test and develop a wide range of configurations much earlier [6,7].

VR improves testing, design, and review capabilities while reducing time-to-market. It also provides a highly collaborative environment that allows academics, designers, and developers to work efficiently. VR enables car safety assessments by simulating specific scenarios, such as crashes during the assembly process in Industry 4.0 [8,9]. As a result, this technique can detect automobile design flaws before the prototype is built. VR is an innovative platform for design teams, manufacturers, and end users to rapidly iterate on ideas, produce the desired product, assess its qualities, and observe how they affect the automobile [10,11].

VR enables the healthcare industry to improve current systems and procedures in various ways, delivering value to consumers. It allows pharmacists to read, learn, and practise more effectively than traditional learning methods, thereby increasing efficiency, efficacy, and productivity. Visualising a new reality might provide essential insights. This enables the firm to provide its employees with a realistic representation of the real world. The training sessions enabled by this technology will help address problems in real time [12–14].

VR can assist in updating the plant and detecting and repairing flaws across the device. It is possible to construct and modify the entire plant from the start. Training professionals will conduct protection and frequent inspections in accordance with virtual production procedures. Operators may use VR to acquire advanced competency levels faster, build skills, and improve team training [15,16]. In the industry, VR technology can operate machinery. The high-tech sector successfully acknowledges the application of this technology. It aids processes ranging from space transportation to rail to hospital infrastructure, making them more competitive, long-lasting, cost-effective, and easier to construct [17,18].

Digital Twin and Industry 4.0. DOI: https://doi.org/10.1016/B978-0-443-54019-6.00001-7

VR significantly impacts workplace safety, health, and environmental sustainability. Plant engineers may use VR to deliver more synchronised, informed projects, whether on-site or at remote locations, by accessing real-time data and analytics. Owners of industries, vendors, and customers can monitor remote operations from anywhere in the world, enabling them to build trust and make decisions based on real-time data [19,20]. Plant operators may be operating complex machinery, debugging problems, installing systems, or conducting safety inspections, and VR may help them improve their overall outcomes and experience. This optimises processes, provides additional knowledge and verification tools, reduces facility downtime, and accelerates development timetables [21–23].

13.2 Working principle of virtual reality

VR employs computer technology to build a virtual 3D environment that users can interact with and explore. To do this, scientists, theorists, and engineers have created many gadgets and applications. The capability to track a user's head and eye movements and appropriately update the visuals on the user's display to reflect changes in perspective [24–26]. The working principle of VR is shown in Fig. 13.1.

This technology uses computers, hardware, software, and input devices to create a 3D virtual object. From a user perspective, 3D graphics are life-sized. A user in a VR environment feels immersed, or as though they are within and a part of that universe. The user is also capable of meaningful interactions with the surrounding environment [27,28].

13.3 Types of virtual reality

VR is a term used to describe a computer-generated experience when a user interacts with a 3D virtual environment while donning electronic devices, such as gloves with sensors or special glasses with screens [28–31]. Here is a list of the many VR formats:

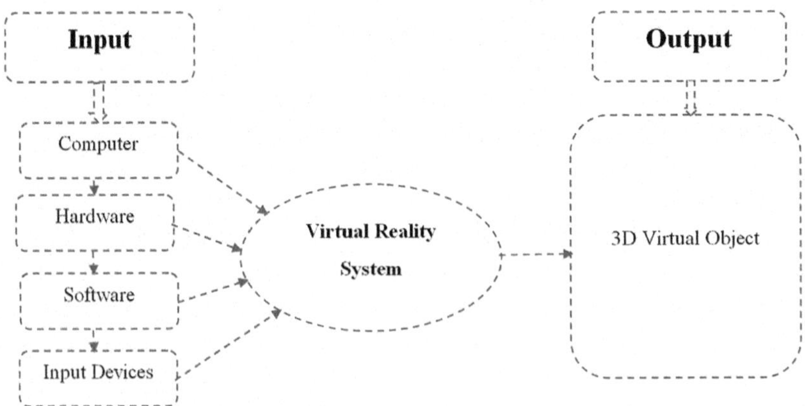

FIGURE. 13.1 Working principle of Virtual Reality.

13.3.1 Nonimmersive virtual reality

Nonimmersive VR allows us to engage with a virtual world, usually via a computer, and control certain characters or activities within the experience, without the virtual environment physically engaging with us.

13.3.2 Fully immersive virtual reality

Fully immersive VR is the antithesis of nonimmersive VR. A realistic virtual experience is ensured. If the virtual world were the focus of its events, we would feel it as though we were physically present. A realistic virtual experience requires specialised equipment, such as body scanners with sense detectors, gloves, and VR eyewear. The computer provides customers with a realistic virtual experience using data from these sensors, and the virtual world responds in real time.

13.3.3 Semi-immersive virtual reality

Semi-immersive VR lies between fully immersive and nonimmersive VR. Using a computer screen or VR glasses, we can move around in a virtual world, but aside from the visual experience, there are no tactile sensations to enhance the immersion. A virtual tour is a great example of semi-immersive virtual technology. It might be device-based or web-based. Because it enables users to navigate a space without being physically present, this technology is widely used by many enterprises.

13.3.4 Augmented reality

Augmented reality, a type of VR, allows users to view the real environment, usually on a phone screen, and alter it virtually. The smartphone game could help us better understand augmented reality. We might turn on our phone's camera and aim it in the approximate direction.

13.3.5 Collaborative virtual reality

Users from different places can engage with 3D projected actors in a fictional environment in this VR. The virtual environment of a mobile game like PlayerUnknown's Battlegrounds, where virtual 3D characters of different players gather and engage with one another, is an example of collaborative VR in action.

13.4 Areas of virtual reality

VR has dominated tech headlines in recent years because it can submerge users in a simulated yet secure environment. One of the most popular uses of VR is for gaming, but there are other applications [32–35]. The significant areas of VR are as follows:

13.4.1 Education

VR has been used in educational settings for both teaching and learning. In a 3D environment, students can interact with one another and with objects. They can also go on virtual field trips to historical sites, museums, and solar system tours. VR may be quite helpful for students with certain challenges, like autism. Children, particularly those with autism spectrum disorders, can practise social skills in a safe, engaging VR environment.

13.4.2 Architecture

Architects may see and feel their creations in VR. This allows them to test the area before creation and make real-time alterations to satisfy customers.

13.4.3 Military

In a culture where technology is accepted from an early age, and children are used to computers and video games, VR training works well. VR has the potential to immerse pupils in a variety of situations, places, and environments for training purposes. Virtual boot camps, medical training, battle simulations, and flight simulators are just a few ways the military employs technology. To train and educate soldiers without endangering them until they are prepared for battle, VR can safely replicate dangerous training scenarios. VR offers a completely immersive visual and aural experience. It can also be used to teach soldiers soft skills, such as how to deal with foreign enemies and local populations on the battlefield.

13.4.4 Sport

VR is transforming sports for players, coaches, and fans. By seeing and reliving certain circumstances, coaches and athletes can use VR to train more efficiently across a variety of sports and improve their performance each time. It is a training tool for analysing athletic performance and technique. It allows athletes to experience game conditions virtually and can enhance players' cognitive abilities in the event of an injury. VR has been used to enhance the experience of watching sports events. Today, broadcasters use VR to show live athletic events. In the future, they plan to offer virtual tickets for live events so that people around the world can virtually 'attend' any athletic event.

13.4.5 Mental health

VR technology is being widely used to treat posttraumatic stress disorder. A patient undergoing VR exposure treatment enters a simulation of a horrifying incident to process the incident and heal. It has also been used to treat depression, anxiety, and phobias. For instance, some anxiety sufferers think that VR meditation is a helpful strategy for enhancing coping skills and controlling stress reactivity. VR technology allows patients to experience their fears in a safe and controlled environment.

13.4.6 Medical training

Students studying medicine and dentistry have used VR to rehearse operations and procedures due to its interactive nature. Because there is no risk of injury or error when practising on real patients, this fosters a learning environment free from negative outcomes. Students can develop skills that they can utilise in the real world by using virtual patients. Using VR technology in the medical field is a great way to improve student training quality and offer a fantastic opportunity to optimise costs, especially given tight budgets and ongoing strain on health services. The most important way that VR is transforming healthcare is through training. It is not stressful for practitioners to become acquainted with the hospital environment. Additionally, the technology is employed in cognitive behaviour therapy, which gives patients with phobias and anxieties a safe environment to work through their issues.

13.4.7 Fashion

One lesser-known use of VR is in the fashion sector, where it has a significant impact. For example, retailers can use virtual simulations of shop environments to design their signs and product displays without fully committing to the build as they would in the real world. The right amount of time and resources can be allocated to the shop layout. Customers may virtually try on clothes and enjoy a 360-degree fashion and show experience via VR.

13.4.8 Social science and psychology

A large segment of this industry uses VR to enable patients to visualise reality from many perspectives and assume multiple personalities. Future social interactions may benefit from immersive environments.

13.4.9 Entertainment

The entertainment industry was among the first to employ VR, and it remains one of its most effective applications to this day. The online and console gaming industries make extensive use of VR. It is used in theatres and theme parks to replicate cinematic experiences and allow guests to see their favourite movie classics.

13.4.10 Automotive

VR devices help cars analyse traffic conditions and vehicle conduct. They can use the simulated scenarios to assess and refine the prototypes before developing a new model. VR is widely used in the development of smart cars that will dominate the market in the future. Driving, turning, and stopping are being taught using AI and VR.

13.4.11 Digital marketing

It can be entertaining and instructive to observe a product in use up close, even though most people do not. VR is useful for digital marketing in several ways. For example, retailers can show potential customers products in their homes.

13.4.12 Occupational safety

Occupational safety and health should be a priority in workplaces with mechanical or natural hazards. To help employees learn how to respond properly, these workplace hazards can be addressed in a controlled simulation environment.

13.5 Advantages of virtual reality

The significant advantages of VR are as follows [36–39]:

- VR delivers immersive experiences that aid in experience-based learning, resulting in enhanced knowledge exchange and retention.
- This technology is used to create digital twins of machinery and assembly lines in a variety of industries.
- R&D teams may visualise assembly line actions, automation techniques, and specialised equipment.
- The virtual environment simulation will help businesses prepare for deployment, testing, and preventive maintenance.
- Enhance cooperation, teamwork, and production.
- Offers a variety of options for improving product and service presentation.
- Aids in the training of workers in real-world scenarios.
- Assist in creating a popular virtual workspace that connects many people working on the same project.
- Enables peer engagement for validation procedures without a physical encounter
- Creates interest and enjoyment.
- Allows doctors to oversee operations from afar and contribute to industrial processes and quality control, among other things.
- Creates virtual signals and directives.
- Allows digital information systems to be integrated with manufacturing and process analyses.

13.6 Roles of virtual reality for developing Industry 4.0

VR may be applied in Industry 4.0 in various ways. VR demonstrates its exceptional capacity to immerse learners in a realistic world while cutting training expenses. It is the most capable method for creating a virtual environment that can imitate the physical world or invent products. It can improve company efficiency or aid in the collection and transmission of information [40–44]. The significant roles of VR for Industry 4.0 are as follows:

13.6.1 Helpful for repairing, servicing, and production

VR is mainly used in the automobile sector for repair, servicing, and production. Engineers and designers may use VR to create and test visual displays of colours, forms, and designs before prototypes are made. This assists workers in maximising assembly and manual manufacturing processes. It is helpful in the early phases when efficiency and productivity are more crucial than in later stages of innovation and actual business change. This strategy may be fully incorporated to reduce the number of mistakes.

13.6.2 Training & capacity building

VR is helpful for training and improving manufacturing efficiency. VR training is used in manufacturing and other sectors. Because students lack practical skills, VR training is essential to enhance their skills and learning.

13.6.3 Production planning

Production planning is used in many businesses, though it is more complex and challenging in some. This technology is beneficial for effective planning. When building space for assembly lines or manufacturing workshops, various factors must be considered, including equipment size, connections between multiple systems, safety distances, power outlets, supply line inputs, and many more.

13.6.4 Designing a manufacturing plant

Manufacturing plant design is significantly more accessible with VR. The manufacturer can ensure that everything is properly positioned and connected by reconstructing the actual regions in VR and placing the equipment models there. They may verify in VR that all equipment is available, all safety distances are maintained, and the manufacturing process is set up correctly. VR in manufacturing allows companies to maximise their floor space before physically shifting equipment.

13.6.5 Enhance the manufacturing procedure

VR enhances manufacturing processes and minimises the overall product cost. 3D virtual models may be used in plant operations to track and analyse the design and to create a virtual mirror of the actual process.

13.6.6 Inspection

VR can expedite inspection and testing by reducing the time required. Companies can now utilise this technology to verify manufactured parts in their plants. VR eliminates the need to assign a manufacturing inspector to the unit.

13.6.7 Improving worker safety

Worker safety is a crucial area in which VR might significantly impact. Health practitioners are utilising this technology to train human bodies. It is used to treat the worker's mental health conditions.

13.6.8 Data collection

VR improves industrial productivity by gathering and analysing data to generate insights that inform better operational decisions. Process simulators will be used to test and confirm that these choices are of high quality. Users can keep track of all aspects of their 3D plants through remote collaboration.

13.6.9 Saves money

To develop items that fulfil specifications, manufacturers frequently create a prototype for testing and retesting. This may be expensive and raise the cost of research and development. Manufacturers employed this technique to minimise the need for prototypes, thereby saving money.

13.7 Roles of digital twin for virtual reality

Digital twins play a crucial role in enhancing VR applications by providing accurate, real-time, data-driven virtual models of physical systems. By integrating live sensor data into VR environments, digital twins enable users to interact with realistic, dynamically updated representations of machines, production lines, or entire facilities [45–48]. This improves visualisation, system understanding, and the accuracy of immersive simulation. Digital twins strengthen VR-based training, maintenance, and design optimisation processes. In training scenarios, the digital twin ensures that the virtual environment reflects actual operational conditions, allowing workers to practice procedures in a realistic yet risk-free setting. Digital twin integration enables what-if analysis and predictive simulations within VR, supporting innovation, improved safety, and enhanced human-machine collaboration in Industry 4.0 environments [49–53]. The major roles of digital twin for VR are as follows:

- Physical assets and VR environments are synchronised in real time.
- Using real-time sensor data, virtual models are updated dynamically.
- Precise representation of machine behaviour in VR platforms.
- Using data from the real world, an immersive visualisation of factory activities.
- Simulation of predictive maintenance in VR settings.
- Risk-free training materials that accurately depict system circumstances.
- Visualising fault detection using interactive VR models.
- Dashboards for performance monitoring are incorporated into VR environments.

- Digital twins based on VR for remote asset assessment.
- Accurate real-time simulation of assembly and disassembly.
- Analysis of workflow optimisation in realistic virtual factories.
- Modelling human-machine interaction to enhance ergonomics.
- Simulation of safety procedures under actual operational circumstances.
- Before physical implementation, design validation and modification testing should be done in VR.
- Cooperation between several users in a shared VR digital twin paradigm.
- Equipment lifecycle tracking in virtual environments.
- Constant feedback integration between VR simulations and physical systems.

13.8 Upcoming goals of virtual reality

The upcoming goals of VR are as follows [54–56]:

- VR is ideally used to examine different features of a product before committing to physical manufacturing in design and prototyping.
- It will train employees with valuable, in-depth knowledge of ergonomic standards for individual parts.
- It will be used as a real-time training tool for manufacturing lines, simulating a wide range of settings and operations.
- This technology will significantly increase capacity, speeding up the entire manufacturing process and enabling even faster product launches.
- By embracing and investigating the potential of interactive VR, manufacturers will place themselves at the forefront of sectors undergoing fundamental upheavals.
- Production, profitability, and the protection system will improve in the industrial sector with proper implementation.

13.9 Disadvantages of virtual reality

The significant disadvantages of VR are as follows [45,57]:

- VR often isolates an individual.
- Lack of flexibility in VR applications.
- VR is slow for demos.
- Harmful to children.
- Effects overall human communication ability.
- Functionality issues.
- Quite expensive to adopt.
- Complex technology is developing.

13.10 Conclusion

VR has the potential to be employed in a broad range of businesses due to its adaptability. This technique offers significant potential to increase industrial flexibility. For Industry 4.0, this will provide several advantages. It might be the most effective technique for technicians and other workers to reflect on critical information and monitor their work in real time. In terms of training and preparing personnel for new product lines and procedures, VR may be a huge help. Employees may train in real time, reducing risk while maximising the benefits of regular, self-directed learning.

VR devices provide training aids in learning specific instructions on where the activities are to be completed. The customer can use a flexible assistance system to execute the instructions. Inspections and remote maintenance may be conducted far more efficiently through this technology. VR combines digitally created data with real-world data. It allows personnel to learn more about a system's problem, obtain a user guide, or even call a specialist for immediate assistance. It is utilised throughout the manufacturing stage of devices or processes, as well as for idea assessment, enabling engineers to track progress.

References

[1] L. Damiani, M. Demartini, G. Guizzi, R. Revetria, F. Tonelli, Augmented and virtual reality applications in industrial systems: a qualitative review towards the Industry 4.0 era, IFAC-PapersOnLine 51 (2018) 624–630.

[2] V. Liagkou, D. Salmas, C. Stylios, Realising virtual reality learning environment for industry 4.0, Procedia CIRP 79 (2019) 712–717.

[3] A.C. Firu, A.I. Tapîrdea, A.I. Feier, G. Drăghici, Virtual reality in the automotive field in Industry 4.0, Materials Today: Proceedings 45 (2021) 4177–4182.

[4] B. Salah, M.H. Abidi, S.H. Mian, M. Krid, H. Alkhalefah, A. Abdo, Virtual reality-based engineering education to enhance manufacturing sustainability in Industry 4.0, Sustainability 11 (2019) 1477.

[5] I.J. Akpan, O.F. Offodile, The role of virtual reality simulation in manufacturing in Industry 4.0, Systems 12 (2024) 26.

[6] F. Bano, M.A. Alomar, F.M. Alotaibi, S.H. Serbaya, A. Rizwan, F. Hasan, Leveraging virtual reality in engineering education to optimise manufacturing sustainability in Industry 4.0, Sustainability 16 (2024).

[7] J.J. Roldán, E. Crespo, A. Martín-Barrio, E. Peña-Tapia, A. Barrientos, A training system for Industry 4.0 operators in complex assemblies based on virtual reality and process mining, Robotics and Computer-Integrated Manufacturing 59 (2019) 305–316.

[8] P.B. Falola, A.E. Adeniyi, J.B. Awotunde, S.O. Akinola, M. Olagunju, O.D. Olanloye, The role of augmented reality, virtual reality, and mixed reality in Industries 4.0 and 5.0, Computational Intelligence in Industry 4.0 and 5.0 Applications, Auerbach Publications, (2025), pp. 329–356.

[9] Y. Qawqzeh, A.A. Shraah, A. Rizwan, M. Sánchez-Chero, L.A.V. More, M. Shabaz, Exploring the effectiveness of virtual reality-based training for sustainable health and occupational safety in Industry 4.0, Scientific Reports 15 (2025) 28930.

[10] M.D.P. Carretero, S. García, A. Moreno, N. Alcain, I. Elorza, Methodology to create virtual reality-assisted training courses within the Industry 4.0 vision, Multimedia Tools and Applications 80 (2021) 29699–29717.

[11] A. Cimino, F. Longo, G. Mirabelli, V. Solina, S. Verteramo, An ontology-based, general-purpose and Industry 4.0-ready architecture for supporting the smart operator (Part II–virtual reality case), Journal of Manufacturing Systems 73 (2024) 52–64.

[12] K. Logeswaran, S. Savitha, P. Suresh, K.R. Prasanna Kumar, M. Gunasekar, R. Rajadevi, et al., Unifying technologies in industry 4.0: harnessing the synergy of internet of things, big data, augmented reality/virtual reality, and blockchain technologies, Topics in Artificial Intelligence Applied to Industry 4.0 (2024) 127–147.

[13] J.E. Naranjo, D.G. Sanchez, A. Robalino-Lopez, P. Robalino-Lopez, A. Alarcon-Ortiz, M.V. Garcia, A scoping review on virtual reality-based industrial training, Applied Sciences 10 (2020) 8224.

[14] H. Schroeder, A. Friedewald, C. Kahlefendt, H. Lödding, Virtual reality for the training of operators in Industry 4.0, IFIP International Conference on Advances in Production Management Systems, Springer International Publishing, Cham, (2017), pp. 330–337. August.

[15] T. Krupnova, O. Rakova, A. Lut, E. Yudina, E. Shefer, A. Bulanova, Virtual reality in environmental education for manufacturing sustainability in Industry 4.0, 2020 Global Smart Industry Conference (GloSIC), IEEE, (2020), pp. 87–91 November.

[16] P.B. Prasad, N. Padmaja, B.S. Kumar, A. Aravind, Industry 4.0: augmented and virtual reality in education, Innovating with Augmented Reality, Auerbach Publications, (2021), pp. 29–52.

[17] Z. Guo, D. Zhou, Q. Zhou, X. Zhang, J. Geng, S. Zeng, et al., Applications of virtual reality in maintenance during the industrial product lifecycle: a systematic review, Journal of Manufacturing Systems 56 (2020) 525–538.

[18] S. Chandra Sekaran, H.J. Yap, S.N. Musa, K.E. Liew, C.H. Tan, A. Aman, The implementation of virtual reality in digital factory—a comprehensive review, The International Journal of Advanced Manufacturing Technology 115 (2021) 1349–1366.

[19] S. Machała, N. Chamier-Gliszczyński, T. Królikowski, Application of AR/VR technology in Industry 4.0, Procedia Computer Science 207 (2022) 2990–2998.

[20] C.J. Turner, W. Hutabarat, J. Oyekan, A. Tiwari, Discrete event simulation and virtual reality use in industry: new opportunities and future trends, IEEE Transactions on Human-Machine Systems 46 (2016) 882–894.

[21] K. Lewczuk, P. Żuchowicz, Virtual reality application for improving safety in intralogistics systems, Sustainability 16 (2024) 6024.

[22] J. Wolfartsberger, J. Zenisek, C. Sievi, Chances and limitations of a virtual reality-supported tool for decision making in industrial engineering, IFAC-PapersOnLine 51 (2018) 637–642.

[23] M. Simonetto, S. Arena, M. Peron, A methodological framework to integrate motion capture systems and virtual reality for assembly system 4.0 workplace design, Safety Science 146 (2022) 105561.

[24] A. Singh, G. Madaan, S. Hr, A. Kumar, Smart manufacturing systems: a futuristics roadmap towards application of Industry 4.0 technologies, International Journal of Computer Integrated Manufacturing 36 (2023) 411–428.

[25] P. Zawadzki, K. Żywicki, P. Buń, F. Górski, Employee training in an intelligent factory using virtual reality, IEEE Access 8 (2020) 135110–135117.

[26] S. Parhi, K. Joshi, T. Wuest, M. Akarte, Factors affecting Industry 4.0 adoption–a hybrid SEM-ANN approach, Computers & Industrial Engineering 168 (2022) 108062.

[27] K. Żywicki, P. Zawadzki, F. Górski, Virtual reality production training system in the scope of intelligent factory, International Conference on Intelligent Systems in Production Engineering and Maintenance, Springer International Publishing, Cham, (2017), pp. 450–458. August.

[28] Different Types of Virtual Reality. https://www.starsandstrikes.com/blog/different-types-of-virtual-reality/.

[29] A.S. Behr, L.M. Neuendorf, P. Sakthithasan, M. Karan, Q. Fang, K.E. Boettcher, et al., Uniting knowledge and application in a hybrid laboratory experiment in virtual reality–a cross-reality laboratory with applications of artificial intelligence for Industry 4.0, International Conference on Remote Engineering and Virtual Instrumentation, Springer Nature Switzerland, Cham, (2023), pp. 287–298 March.

[30] F. Caputo, A. Greco, E. D'Amato, I. Notaro, S. Spada, On the use of virtual reality for a human-centered workplace design, Procedia Structural Integrity 8 (2018) 297–308.

[31] P. Fraga-Lamas, T.M. Fernandez-Carames, O. Blanco-Novoa, M.A. Vilar-Montesinos, A review on industrial augmented reality systems for the Industry 4.0 shipyard, IEEE Access 6 (2018) 13358–13375.

[32] 5 Exciting Uses for Virtual Reality. https://www.fdmgroup.com/blog/5-exciting-uses-for-virtual-reality/.

[33] A. Cotrino, M.A. Sebastián, C. González-Gaya, Industry 4.0 roadmap: Implementation for small and medium-sized enterprises, Applied Sciences 10 (2020).

[34] L. Pérez, E. Diez, R. Usamentiaga, D.F. García, Industrial robot control and operator training using virtual reality interfaces, Computers in Industry 109 (2019) 114–120.

[35] A. Burova, J. Mäkelä, H. Heinonen, P.B. Palma, J. Hakulinen, V. Opas, et al., Asynchronous industrial collaboration: How virtual reality and virtual tools aid the process of maintenance method development and documentation creation, Computers in Industry 140 (2022) 103663.

[36] M. Javaid, A. Haleem, R.P. Singh, R. Suman, Virtual reality applications in the manufacturing field. In modern manufacturing systems, Apple Academic Press (2022) (107–122).

[37] D. Patel, R. Manickam, Enhancing maintenance learning in the industrial metaverse: integrating virtual reality and machine learning, International Conference on Industry 4.0 and Advanced Manufacturing, Springer Nature Singapore, Singapore, (2024), pp. 153–163 January.

[38] V.A. Huerta-Torruco, O. Hernandez-Uribe, L.A. Cárdenas-Robledo, N.A. Rodríguez-Olivares, Effectiveness of virtual reality in discrete event simulation models for manufacturing systems, Computers & Industrial Engineering 168 (2022) 108079.

[39] A. Al Mansoori, S. Ali, S.A. Pasha, M. Alghizzawi, M. Elareshi, A.D. Ziani, et al., Technology enhanced learning through learning management system and virtual reality googles: a critical review, in: A. Hamdan, A. Harraf, A. Buallay, P. Arora, H. Alsabatin (Eds.), From Industry 4.0 to Industry 5.0. Studies in Systems, Decision and Control, vol 470, Springer, Cham, (2023), pp. 557–564. Available from https://doi.org/10.1007/978-3-031-28314-7_48.

[40] S.R. Howie, S. Liang, A. Grech, S.A. Munawar, M. Gilardi, Industry 4.0 design reviews: a conceptual virtual reality interface, ICIEA-EU 2023: 2023 The 10th International Conference on Industrial Engineering and Applications Rome Italy, Association for Computing Machinery, New York, NY, (2023), pp. 273–282.

[41] K. Gkagiannis, A. Komninos, Spatial multimodal alert cues for virtual reality control environments in Industry 4.0, CHIGREECE '23: Proceedings of the 2nd International Conference of the ACM Greek SIGCHI Chapter, Association for Computing Machinery, New York, NY, (2023), pp. 1–6. Available from https://doi.org/10.1145/3609987.3610003.

[42] L.N. Tovar, E. Castañeda, V.R. Leyva, D. Leal, Work-in-progress—a proposal to design of virtual reality tool for learning mechatronics as a smart industry trainer education, 2020 6th International Conference of the Immersive Learning Research Network (iLRN), IEEE, (2020), pp. 381–384 June.

[43] V. Havard, B. Jeanne, M. Lacomblez, D. Baudry, Digital twin and virtual reality: a co-simulation environment for design and assessment of industrial workstations, Production & Manufacturing Research 7 (2019) 472–489.

[44] Z. Zhang, F. Wen, Z. Sun, X. Guo, T. He, C. Lee, Artificial intelligence-enabled sensing technologies in the 5G/internet of things era: from virtual reality/augmented reality to the digital twin, Advanced Intelligent Systems 4 (2022) 2100228.

[45] A. Bucchiarone, Gamification and virtual reality for digital twin learning and training: architecture and challenges, Virtual Reality & Intelligent Hardware 4 (2022) 471–486.

[46] L. Pérez, S. Rodríguez-Jiménez, N. Rodríguez, R. Usamentiaga, D.F. García, Digital twin and virtual reality based methodology for multi-robot manufacturing cell commissioning, Applied Sciences 10 (2020) 3633.

[47] R. Rocca, P. Rosa, C. Sassanelli, L. Fumagalli, S. Terzi, Integrating virtual reality and digital twin in circular economy practices: a laboratory application case, Sustainability 12 (2020) 2286.

[48] V. Kuts, T. Otto, T. Tähemaa, Y. Bondarenko, Digital-twin-based synchronised control and simulation of the industrial robotic cell using virtual reality, Journal of Machine Engineering 19 (2019).

[49] J. Pirker, E. Loria, S. Safikhani, A. Künz, S. Rosmann, Immersive virtual reality for virtual and digital twins: a literature review to identify state of the art and perspectives, 2022 IEEE Conference on Virtual Reality and 3D User Interfaces Abstracts and Workshops (VRW), IEEE, (2022), pp. 114–115 March.

[50] P.K. Pandey, P.K. Pandey, S. Mahajan, J. Paul, S. Iyer, Digital twin and virtual reality, augmented reality, and mixed reality, Digital Twins for Smart Cities and Villages, Elsevier, (2025), pp. 273–293.

[51] M. Li, S. Wenjie, Application of virtual reality technology and digital twin in digital media communication, Journal of Intelligent & Fuzzy Systems 40 (2021) 6655–6667.

[52] S. Brettschuh, M. Holly, M. Hulla, P. Herstätter, J. Pirker, Virtual reality in small and medium-sized enterprises: a systematic literature review, 2022 IEEE Conference on Virtual Reality and 3D User Interfaces Abstracts and Workshops (VRW), IEEE, (2022), pp. 502–507 March.

[53] V.M. Manghisi, A. Evangelista, A.E. Uva, A virtual reality approach for assisting sustainable human-centred ergonomic design: the ErgoVR tool, Procedia Computer Science 200 (2022) 1338–1346.

[54] A. Richert, M. Shehadeh, L. Plumanns, K. Groß, K. Schuster, S. Jeschke, Educating engineers for Industry 4.0: virtual worlds and human-robot teams: empirical studies towards a new educational age, 2016, IEEE Global Engineering Education Conference (EDUCON), IEEE, (2016), pp. 142–149. April.

[55] M. Fallaha, O. Korhan, Q. Zeeshan, Virtual reality: a possibility for training operator 4.0, Global Joint Conference on Industrial Engineering and Its Application Areas, Springer International Publishing, Cham, (2020), pp. 335–347.

[56] N. Kraus, K. Kraus, Digitalisation of business processes of enterprises of the ecosystem of Industry 4.0: virtual-real aspect of economic growth reserves, WSEAS Transactions on Business and Economics 18 (2021) 569–580.

[57] D.A.N. Venkatesh, Industry 4.0: reimagining the future of workplace (five business case applications of artificial intelligence, machine learning, robots, virtual reality in five different industries), International Journal of Engineering, Business and Enterprise Applications (IJEBEA) 26 (2018) 05–08.

14

Holography for Industry 4.0

14.1 Introduction to Holography

The word 'hologram' is derived from a combination of two Greek words: 'holos', meaning 'complete', and 'gramma', meaning 'message'. These two words combine to form a complete message or a complete picture. Unlike typical photography, a hologram is a three-dimensional (3D) image. Holographic techniques employ light, often in the form of lasers, to create diverse effects [1,2]. Thus an angled mirror separates a laser beam into two beams to create these holograms. The holographic image appears 3D because it provides depth-perception cues to the spectator. The object disperses the entire light field, which is then reconstructed. The observer may roam around the room, gazing behind an object in front of them and analysing it from various angles [3–5].

The interference of light beams reflecting genuine physical objects in the real world produces holograms, which are virtual 3D representations. The programme allows users to import models from various sources or create 3D objects from scratch. It is a one-of-a-kind photographic process in which 3D objects are photographed with a laser and recreated as closely as possible to resemble the original item. Holograms can create an exact 3D replica of an item and replicate its characteristics when illuminated by a laser. To accurately display a hologram at a given position in space, two light waves must move simultaneously [6–8].

In addition to the other essential technologies, Holography is a breakthrough technology with the potential to transform Industry 4.0's vision radically. Hologram technology is a new industry trend with the potential to impact numerous fields of Industry 4.0. It is used to create virtual overlays of real-world environments and can be a game-changer due to its diverse applications [9,10].

Holography technology also creates virtual, real-time training, teaching, and visualisation environments. Holographic recordings can be made in a variety of methods. This technology is also used on screens that modify the phase of light waves. 3D hologram is a novel, pioneering approach that presents content in a high-quality, exclusive, and unique manner, with the needs of the respective target groups carefully considered. It is also good to use an acoustic backdrop compatible with the 3D hologram, including the background. Developing holographic images in the Industry 4.0 culture entails a range of progressive processes to drive change in holographic applications [11–13].

Digital holographic technology produces fantastic visual effects for performances, live events, product launches, advertising, and promotions, using digital signage and interactive kiosks, as well as exhibitions, seminars, and conferences. This method may serve various industries, including automobiles, medical devices, materials, and design. It has also offered advantages for exploring Industry 4.0 culture by enabling improved quality and impactful services, with a vision for data cloud storage, cost-effective solution creation, and the use of software-based, innovative tools, among other things [14,15].

Digital Twin and Industry 4.0. DOI: https://doi.org/10.1016/B978-0-443-54019-6.00012-1

A holographic image can be viewed in various ways. The most prevalent methods are staring into an illuminated holographic print, shining a laser through a hologram, and projecting the picture onto a screen. By directing light into a picture, this technology creates virtual holograms. Using flexible holographic training scenarios, employees may learn in a simulated real-world setting without the risks or costs associated with many traditional industrial training scenarios [16,17].

14.2 Types of holograms

Reflection and transmission holograms are the two categories into which holograms fall as discussed below [10,18–20]:-

14.2.1 Reflection holograms

A reflection hologram allows one to see a 3D image of an item up close. Angles are placed on the other side of the hologram, which is created by igniting a point of white light from an incandescent source that is held at a certain distance. The image is created by light reflected from the hologram.

14.2.2 Transmission holograms

A type of hologram known as a transmission hologram uses laser light from the same source as the hologram's recording to create a 3D image of an item. A laser light directed from behind the hologram conveys this image to the observer.

14.3 Working principle of Holography

Dennis Gabor invented a method in 1948 to capture the amplitude and phase of a light wave, thereby creating a 3D picture of an object. Holography is a method for creating a 3D picture of an item [11,21–23]. Fig. 14.1 shows the working principle of Holography.

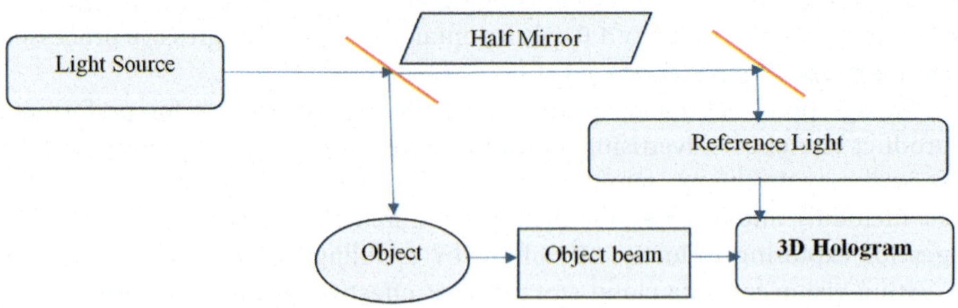

FIGURE 14.1 Working principle of Holography.

3D scenes are captured as a two-dimensional (2D) scene in a regular snapshot of an item, since the photographic plate's emulsion is solely responsible for intensity fluctuations. As a result, just the intensity distribution in a particular plane is captured in the snapshot [10]. Phase dispersion at the photograph's plane is common in typical photos. As a result, the snapshot merely captures a 2D representation of the 3D scene, losing the object's 3D. The principle of Holography is explained in two steps: recording the hologram and reconstructing the image. These are discussed below [24,25].

14.3.1 Recording of the holograph

The laser beam is divided in half using a plane mirror. One part of the laser beam illuminates the object whose hologram is to be produced. A beam or wave is obtained and recorded on the photographic plate after the laser beam's extra component, referred to as a reference, is reflected from the plane mirror. As a result, the photographic plate is simultaneously exposed to the reference beam and the light reflected from the object. These two beams of light are coherent waves. Thus by superimposing the two coherent light beams, an interference pattern is produced on the holographic plate. After the photographic plate or film is developed, we get a hologram. Both the phase and the amplitude of the object wave are described in the hologram.

14.3.2 Reconstructing the image

The process of recreating an object's 3D picture is called reconstruction. The reconstruction wave, a single laser beam, illuminates the hologram to examine an item's image continuously. This reconstruction wave is identical to the reference wave that is used to record the hologram. The object is represented both physically and virtually. When the observer moves his gaze from one side to the other, the near portion of the object appears to rotate around the distant portion. Consequently, the viewer may perceive the image's depth and see its three-dimensionality.

14.4 Areas of Holography

Holography is not limited to the visible spectrum and is used for more than creating 3D images [10,24]. Through obstacles that would otherwise be impenetrable, microwaves detect objects. Holography is also used to find material stress. When a material is under stress, it will deform, often inadvertently and unnaturally. A hologram can amplify this change, since the light reflected from the material will now be at a different angle than it was initially [4,25]. The significant areas of Holography are as follows:

14.4.1 Holographic microscopy

The microscopic objects or particles that float in a liquid are always moving. When a specimen or particle goes out of the range of view of a standard microscope, it becomes difficult to locate it in time to study its size or another feature. Nevertheless, the entire volume that contains these particles or specimens is recorded holographically. In that scenario, a particle or a moving specimen can be frozen in a hologram. As a result, by examining the reconstructed 3D image, the characteristics of a particle or specimen can be easily determined.

14.4.2 Holography interferometry

In a double-exposure photographic interferometry approach, an object is first exposed to a photographic plate using a reference wave. A reconstruction wave illuminates the resulting hologram. These interference fringes reveal the object's strain characteristics. This makes it feasible to investigate the distribution of strain inside the strained object.

14.4.3 Data storage

A media's ability to store data or information is essential. Data or information is stored at a high density in a crystal or photopolymer via holographic data storage. The main advantage of holographic data storage is that, unlike traditional recording media, it uses the entire volume of the medium rather than just its surface. As a result, a hologram has an enormous data storage capacity.

14.4.4 Cinematography

Its use to produce 3D images for television and film is growing. Filmmakers want holograms in films almost as much as audiences do. One remarkable technological advancement that grabs our interest and imagination is holograms. They expand our potential horizons with their really cool design, 3D images etched in light, and seamless interactivity that lets on-screen users zoom in, out, and around with intuitive ease. Eventually, holograms in the real world will become so successful that holographic films will be feasible.

14.4.5 Security

Holography is used to make security holograms. They are difficult to fake because they are replicas of master holograms that need costly, sophisticated equipment. They are used in publications, bank and credit cards, passports, identification cards, and currencies of various countries.

14.4.6 Three-dimensional volumetric

The most visually appealing holographic technology on the market right now is volumetric holograms. Every second, they project more than half a billion points of light into a volumetric world using an extremely fast digital light engine and a highly tuned volume rendering engine. Hundreds of digitised cross-sections of an image are shown simultaneously with a perfectly constructed high-speed reciprocating screen. The photons of light strike the screen and diffuse into a physical cross-sectional image at just the right point in physical space, tricking our eyes into combining layers into a single, 3D image.

14.4.7 Military mapping

Geographic information is a crucial part of military strategy, and fully dimensional holographic images are being used to improve reconnaissance. This helps soldiers study 3D terrain and peep 'around' corners during mission training. The company does this by creating a holographic sheet using intricate computational image data. In addition to allowing 'see into' the intricate 3D terrain image captured in the holographic sheet, the technology is portable, easy to operate, and foldable for storage. When it comes to rescue and evacuation after a disaster, the maps are useful.

14.4.8 Healthcare

As a tool for teaching students and medical professionals how to display patient data, Holography has the potential to revolutionise medicine. Many medical systems use state-of-the-art imaging methods, such as ultrasound and magnetic resonance imaging, to provide comprehensive data. Although electrical data is frequently used to display a flat image on a computer screen, it may also be used to produce full-colour, computer-generated 3D holographic visuals. The 3D visuals are freely accessible to medical practitioners and students. Multiple images are stored in a single hologram, allowing the user to explore the display and see different organs or bodily parts. The brain, liver, lungs, heart, skeleton, vascular system, nerves, and muscles are just a few structures rendered in 3D holographic representations.

14.5 Advantages of Holography

The significant advantages of Holography are as follows [25,26]:

- Increases object feasibility.
- Allows for many pictures in 3D.
- Compatibility with other digital technologies.
- Cost-effective approach.
- In comparison to other approaches, it has a larger storage capacity.
- Provide better security.
- The product can be viewed from any angle and does not require special glasses.
- It does not necessitate the use of a projection screen.
- Sensing heads are used to detect local values in the measuring space.
- Allows for recording real-time occurrences.
- Natural intangible interaction with pictures.
- Simpler interfaces and intuitive usage daily.

14.6 Roles of Holography for Industry 4.0

For Industry 4.0, Holography can store large volumes of data. A hologram is a 3D image that can be viewed without special equipment, such as cameras or glasses. The object moves and shifts realistically as the spectator travels around the display since the picture can be viewed from any angle. Holographic images can be static, such as commercial images, or dynamic sequences that can be viewed from any angle by many people [27–30]. The significant roles of Holography for Industry 4.0 are as follows:

14.6.1 Visualise product

Holographic technology helps visualise products in real time in a virtual environment. Real-time holographic communications may become the most extensively utilised kind of communication. Companies may use this technology to deliver solutions to several places in various ways simultaneously.

14.6.2 Customisable holographic training

Employees may learn in a simulated real-world context utilising customisable holographic training scenarios. This creates real-time 3D models that can be updated and used to identify issues earlier in the product development process.

14.6.3 Explain a complex scenario

Holography uses real-time data visualisation to explain complex scenarios and uncover design flaws before they are manufactured graphically. It can produce, position, organise, and re-order holographic elements in the environment, speeding up and simplifying the planning process. Using 3D Holography, organisations and enterprises can better grasp realistic production.

14.6.4 Streamline the product development process

Holography is a technique for creating realistic digital representations in simulations, helping streamline product development. Perfectly scaled 3D holograms accelerate product design and reduce iteration. Simulations are widely used throughout plant operations in Industry 4.0 to utilise real-time data and simulate the actual environment.

14.6.5 Visualise materials in three-dimensional

It developed an informative approach for visualising 3D material with higher data transmission rates than previous methods. It offers a different data storage record and reads many bits in parallel. Hologram technology enters the picture, separating itself from earlier 3D image breakthroughs.

14.6.6 Improved management resources

Holography enables better resource management and prioritises containment decisions. This technology employs real-time overlays to support emergency services. Holograms may also create virtual environments where public safety workers can visualise and respond to disasters.

14.6.7 Products assessment

Holography is a valuable technology that enables more effective product assessment. This method enables precise measurement of 3D geometry in a short time. Holography inspection is increasingly used to evaluate components that improve customer satisfaction.

14.6.8 Meeting through the Internet

HR may use holographic technology to hold a successful virtual meeting. Virtual meetings help businesses in many ways. This eliminates the need for attendees to travel to a conference and lets them attend from anywhere, without transportation costs.

14.6.9 Improvements in security

Over time, a variety of holograms have evolved, helping improve workplace security. Banknotes, identity cards, and credit cards now feature security holograms, a 3D holographic display that is more human-friendly digital content.

14.6.10 Research and development

Research and development using this technology can help companies improve their products, bringing science fiction into our future and transforming the world. Smart holographic sensors make diagnostic testing accessible by changing their optical properties in response to external inputs. Holograms are cost-effective because they can be made with a well-established research and development process.

14.7 Roles of digital twin for Holography

By offering precise, real-time, data driven 3D models for holographic visualisation, digital twins significantly advance Holography. The digital twin ensures that holographic projections accurately depict the performance, status, and structural aspects of equipment or systems by continuously synchronising physical assets with their virtual counterparts. This enables engineers and operators to view system behaviour, internal components, and real-time operational data in an interactive holographic format [31–34]. Digital twins strengthen Holography-based applications in predictive analysis, maintenance, and training. Digital twins provide dynamic data that may be holographically projected to simulate fault states and maintenance processes. Before adjusting in the actual environment, this enables personnel to engage with holographic models for system optimisation and troubleshooting. In Industry 4.0

ecosystems, industries can improve human-machine collaboration, reduce downtime, and enhance decision-making accuracy by integrating immersive holographic displays with digital twin intelligence [35–39]. The major roles of digital twin for Holography are as follows:

- Using live sensor data, holographic representations can be updated in real time.
- Precise 3D asset depiction for holographic display.
- Machine performance is dynamically visualised via holographic screens.
- Internal component structures projected without actual disassembly.
- Predictive analytics-based holographic defect simulation.
- Data-driven holograms provide interactive maintenance recommendations.
- Holographic dashboards for real-time production monitoring.
- Using holographic digital prototypes for design validation.
- Immersion 3D visualisation of operational parameters.
- The use of shared holographic twins for remote expert collaboration.
- Emergency scenario simulation in holographic settings.
- Training for assembly and disassembly with real-time digital twin data.
- Visualisation of performance trends in 3D holographic space.
- Holographic analytics for representing energy consumption.
- Modelling workflow optimisation in holographic simulations.
- Equipment lifecycle tracking with holographic interfaces.
- Holographic deviation mapping for visualisation of quality inspection.
- Immersion 3D twins to improve human-machine interaction.
- Constant feedback integration between holographic displays and physical systems.

14.8 Upcoming goals of Holography

The upcoming goals of Holography are as follows [40–42]:

- Unlike traditional optical storage media, which only employ a thin 2D layer to store data, Holography will store data in 3D, allowing for massive storage capacity in small spaces.
- People will focus on the road and its surroundings instead of focusing on the inside of a vehicle and its comfort features.
- These technologies will provide users with an interactive experience.
- New product holograms can be easily created using this technology.

14.9 Disadvantages of deploying Holography

The major disadvantages of deploying Holography are as follows [43–44]:

- Only static images are created.
- Significant computing power, memory, and storage are required.
- Does not make photographs with much movement.

- Requires high-density data storage.
- Requires precise apparatus to create and display a picture.
- Collinearity of reference illumination beams.
- Holographic interferometry equipment is sophisticated and costly.
- Higher computer cost of processing holograms.

14.10 Conclusion

Holograms are 3D images made by interfering light beams that reflect real-world objects. In Holography, light diffraction is used to create images and is now being utilised in popular culture to produce 3D images and holograms, which might be helpful in Industry 4.0. Holograms may be seen with the naked eye. The depth, parallax, and other features of the original item are preserved in holograms. They are great for showcasing aesthetically pleasing products and explaining complex technological issues. Holograms offer significant benefits for commercial security. By utilising light from diverse viewpoints, the impacts of virtual holographic technology create a particular procedure. Holograms make it easier to reproduce designs in a 3D virtual environment for Industry 4.0. Product designers can use this technology to visualise and discuss designs, enabling quick decision-making. On a specially developed high-speed reciprocating screen, thousands of digitised cross-sections of an image are projected concurrently. In the future, this technology will enable innovative applications in engineering, healthcare, and related areas.

References

[1] M. Fratz, T. Seyler, A. Bertz, D. Carl, Digital Holography in production: an overview, Light: Advanced Manufacturing 2 (3) (2021) 283–295. Available from https://doi.org/10.37188/lam.2021.015.

[2] R. Kumar, G. Dwivedi, Emerging scientific and industrial applications of digital Hologaphy: an overview, Engineering Research Express 5 (3) (2023) 032005. Available from https://doi.org/10.1088/2631-8695/acf97e.

[3] D. Khodadad, Digital Holography and its application, Applied Sciences 14 (23) (2024) 11254. Available from https://doi.org/10.3390/app142311254.

[4] A. Haleem, M. Javaid, R.P. Singh, R. Suman, S. Rab, Holography and its applications for Industry 4.0: an overview, Internet of Things and Cyber-Physical Systems 2 (2022) 42–48. Available from https://doi.org/10.1016/j.iotcps.2022.05.004.

[5] V. Jokanović, The extraordinary importance of 6G network development and 3D Holography in future healthcare, Deep Learning in Internet of Things for Next Generation Healthcare, Chapman and Hall/CRC, (2024), pp. 224–241. Available from https://doi.org/10.1201/9781003451846-18.

[6] N. Schloer, S. Boos, F. Harst, C. Lanquillon, M. Ohrnberger, F. Schoch, et al., Optimising the industrial human-machine interface: holographic surfaces and their role in anomaly detection, International Conference on Human-Computer Interaction, Springer Nature Switzerland, Cham, (2024), pp. 430–435.

[7] M.K. Ramlie, M.N.A. Aziz, N.H. Ibrahim, F. Khalid, Hologram tutor: the 3D visualization technology revolution in education, International Journal of Academic Research in Progressive Education and Development 11 (2022) 98–116.

[8] A. Jaiswal, A. Pandey, V.A. Avinash, S.L. Shiva Darshan, M.V. Manoj Kumar, B.S. Prashanth, et al., 3D holography in the air using artificial intelligence and unity without 3D glasses, 2022 IEEE 2nd Mysore Sub

Section International Conference (MysuruCon), Mysuru, India, IEEE, (2022), pp. 1–6. Available from 10.1109/MysuruCon55714.2022.9972414.

[9] R. Ahmed, K. Hidayat, S. Ahmed, R. Ahmed, Revolutionary Hologram systems: pioneering a new frontier in visual technology, International Journal of Innovations in Science & Technology 6 (2024) 1943–1955.

[10] Shabbusharma, What is Holography? |How does Holography Work | 5 Application of Holography?, 2020. https://physicswave.com/how-does-holography-work/.

[11] J.H. Park, B. Lee, Holographic techniques for augmented reality and virtual reality near-eye displays, Light: Advanced Manufacturing 3 (1) (2022) 137–150. Available from https://doi.org/10.37188/lam.2022.009.

[12] M.Z. Win, Z. Wang, Z. Liu, Y. Shen, A. Conti, Location awareness via intelligent surfaces: a path toward holographic NLN, IEEE Vehicular Technology Magazine 17 (2022) 37–45.

[13] L.N. Fewella, A developed framework for utilising holograms and artificial intelligence in customised furniture design, Ain Shams Engineering Journal 16 (2025) 103747.

[14] T.M. Fernández-Caramés, P. Fraga-Lamas, M. Suárez-Albela, M. Vilar-Montesinos, A fog computing and cloudlet-based augmented reality system for the Industry 4.0 shipyard, Sensors 18 (2018) 1798.

[15] C. Marinagi, P. Reklitis, P. Trivellas, D. Sakas, The impact of Industry 4.0 technologies on key performance indicators for a resilient Supply Chain 4.0, Sustainability 15 (2023) 5185.

[16] G. Ekren, B. Oberer, A. Erkollar, Augmented reality in Industry 4.0: enabling technologies and the potential for SMEs, International Symposium for Production Research 9 (17) (2017) 1–3.

[17] B. Javidi, A. Carnicer, A. Anand, G. Barbastathis, W. Chen, P. Ferraro, et al., Roadmap on digital Holography, Optics Express 29 (2021) 35078–35118.

[18] A. Haleem, M. Javaid, Additive manufacturing applications in Industry 4.0: a review, Journal of Industrial Integration and Management 4 (2019) 1930001.

[19] V. Reljić, I. Milenković, S. Dudić, J. Šulc, B. Bajči, Augmented reality applications in Industry 4.0 environment, Applied Sciences 11 (2021) 5592.

[20] S. Ivashov, N. Lichkov, A. Tumbarska, Potential of using holographic subsurface radars for mine detection, Security & Future 4 (2020) 43–46.

[21] J. Butt, A strategic roadmap for the manufacturing industry to implement Industry 4.0, Designs 4 (2) (2020) 11. Available from https://doi.org/10.3390/designs4020011.

[22] K. Nosalska, Z.M. Piątek, G. Mazurek, R. Rządca, Industry 4.0: coherent definition framework with technological and organisational interdependencies, Journal of Manufacturing Technology Management 31 (5) (2020) 837–862. Available from https://doi.org/10.1108/JMTM-08-2018-0238.

[23] F.A. Bhat, S. Parvez, Emerging challenges in the sustainable manufacturing system: from Industry 4.0 to Industry 5.0, Journal of The Institution of Engineers (India): Series C 105 (2024) 1385–1399.

[24] Five surprising ways holograms are revolutionising the world. https://theconversation.com/five-surprising-ways-holograms-are-revolutionising-the-world-77886

[25] A. Forcina, D. Falcone, The role of Industry 4.0 enabling technologies for safety management: a systematic literature review, Procedia Computer Science 180 (2021) 436–445.

[26] L.N. Hoon, S.S. Shaharuddin, Learning effectiveness of 3D hologram animation on primary school learners, Journal of Visual Art and Design 11 (2019) 93–104.

[27] G. Di Bona, V. Cesarotti, G. Arcese, T. Gallo, Implementation of Industry 4.0 technology: new opportunities and challenges for maintenance strategy, Procedia Computer Science 180 (2021) 424–429.

[28] J. Birdi, S. Rajora, M. Butola, K. Khare, True 3D reconstruction in digital Holography, Journal of Physics: Photonics 2 (2020) 044004.

[29] B. Wang, F. Dong, D. Yang, Z. Song, L. Xu, W. Chu, et al., Polarisation-controlled colour-tunable holograms with dielectric metasurfaces, Optica 4 (2017) 1368–1371.

[30] J. Wu, J. Zhu, J. Zhang, P. Dang, W. Li, Y. Guo, et al., A dynamic holographic modelling method of digital twin scenes for bridge construction, International Journal of Digital Earth 16 (2023) 2404–2425.

[31] L. Shuguang, B. Lin, Holographic Classroom Based on Digital Twin and Its Application Prospect, 2020 IEEE 3rd International Conference on Electronics and Communication Engineering (ICECE), Xi'An, China, IEEE, (2020), pp. 122–126. Available from 10.1109/ICECE51594.2020.9352884.

[32] Y. Shi, X. Deng, Y. Tong, R. Li, Y. Zhang, L. Ren, et al., Synergistic digital twin and holographic augmented-reality-guided percutaneous puncture of respiratory liver tumor, IEEE Transactions on Human-Machine Systems 52 (2022) 1364–1374.

[33] T. Fushimi, D. Tagami, K. Yamamoto, Y. Ochiai, A digital twin approach for experimental acoustic hologram optimisation, Communications Engineering 3 (2024) 12.

[34] T.L.P. Do, J. Choi, V.Q. Le, P. Gentet, L. Hwang, S. Lee, HoloGaussian digital twin: reconstructing 3D scenes with Gaussian splatting for tabletop hologram visualisation of real environments, Remote Sensing 16 (2024) 4591.

[35] P. Durana, M. Vochozka, A.I. Tucmeanu, P. Tudosă, Digital twin-enabled industrial internet of things, multi-sensory extended reality and immersive holographic imaging technologies, and 3D modelling and virtual simulation tools across blockchain-based decentralised metaverse systems, Economics, Management and Financial Markets 17 (2022) 43–57.

[36] X. Zhang, M. Xu, R. Tan, D. Niyato, Learning-based auction for matching demand and supply of holographic digital twin over immersive communications, IEEE Transactions on Multimedia 26 (2023) 5884–5896.

[37] K.F. Michalikova, R. Ondrejka, E. Johnson, Deep learning-based object detection and holographic virtual imaging technologies, mobile biometric and sentiment data, and digital twin simulation and modelling tools in the metaverse economy, Linguistic and Philosophical Investigations 22 (2023) 247–263.

[38] W. Chai, Q. Ma, Application of digital twin and hologram technology to achieve distribution network reliability forecast, 2022 7th Asia Conference on Power and Electrical Engineering (ACPEE), Hangzhou, China, IEEE, (2022), pp. 783–787. Available from 10.1109/ACPEE53904.2022.9784043.

[39] Z. Wang, L. Miccio, S. Coppola, V. Bianco, P. Memmolo, V. Tkachenko, et al., New needs to measure soft matter at micro-/nanoscale: a new possibility thanks to digital Holography, Optics and Photonics for Advanced Dimensional Metrology II, SPIE, (2022), pp. 121–125 vol. 37121.

[40] A. Cotrino, M.A. Sebastián, C. González-Gaya, Industry 4.0 roadmap: implementation for small and medium-sized enterprises, Applied Sciences 10 (2020) 8566.

[41] P. Fraga-Lamas, T.M. Fernández-Caramés, O. Blanco-Novoa, M.A. Vilar-Montesinos, A review on industrial augmented reality systems for the Industry 4.0 shipyard, IEEE Access 6 (2018) 13358–13375.

[42] E.C. Strinati, S. Barbarossa, J.L. Gonzalez-Jimenez, D. Ktenas, N. Cassiau, L. Maret, et al., 6G: the next frontier: from holographic messaging to artificial Intelligence using Subterahertz and visible light communication, IEEE Vehicular Technology Magazine 14 (2019) 42–50.

[43] C.M. Bolognesi, S. Teruggi, F. Fiorillo, Holographic visualisation and management of big point clouds, The International Archives of the Photogrammetry, Remote Sensing and Spatial Information Sciences 46 (2021) 71–78.

[44] R. Tayeh, R.R. Issa, Interactive holograms for collaborative visualisation of building information models, Research Companion to Building Information Modelling, Edward Elgar Publishing, (2022), pp. 268–293.

15

Cybersecurity for Industry 4.0

15.1 Introduction to cybersecurity

Cybersecurity technology is essential for protecting systems, networks, programmes, devices, and data from cyberattacks and related threats. Its purpose is to reduce the risk while protecting systems, networks, and technology from unauthorised access. The importance of cybersecurity is growing as the number of people, devices, and programmes in contemporary businesses grows. The growing numbers, skills, and attack techniques of cyber attackers exacerbate the situation [1–3].

Industry 4.0 integrates cyber-physical systems, thereby connecting the physical and digital worlds by gathering data from physical objects and processes to drive automation and optimisation. This digital technology revolution collects and processes large volumes of high-velocity data while automating field operations and supply chain activities [4,5]. Cybersecurity is a complex process that helps address the numerous hacking challenges in Industry 4.0. Cybersecurity is critical for excellent data protection in many enterprises and industrial control systems [6,7].

For cybersecurity, earning and retaining clients has emerged as a significant challenge. As data breaches continue at an alarming rate, companies want to know they are doing business with a security company they can trust to keep their data safe. It is a new generation of physical-process computing and networking [8,9]. The primary benefit of this technology is to improve industrial efficiency and flexibility. It enables producers to make last-minute changes at various stages of the manufacturing process, resulting in unprecedented flexibility. The decision-making and transparency of the production process are improved [10–12].

As part of Industry 4.0, it exchanges data and intellectual property across supply chains and many stakeholders. Consumers and providers use cybersecurity to connect their systems. Employees often confuse a computer attack with a production system failure, underscoring the criticality of cybersecurity for small- and medium-sized manufacturing enterprises. Small and medium-sized firms are the most appealing targets for hackers, even though the amounts and relevance of data stolen are often less than those of large international corporations. This technology can quickly fulfil this challenge [13–15].

15.2 Technologies used in cybersecurity

To improve cybersecurity, researchers and information security experts continually develop new technologies and proactive approaches. Ransomware attacks and vulnerabilities arising from the increased use of cloud services are among the emerging risks [14]. The growth of the Internet of Things (IoT), including smart home devices, and the potential vulnerabilities in 5G

Digital Twin and Industry 4.0. DOI: https://doi.org/10.1016/B978-0-443-54019-6.00007-8

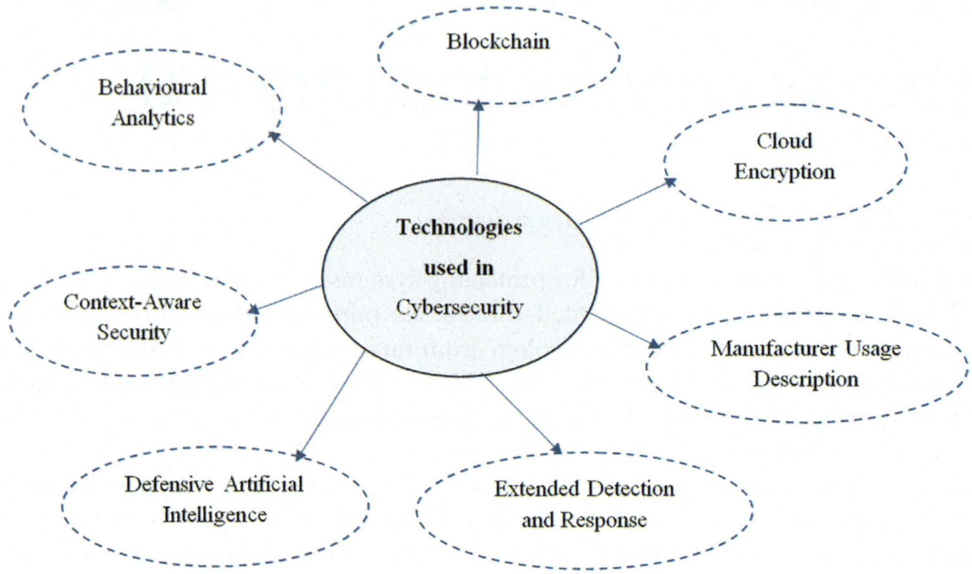

FIGURE 15.1 Technologies used in Cybersecurity.

technology, also raises security concerns. The development of improved cybersecurity technologies is stimulated by cybersecurity threats [16–19]. Fig. 15.1 displays the main technologies utilised in cybersecurity.

15.2.1 Behavioural analytics

To study user behaviour on websites, mobile applications, systems, and networks, data is analysed using behavioural analytics. Cybersecurity specialists can use behavioural analytics platforms to identify potential risks and vulnerabilities. Analysing activity patterns can help identify unusual events and behaviours that may indicate cybersecurity weaknesses. Behavioural analytics may find that one gadget is generating unusually large amounts of data.

15.2.2 Blockchain

Data can be safely stored in blocks using a blockchain database. It connects the blocks via cryptography. Blockchain enables data acquisition but not modification or deletion. Cybersecurity professionals can use blockchain to safeguard systems or devices, create industry-standard security protocols, and make it nearly impossible for hackers to access databases. Blockchain technology reduces human error, improves user privacy, increases transparency, and reduces costs. The security problem with centralised data storage is also resolved by blockchain. Data storage across networks results in a decentralised system that is less vulnerable to hackers.

15.2.3 Cloud encryption

Cloud services help businesses save money, increase productivity, and enhance remote services. However, the cloud's remote data storage may make data more sensitive. Before being saved to the cloud, data is encrypted using cloud encryption technology to render it unreadable. Only approved users with the encryption key can decrypt and read the data again. This restricted access reduces the risk of data breaches by unauthorised attackers. One excellent cybersecurity option for data protection is cloud encryption. Additionally, cloud encryption helps prevent unauthorised access to valuable data. Additionally, cloud encryption can increase consumer confidence in cloud services and facilitate regulatory compliance for businesses.

15.2.4 Context-aware security

One type of cybersecurity technology that helps companies make better security decisions in real time is context-aware security. Thus classic cybersecurity solutions use yes/no questions to determine whether to grant someone access to a system or data. Some legitimate users might be rejected by this straightforward procedure, reducing productivity. Context-aware security reduces the likelihood that an authorised user will be denied access. Context-aware security simplifies data access procedures and facilitates approved users' work.

15.2.5 Defensive artificial intelligence

To identify and stop cyberattacks, cybersecurity professionals may employ artificial intelligence (AI) for defence. Cunning thieves exploit technologies such as adversarial machine learning and hostile AI because they are harder for traditional cybersecurity methods to detect. Examples of offensive AI include deepfakes, manufactured images, personas, and movies that accurately depict people or events that never happened or do not exist. Malicious actors can use adversarial machine learning to trick computers into malfunctioning by providing them with misleading information. Cybersecurity experts can utilise defensive AI to spot and stop offensive AI's attempts to monitor, test, and comprehend how a system or network functions. Defensive AI can improve algorithms, making them more difficult to compromise. Researchers in cybersecurity can subject machine learning models to more rigorous vulnerability evaluations.

15.2.6 Extended detection and response

Advanced cybersecurity technology known as extended detection and response (XDR) is utilised to identify security risks and concerns and take appropriate action. XDR reacts to endpoints, networks, and the cloud. It evolved from the simple, traditional endpoint-detection-and-response. XDR connects data from multiple sources to provide a more complete picture. By using this technology, cybersecurity professionals can now automatically identify and assess threats at a higher level. Throughout a company's whole asset ecosystem, it can help prevent or minimise current and impending data breaches. With XDR, cybersecurity professionals can detect and react to targeted attacks. Additionally, they can generate comprehensive analytics and automatically verify and correlate warnings. Automation of tedious operations, robust automatic detection, and decreased occurrences requiring enquiry are all advantages of XDR.

15.2.7 Manufacturer usage description

The internet engineering task force created the manufacturer-use description standard to enhance the security of IoT devices on household and small-business networks. IoT devices are vulnerable to network-based attacks. These attacks could cause a computer to malfunction or result in the loss of private data. IoT gear needs to be secure without being too costly or complicated. By utilising manufacturer usage description (MUD), IoT device security can be rapidly and economically improved. Cybersecurity professionals can use MUD to improve device defences against distributed denial-of-service (DDoS) attacks. If an attack is successful, MUD can reduce damage and data loss.

15.3 The human factor which increases cybersecurity risk

The significant ways that human factors can increase cybersecurity risk are as follows [20–25]:

15.3.1 Suspicious URLs and emails

If something seems strange, let the personnel know. Encourage staff members to carefully review URLs, delete emails that appear to be from a phoney address or have no content, and stress the need to protect personal data. As IT professionals, it is our responsibility to raise awareness of potential cybersecurity risks.

15.3.2 Password idleness

We know it is not a good idea to reuse passwords. Teach employees the importance of changing passwords frequently and using specific combinations. Some of us need to jot down our various passwords because we all carry them around.

15.3.3 Personally identifiable information

The majority of employees should understand that their gadgets should be used for personal browsing, purchasing, and financial tasks. Emphasise the importance of being mindful of websites that may lead to others. That also includes social media. Customer support may need to be informed that hackers often obtain information by posting excessive amounts of personally identifiable information on social media platforms like Facebook, Twitter, and Instagram.

15.3.4 Backups and updates

Uninformed IT users can continue with their daily lives without regularly updating their antivirus software or backing up their data. The IT department should handle this. The most difficult element is letting the personnel know when they need help.

15.3.5 Physical security for devices

Think about how many workers get up from their workstations to attend meals, meetings, and get-togethers. Stress the significance of safeguarding data whenever a device is left unsupervised. Encourage employees to handle their devices as carefully as they do their bags.

15.4 Types of cybersecurity threats

Keeping ahead of cybersecurity threats is difficult. There are many hazards that IT professionals are aware of, but the list keeps growing. These days, cyberattacks happen frequently. While some attacks are small and may be prevented, others spread quickly and wreak havoc [6,26-30]. The following are the main categories of cybersecurity risks:

15.4.1 Malware

Malware is software created expressly to cause harm. Only by opening the wrong attachment or clicking on the wrong link can malware, also known as a virus, infect a computer.

15.4.2 Illegal websites

On questionable websites that promote free antivirus software, other applications, or data, hackers often include malware, spyware, and worms. It contains viruses and ransomware that might destroy the user's computer system. These websites use URLs and links that redirect visitors to malicious sites, infecting computers with viruses and malware.

15.4.3 Ransomware

Ransomware is actually a type of malware. After infiltrating a network or stealing personal information, ransomware encrypts our systems in exchange for a ransom. This is a crucial difference.

15.4.4 Phishing attacks

Phishing is a swindle, just as it sounds. Hackers put out a hook in the hopes that we will bite, and when we do, they use our behaviour to steal personal information, including credit card numbers and passwords. Phishing efforts usually take the form of official-looking emails that ask us to respond.

15.4.5 File sharing

File-sharing systems transfer data, including files, folders, photos, programmes, and more, over the Internet. However, hackers exploit this to disseminate malware and spyware alongside these files, exposing users' private information.

15.4.6 Social engineering

Deceptive human contact is a component of social engineering. This is an example of someone lying and forcing someone to reveal personal information. These people usually rely on social media posts and profiles for their information.

15.4.7 Cybersecurity careers

Given the rising number of cybersecurity dangers, there is a growing need for cybersecurity jobs.

15.4.8 Emails

The majority of emails infect our systems with malware, including viruses, ransomware, and spyware. Links to questionable websites and the majority of emails from unidentified sources may also conceal malicious downloads.

15.4.9 Phishing

Phishing is the practice of sending phoney emails and messages with malicious URLs to a large number of recipients. When the receiver opens the emails, malware, spyware, or ransomware is downloaded in the background onto the recipient's computer.

15.5 Types of cybersecurity

The major types of cybersecurity are as follows [4,5,10]:

15.5.1 Information security

Information security primarily concerns safeguarding user privacy from hackers and other cybercriminals. It prevents data theft and the scenario in which hackers demand a ransom from customers in return for their personal information. It discusses authentication, user authorisation, and cryptography.

15.5.2 Application security

Application security focuses on defending computer programmes and apps against potential hacking threats and vulnerabilities arising from various weaknesses that hackers exploit.

15.5.3 Network security

The main goals of network security are to prevent denial-of-service attacks, which prevent authorised users from accessing the computer network, and to protect the integrity and security of the network components that users utilise. Firewalls, antivirus software, and virtual private networks are among the items it sells.

15.6 Areas of cybersecurity

The significant areas of cybersecurity are as follows [10,13,31–34]:

15.6.1 Distributed denial-of-service security

To keep the web server busy processing requests from different devices, the attacker uses multiple devices in this cyberattack. It creates fake website traffic on the server. By rerouting traffic to alternative cloud-based servers, cybersecurity's DDoS mitigation solution helps manage and resolve the issue.

15.6.2 Business continuity and disaster recovery planning

Data recovery enables businesses to continue in the event of data loss, threats, or disasters. By regularly backing up data and investing in a system that enables corporate activities to continue, this application offers models and tactics to help firms handle major data loss.

15.6.3 Web firewall

A web application server-based firewall is set up on a wide-area network. It automatically identifies and removes harmful website traffic by analysing all incoming and outgoing traffic on the server. This cybersecurity method helps identify and enable automated traffic monitoring by reducing the risk of attack.

15.6.4 Bots

To shut down a server, many hackers and attackers now use bots to overload it with traffic from multiple devices. Cybersecurity helps combat fraudulent users, often known as bots, by identifying them and requiring them to end their sessions so they don't disrupt the experiences of legitimate users.

15.6.5 Software development security

Identifying software defects throughout development and making sure rules and guidelines are adhered to. To identify any flaws, vulnerabilities, or holes that hackers or rival companies might exploit, this programme thoroughly cheques, analyses, and evaluates the software.

15.6.6 Antivirus and antimalware

Antivirus and antimalware programmes that protect computers and other electronic devices from online threats and malicious attacks are developed in accordance with cybersecurity principles. It also helps with network security and firewall management for all linked devices.

15.6.7 Threat management systems

Cybersecurity helps protect against online threats and attacks on computer systems. It identifies several vulnerabilities and flaws in the system that attackers could exploit and automatically resolves all defects that could worsen performance issues. It also makes it easier to stop a cyberattack immediately and gives customers more efficient control over security vulnerabilities.

15.6.8 Critical systems

One serious issue that cybersecurity helps address is attacks on large servers connected to wide-area networks. It maintains stringent, industry-standard safety requirements for users to follow to secure devices through cybersecurity measures. It monitors every application in real time and regularly assesses the security of the network, servers, and users.

15.6.9 Operations security

This entails vulnerability scanning, system backups, employee background cheques, vulnerability analysis, and penetration testing.

15.6.10 Rules and regulations

Cybersecurity helps develop new norms and standards that users, attackers, and other network users must follow to use the Internet safely. It allows us to investigate security concerns and adjust the network as necessary.

15.6.11 Access control

The three primary focuses of this application are monitoring, authorisation, and authentication. Experts must utilise cybersecurity frameworks and techniques to limit users and delegate tasks in this scenario. To prevent unauthorised access or activity, it is also crucial to track down and closely monitor any questionable activity.

15.6.12 Telecommunications and network security

Confidential information and conversations are secured. This application's significant goals are safeguarding data and guaranteeing its continuous availability. Networks are strengthened and guarded while telecommunications security is ensured.

15.6.13 Information security governance and risk management

This application of cybersecurity covers risk management, data integrity, security governance, risk analysis, and security education. Essential elements of risk management include assessing risks and controlling potential harm. Information security also addresses the integrity of sensitive data.

15.6.14 Cryptography

Sensitive information is encrypted, and only those with the proper access may decode it. By using cryptography, only the intended recipients can see a particular piece of data. This helps minimise data breaches or leaks.

15.6.15 Security architecture and design

Authorising protocols, services, and connectors for the entire infrastructure. This module covers identifying flaws in the current architecture and building more resilient IT systems. Since a leaky defence is easily compromised, the security architecture needs to be safeguarded.

15.7 Advantages of cybersecurity

The significant advantages of cybersecurity are as follows [35–39]:

- Antitheft protection for data aims to make other equipment safer and protected from hackers.
- Prevention of Financial fraud, embezzlement, other financial activities, and businesses, and Intellectual property protection.
- Increases consumer trust in different services and even in automated and nonautomated production environments.
- Defends the computer against viruses and prevents hackers from gaining access to the computer.
- Reduces the chances of the computer stalling or crashing and maintains data and system confidentiality.
- Prevents data theft by creating a high-security protocol and making the device secure.
- Allows people to work in a comfortable setting and keeps everything running smoothly.
- Implement security protocols and initiatives to safeguard our data and equipment.
- Guarding against unauthorised and illegal access to our network, devices, and data.
- Guard against malicious malware, compromised files, and other online threats.
- Cyber espionage prevention.

15.8 Roles of cybersecurity for Industry 4.0

Organisations must perform frequent software patching and vulnerability assessments to prevent hacker attacks. Advances in cybersecurity technology, including threat segmentation, risk containerisation, network zoning, and profiling, have led to widespread adoption in bolstering a company's cyber posture. Any device, application, or programme with direct, remote, or even indirect access to an organisation's networks must be monitored using cybersecurity [40–48]. The significant roles of cybersecurity for Industry 4.0 are as follows:

15.8.1 Protecting the computer system

Cybersecurity is a set of safeguards meant to protect computer systems against cyberattacks. This technology considers many technological, organisational, legal, and human factors. These capabilities aid in evaluating, implementing, and maintaining data integrity, whether the data is provided externally or internally.

15.8.2 Distribute information

Cybersecurity systems help physical components and digital networks automate processes and safely transmit information. Organisations carefully analyse the security implications to achieve a thriving Industry 4.0 scenario with this technology. This enhances IoT by enabling it to connect with machines on the production line to distribute information.

15.8.3 Tackle risky circumstances

Cybersecurity risks are crucial for all businesses, and addressing them with this technology is a top priority. Businesses and individuals have migrated in droves to the online world. As the number of online transactions grew, businesses were obliged to expand their online presence and to handle large volumes of data. This technology can handle massive amounts of data with intelligent devices.

15.8.4 Controlling cyber risk

Controlling cyber risk in the hyperconnected environment of Industry 4.0 necessitates immediate action to mitigate risks. This technology has significant potential to mitigate cyber risk. This can continuously change and update the danger landscape.

15.8.5 Developing digital businesses

Preventing and improving cybersecurity have always been top goals for digital businesses. With the rising adoption of Industry 4.0, the risks and stakes have grown as well. This technology's potential lies in developing digital firms by enhancing processes and resolving ongoing production issues.

15.8.6 Analysis of large amounts of data

Cybersecurity applications can analyse large volumes of data, detect signs of hostile behaviour, and recommend actions. This new technology presents opportunities and responsibilities that subject-matter experts will confront. Some forward-thinking companies are altering their work processes to maintain a strong focus on cybersecurity.

15.8.7 Improving output

Machines may communicate with workers and interact remotely, enabling remote control and maintenance as this technology evolves. This technology will be entirely automated, from design to manufacturing to the supply chain, increasing output. Network and IT system administrators will find themselves managing the proper complex systems.

15.8.8 Sufficient level of protection

In cybersecurity, manufacturers are expected to provide features in their products that give a sufficient level of protection. Similarly, Industry 4.0 operators must implement the manufacturer's security improvements. Cybersecurity is used to organise related tasks, addressing substantial gaps and overlaps to overcome various problems.

15.8.9 Improved business

Businesses securely benefit from cybersecurity. It allows them to safeguard both material and intangible assets and their reputation. It adds to increased client trust, enabling greater market competitiveness. Cybersecurity will play an increasingly important role in the future in improving business.

15.8.10 Transfer of data across devices

A factory equipped with this advanced technology may achieve greater production efficiency, produce higher-quality items with fewer defects, and offer greater operational flexibility, greatly benefiting the manufacturing industry. Cybersecurity aids in the seamless transfer of data across devices. The information may be used to adjust and improve the equipment's performance, achieving optimal efficiency while reducing waste.

15.8.11 Provide dynamic changes

Cybersecurity enables swift responses to dynamic changes, such as shifts in demand and supply, as well as the possibility of human error. Resource efficiency is the main focus, particularly regarding materials, energy, and human resources. Using this technology, businesses can embrace the advancements brought about by automation and digitalisation while carefully managing a new generation of risks.

15.9 Roles of digital twin for cybersecurity

In Industry 4.0, digital twin technology is crucial to bolstering cybersecurity. When it comes to cybersecurity, digital twins allow businesses to test vulnerabilities, model cyberattacks, and assess the impact of potential threats without disrupting regular operations. Digital twins also facilitate anticipatory cybersecurity management and ongoing monitoring. The digital twin can identify anomalies that may indicate malware infections, unauthorised access, or network breaches by analysing behavioural patterns and system performance data [49–52]. Rapid response planning and automated threat modelling are made possible by integration with machine learning and AI technologies.

Additionally, by modelling recovery plans and reducing downtime during cyber emergencies, digital twins help to enhance incident response. As a result, in addition to improving operational efficiency, digital twin technology serves as a strategic cybersecurity tool, ensuring resilience, reliability, and a safe digital transformation in smart industrial ecosystems [53–56]. The significant roles of digital twins for cybersecurity are as follows:

- Simulating the routes taken by ransomware as it spreads throughout linked production lines.
- Network traffic is mirrored in real time for anomaly-based intrusion detection and further prevention.
- Using a regulated virtual factory model to simulate zero-day attack scenarios.
- Examining how well firewall rules defend against attacks that mimic lateral movement.
- Testing multifactor authentication systems in the event of a simulated breach.
- Verification of safe firmware upgrades before applying patches to actual computers.
- Cyber-physical effect study using machine behaviour correlation with cyber events.
- Finding unusual machine command sequences that point to malevolent control
- Simulating how industrial gateways are affected by DDoS.
- Modelling the abuse of insider credentials in role-based access systems.
- Evaluating machine-to-machine communication channels' encryption strength.
- Evaluation of smart factory edge computing node vulnerabilities.
- Cyber incident reconstruction for root cause analysis and forensic enquiry.
- Ensuring the accuracy of data transfers in the digital supply chain and similar environments.
- Predictive detection of industrial network topologies that unauthorised users can exploit.
- Coordinated cyber-physical attacks are simulated to assess disaster recovery procedures.

15.10 Upcoming goals of cybersecurity

The primary upcoming goals of cybersecurity are as follows [57–62]:

- Cybersecurity will allow objects in an intelligent network to interact with humans in a seamless manner.
- In this new world of accessibility and technology, secure networks are necessary due to the implications they entail.
- It will provide viable solutions and preventative measures against cybersecurity threats.

- Making cybersecurity an effective solution to numerous challenges in the future, protecting all information and data.

15.11 Challenges with cybersecurity

The significant disadvantages of cybersecurity are as follows [63,64]:

- New software must be updated regularly to maintain security.
- Firewalls can be tricky to set up properly for general users.
- Cybersecurity can be too expensive for firms and even individuals.
- It slows the system down even more than it previously did.
- It might be expensive for the typical user.
- Firewall rules are challenging to set up properly.
- The perplexing nature of cybersecurity.
- Constant monitoring is required.

15.12 Conclusion

Potential cybersecurity applications of Industry 4.0 are beneficial for data protection and security. By utilising cybersecurity, firms can gain a competitive edge by deriving intelligent insights from data. This technology preserves information and other critical digital systems for Industry 4.0. This increases profitability and allows them to create new intellectual property. Cybersecurity strategies and solutions can provide partial protection for manufacturing organisations as factories become digitally linked and value chains become digitally integrated.

By supporting effective decision-making, cybersecurity will help increase efficiency and productivity in the future. Various high-risk scenarios can be handled quickly, and cyber risk can be readily minimised through cybersecurity. This technology will serve as a precaution against various technologies that might be valuable targets for hackers and unscrupulous individuals. This helps with safe, efficient communication. Cybersecurity for Industry 4.0 helps increase cross-departmental collaboration by making the right information available to the right people in real time.

References

[1] M. Lezzi, M. Lazoi, A. Corallo, Cybersecurity for Industry 4.0 in the current literature: a reference framework, Computers in Industry 103 (2018) 97–110.

[2] G. Culot, F. Fattori, M. Podrecca, M. Sartor, Addressing Industry 4.0 cybersecurity challenges, IEEE Engineering Management Review 47 (2019) 79–86. Available from https://doi.org/10.1109/emr.2019.2927559.

[3] G. Czeczot, I. Rojek, D. Mikołajewski, B. Sangho, AI in IIoT management of cybersecurity for Industry 4.0 and Industry 5.0 purposes, Electronics 12 (2023) 3800.

[4] V. Mullet, P. Sondi, E. Ramat, A review of cybersecurity guidelines for manufacturing factories in Industry 4.0, IEEE Access 9 (2021) 23235–23263.

[5] A. Corallo, M. Lazoi, M. Lezzi, Cybersecurity in the context of Industry 4.0: a structured classification of critical assets and business impacts, Computers in Industry 114 (2020) 103165.

[6] Applications of Cybersecurity. Available from https://www.geeksforgeeks.org/applications-of-cybersecurity/.

[7] L. Thames, D. Schaefer, Cybersecurity for Industry 4.0 and advanced manufacturing environments with ensemble intelligence, Cybersecurity for Industry 4.0: Analysis for Design and Manufacturing, Springer International Publishing, Cham, (2017), pp. 243–265.

[8] H. Khalid, S.J. Hashim, S. Ahmad, F. Hashim, M.A. Chaudary, Cybersecurity in Industry 4.0 context: background, issues, and future directions, The Nine Pillars of Technologies for Industry 4 (2020) 263–307.

[9] B.C. Ervural, B. Ervural, Overview of cybersecurity in the Industry 4.0 era, Industry 4.0: Managing the Digital Transformation, Springer International Publishing, Cham, (2017), pp. 267–284.

[10] Anand, S., Important applications of cyber security. Available from https://www.analytixlabs.co.in/blog/important-applications-of-cyber-security/, 2022.

[11] M. Humayun, N.Z. Jhanjhi, M.N. Talib, M.H. Shah, G. Sussendran, Industry 4.0 and cybersecurity issues and challenges, Turkish Journal of Computer and Mathematics Education 12 (2021) 2957–2971.

[12] A. Haleem, M. Javaid, R.P. Singh, S. Rab, R. Suman, Perspectives of cybersecurity for an ameliorative Industry 4.0 era: a review-based framework, Industrial Robot: The International Journal of Robotics Research and Application 49 (2022) 582–597.

[13] M. Dawson, Cybersecurity in Industry 4.0: the pitfalls of having hyperconnected systems, Journal of Strategic Management Studies 10 (2018) 19–28.

[14] L. Simmons, New technologies in cybersecurity. Available from https://www.cyberdegrees.org/resources/hot-technologies-cyber-security/, 2022.

[15] M. Toussaint, S. Krima, H. Panetto, Industry 4.0 data security: a cybersecurity framework review, Journal of Industrial Information Integration 39 (2024) 100604.

[16] R. Kour, Cybersecurity issues and challenges in Industry 4.0, Applications and Challenges of Maintenance and Safety Engineering in Industry 4.0, (2020), pp. 84–101.

[17] E. Avdibasic, A.S. Toksanovna, B. Durakovic, Cybersecurity challenges in Industry 4.0: a state of the art review, Defense and Security Studies 3 (2022) 32–49.

[18] K. Karampidis, S. Panagiotakis, M. Vasilakis, E.K. Markakis, G. Papadourakis, Industrial cybersecurity 4.0: preparing the operational technicians for Industry 4.0, 2019 IEEE 24th International Workshop on Computer Aided Modeling and Design of Communication Links and Networks (CAMAD), IEEE, (2019), pp. 1–6 September.

[19] T. Sawik, A linear model for optimal cybersecurity investment in Industry 4.0 supply chains, International Journal of Production Research 60 (2022) 1368–1385.

[20] W. Torbacki, A hybrid MCDM model combining DANP and PROMETHEE II methods for the assessment of cybersecurity in Industry 4.0, Sustainability 13 (2021) 8833.

[21] J.E. Rubio, R. Roman, J. Lopez, Analysis of cybersecurity threats in Industry 4.0: the case of intrusion detection, International Conference on Critical Information Infrastructures Security, Springer International Publishing, Cham, (2017), pp. 119–130 October.

[22] N.H. Ghoson, V. Meyrueis, K. Benfriha, T. Guiltat, S. Loubere, A review on the static and dynamic risk assessment methods for OT cybersecurity in Industry 4.0, Computers & Security 150 (2025) 104295.

[23] J. Yu, A.V. Shvetsov, S.H. Alsamhi, Leveraging machine learning for cybersecurity resilience in Industry 4.0: challenges and future directions, IEEE Access 12 (2024) 159579–159596.

[24] R. Rudenko, I.M. Pires, P. Oliveira, J. Barroso, A. Reis, A brief review on internet of things, Industry 4.0 and cybersecurity, Electronics 11 (2022) 1742.

[25] F.S. Khoshaba, S. Askar, S. Hamad, S. Maghdid, Cyber security challenges in Industry 4.0: a review, Indonesian Journal of Computer Science 13 (2) (2024) 3848. Available from https://doi.org/10.33022/ijcs.v13i2.3848.

[26] A. Deshmukh, D.S. Patil, G. Soni, A.K. Tyagi, Cybersecurity: new realities for Industry 4.0 and society 5.0, Handbook of Research on Quantum Computing for Smart Environments, IGI Global Scientific Publishing, (2023), pp. 299–325.

[27] R.M. Barrios, D. Schippers, C. Heiden, G. Pappas, A cybersecurity strategy for Industry 4.0, Proc. SPIE 11009, Autonomous Systems: Sensors, Processing, and Security for Vehicles and Infrastructure 2019, SPIE, (2019), pp. 62–78. Available from https://doi.org/10.1117/12.2524667.

[28] M. Gombár, A. Vagaská, A. Korauš, P. Račková, Application of structural equation modelling to cybersecurity risk analysis in the era of Industry 4.0, Mathematics 12 (2024) 343.

[29] A. Alqudhaibi, M. Albarrak, A. Aloseel, S. Jagtap, K. Salonitis, Predicting cybersecurity threats in critical infrastructure for Industry 4.0: a proactive approach based on attacker motivations, Sensors 23 (2023) 4539.

[30] P. Kotuszewski, K. Kukielka, P. Kluk, A. Ordys, K. Bieńkowski, J.M. Kościelny, et al., Cyber-security assessment of an Industry 4.0-enabled mechatronic system, Complexity 2021 (2021) 6670625.

[31] E. Wai, C.K.M. Lee, Seamless Industry 4.0 integration: a multilayered cyber-security framework for resilient scada deployments in cpps, Applied Sciences 13 (2023) 12008.

[32] R.J. Raimundo, A.T. Rosário, Cybersecurity in the internet of things in industrial management, Applied Sciences 12 (2022) 1598.

[33] G. Sriram, S. Rambabu, Importance of cybersecurity in additive manufacturing in the era of Industry 4.0, International Journal of Manufacturing, Materials, and Mechanical Engineering (IJMMME) 2940 (2024) 6442.

[34] N. Benias, A.P. Markopoulos, A review of the readiness level and cybersecurity challenges in Industry 4.0, 2017, South Eastern European Design Automation, Computer Engineering, Computer Networks and Social Media Conference (SEEDA-CECNSM), IEEE, (2017), pp. 1–5 September.

[35] D.P. Möller, Cybersecurity in digital transformation, Guide to Cybersecurity in Digital Transformation: Trends, Methods, Technologies, Applications and Best Practices, Springer Nature Switzerland, Cham, (2023), pp. 1–70.

[36] T.M. Fernández-Caramés, P. Fraga-Lamas, Use-case based blended teaching of IIoT cybersecurity in the Industry 4.0 era, Applied Sciences 10 (2020) 5607.

[37] H. Kayan, M. Nunes, O. Rana, P. Burnap, C. Perera, Cybersecurity of industrial cyber-physical systems: a review, ACM Computing Surveys (CSUR) 54 (11s) (2022) 1–35.

[38] R. Trifonov, E. Sabev, G. Pavlova, K. Raynova, Analysis of deep learning methods for cybersecurity in Industry 4.0, AIP Conference Proceedings 3084 (2024) 060003. Available from https://doi.org/10.1063/5.0193712.

[39] A. Alqudhaibi, M. Albarrak, S. Jagtap, N. Williams, K. Salonitis, Securing industry 4.0: assessing cybersecurity challenges and proposing strategies for manufacturing management, Cyber Security and Applications 3 (2025) 100067.

[40] M. Shukla, S.P. Sarmah, M.K. Tiwari, A multi-objective framework for the identification and optimisation of factors affecting cybersecurity in the Industry 4.0 supply chain, International Journal of Production Research 61 (2023) 5266–5281.

[41] L. Shi, X. Chen, S. Wen, Y. Xiang, Main enabling technologies in Industry 4.0 and cybersecurity threats, International Symposium on Cyberspace Safety and Security, Springer International Publishing, Cham, (2019), pp. 588–597 December.

[42] V. Sklyar, V. Kharchenko, ENISA documents in cybersecurity assurance for Industry 4.0: IIoT threats and attack scenarios, 10th International Conference on Intelligent Data Acquisition and Advanced Computing Systems: Technology and Applications (IDAACS), IEEE, (2019), pp. 1046–1049.

[43] U.A. Usmani, A. Happonen, J. Watada, Advancements in Industry 4.0 asset management: interoperability, cybersecurity challenges, and opportunities, Proceedings of the Future Technologies Conference, Springer Nature Switzerland, Cham, (2023), pp. 468–488 October.

[44] P.R. Brandao, Bases, challenges, and main dangers for deploying cybersecurity in Industry 4.0, Advances in Wireless Communications and Networks 5 (2019) 5–33.

[45] F. Böhm, M. Dietz, T. Preindl, G. Pernul, Augmented reality and the digital twin: state-of-the-art and perspectives for cybersecurity, Journal of Cybersecurity and Privacy 1 (2021) 519–538.

[46] M. Homaei, Ó. Mogollón-Gutiérrez, J.C. Sancho, M. Ávila, A. Caro, A review of digital twins and their application in cybersecurity based on artificial intelligence, Artificial Intelligence Review 57 (2024) 201.

[47] M. Masi, G.P. Sellitto, H. Aranha, T. Pavleska, Securing critical infrastructures with a cybersecurity digital twin, Software and Systems Modelling 22 (2023) 689–707.

[48] D. Holmes, M. Papathanasaki, L. Maglaras, M.A. Ferrag, S. Nepal, H. Janicke, Digital twins and cyber security–solution or challenge? 2021 6th South-East Europe Design Automation, Computer Engineering, Computer Networks and Social Media Conference (SEEDA-CECNSM), IEEE, (2021), pp. 1–8 September.

[49] R. Faleiro, L. Pan, S.R. Pokhrel, R. Doss, Digital twin for cybersecurity: towards enhancing cyber resilience, International Conference on Broadband Communications, Networks and Systems, Springer International Publishing, Cham, (2021), pp. 57–76 October.

[50] A. Srivastava, C.C. Liu, A. Stefanov, S. Basumallik, M.M. Hussain, B. Somda, et al., Digital twins serving cybersecurity: more than a model: cybersecurity as a future benefit of digital twins 2, IEEE Power and Energy Magazine 22 (2024) 61–71.

[51] K. Alshammari, T. Beach, Y. Rezgui, Cybersecurity for digital twins in the built environment: current research and future directions, Journal of Information Technology in Construction 26 (2021) 159–173.

[52] K. Achuthan, B.B. Gupta, R. Raman, Bridging cybersecurity with digital twin technology: a thematic analysis, International Journal of Information Security 24 (2025) 207.

[53] S.B. Goyal, A.S. Rajawat, R.K. Solanki, M.A.M. Zaaba, Z.A. Long, Integrating AI with cybersecurity for smart Industry 4.0 application, 2023 International Conference on Inventive Computation Technologies (ICICT), IEEE, (2023), pp. 1223–1232. April.

[54] A. Mpatziakas, I. Schoinas, A. Lalas, A. Drosou, N. Chatzidiamantis, D. Tzovaras, Deciphering standards for cybersecurity in Industry 4.0: advisory AI for cybersecure IIoT, 2025 IEEE International Conference on Cyber Security and Resilience (CSR), IEEE, (2025), pp. 887–892. August.

[55] M. Dawson, Cybersecurity implications of the use of artificial intelligence in Industry 4.0, Scientific Bulletin 26 (2021) 24–31.

[56] A. Chaudhuri, R.K. Behera, P.K. Bala, Factors impacting cybersecurity transformation: an Industry 5.0 perspective, Computers & Security 150 (2025) 104267.

[57] A. Alqudhaibi, M. Albarrak, A. Aloseel, A. Munshi, T. Alsharif, S. Jagtap, et al., Proactive cybersecurity in Industry 4.0: a survey of cybersecurity threat prediction approaches in manufacturing systems, International Journal of Information Security 25 (2026) 14. Available from https://doi.org/10.1007/s10207-025-01188-9.

[58] M. Zajíček, T. Tichý, Z. Lokaj, M. Šrotýř, Cybersecurity challenges and solutions in Industry 4.0, telematics and smart cities, 2024 Smart City Symposium Prague (SCSP), Prague, Czech Republic, IEEE, (2024), pp. 1–7. Available from 10.1109/SCSP61506.2024.10552726.

[59] N. Halawi Ghoson, K. Benfriha, V. Meyrueis, T. Guiltat, C. El Zant, ShieldFlow: a novel framework for OT cybersecurity in the context of Industry 4.0, International Journal of Information Security 24 (2025) 162. Available from https://doi.org/10.1007/s10207-025-01083-3.

[60] F. Hajjej, N. Sarwar, A. Irshad, Integrating machine intelligence and data analytics for enhanced cybersecurity in Industry 4.0 manufacturing environments, The International Journal of Advanced Manufacturing Technology 142 (2026) 2813–2825. Available from https://doi.org/10.1007/s00170-025-17244-6.

[61] A. Corallo, M. Lazoi, M. Lezzi, P. Pontrandolfo, Cybersecurity challenges for manufacturing systems 4.0: assessment of the business impact level, IEEE Transactions on Engineering Management 70 (2021) 3745–3765. Available from 10.1109/TEM.2021.3084687.

[62] G. Bravos, A.J. Cabrera, C. Correa, D. Danilović, N. Evangeliou, G. Ezov, et al., Cybersecurity for industrial internet of things: architecture, models and lessons learned, IEEE Access 10 (2022) 124747–124765.

[63] O. Morozova, A. Nicheporuk, A. Tetskyi, V. Tkachov, Methods and technologies for ensuring the cybersecurity of industrial and web-oriented systems and networks, Radioelectronic and Computer Systems (4) (2021) 145–156. Available from https://doi.org/10.32620/reks.2021.4.12.

[64] A. Efe, A. Isik, A general view of Industry 4.0 revolution from cybersecurity perspective, International Journal of Intelligent Systems and Applications in Engineering 8 (2020) 11–20. Available from https://doi.org/10.18201/ijisae.2020158884.

16 Machine vision for Industry 4.0

16.1 Introduction to machine vision

Machine vision (MV) is an industry term for the technology and methodology used to enable imaging-based autonomous inspection and analysis. Material inspection, signature recognition, component analysis, electronic object recognition, pattern recognition, optical character recognition (OCR), and money recognition are among the applications where MV is employed. Its goal is to discover innovative methods to apply current technology to real-world challenges. The MV process begins with thoroughly examining the needs and project, followed by developing a solution [1–3].

In its broadest sense, MV describes various approaches for extracting visual data from hardware and processing and analysing data. Like other digital technologies, MV is essential for Industry 4.0. Because optical equipment can access large amounts of data, it can immediately detect and flag incorrect items while also recognising their flaws, enabling quick and efficient intervention in Industry 4.0. For applications in inventory management, quality control, and enforcement, MV versions are necessary for large-scale manufacturing [4–6].

One of the goals of MV is to see, communicate, and operate with greater precision than humans can. This allows the robot to perceive and assist people in dynamic systems, opening a world of possibilities [7,8]. MV will play a crucial role in the smart plant of the future, where automated manufacturing lines will adapt to optimise efficiency, performance, and profitability. MV's use of cutting-edge technologies such as AI, deep learning, and robust data analytics is driving rapid change. MV systems easily meet and, in many cases, surpass market requirements [9,10].

MV is significantly user-friendly and straightforward. It collects data on production cycles and other process variances as new technologies emerge. This system uses image processing to automate inspection and evaluation, including automated inspection, process monitoring, and robotic guidance [11,12]. MV is a processing capability that can be used in novel ways to address real-world issues with current technology. The scanned data is analysed using this technology to maximise performance, a process that previously relied solely on human review. MV enables a business to gain enhanced production precision and long-term benefits [13,14].

The techniques used by MV are multidisciplinary, combining cameras, sensors, and computational power to analyse images and enable robots to perform industrial activities such as manufacturing and quality assurance. For Industry 4.0, MV is a critical component that supports industrial automation technologies in various ways, including improving performance through stock management, detecting damaged components, and improving product quality [15,16]. This is useful for imaging and analysis in automated inspection, process control, and robot guidance, among other applications. In industry, MV is utilised for

Digital Twin and Industry 4.0. DOI: https://doi.org/10.1016/B978-0-443-54019-6.00019-4

automated machine control. It necessitates the use of computer vision to analyse and comprehend images [17,18].

Many industries benefit from using MV because it enables precise production, thereby decreasing emissions and reducing worldwide demand. By providing automatic inspection capacity under quality control methods, MV substantially contributes to the industry. From MV in warehouses to sophisticated robots in R&D labs, this technology has a substantial impact at every stage of the manufacturing process. Computers may use MV to recognise items and analyse real-life scenarios and events. MV systems are most successful when strategically integrated into smart manufacturing because they boost human and digital performance [19–21].

16.2 Major parts of the machine vision system

MV systems comprise cameras, lenses, lighting, and image-processing tools [10,22–24]. Fig. 16.1 depicts the significant components of the MV system.

16.2.1 Light

An MV system cannot study anything that the camera cannot see. Therefore, every MV configuration must choose the appropriate illumination. The distance and angle of the installation, along with the form, size, and colour of the illumination, can be adjusted to highlight the objects being examined without allowing the surroundings to intrude.

16.2.2 Camera

Light is converted into digital image data by camera image sensors before being transmitted to the controller.

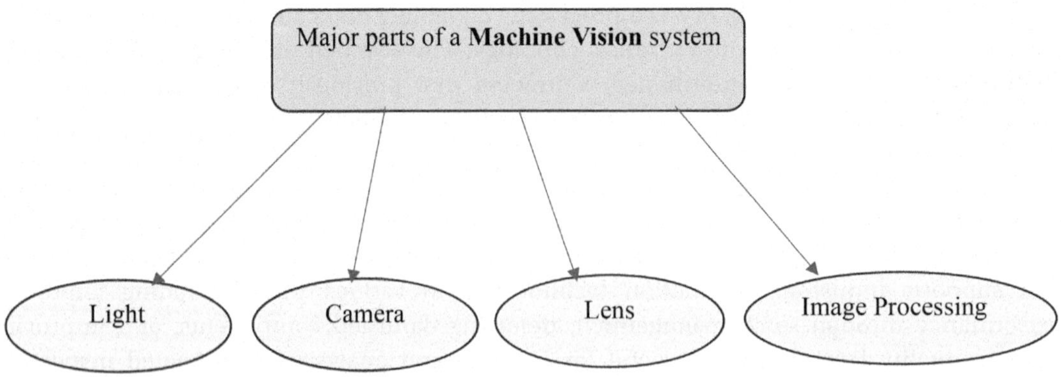

FIGURE 16.1 Major parts of the machine vision system.

16.2.3 Lens

Light is focused onto the camera sensor using lenses.

16.2.4 Image processing

Image processing units, sometimes called controllers, use established algorithms to interpret image input and extract meaningful information.

16.3 Typical areas of machine vision

Any industrial or nonindustrial application where a combination of hardware and software provides operational guidance to equipment that is operating and processing images. While many of the approaches and techniques used in academic and military computer vision applications are similar, the limitations are usually very different. Industrial vision systems need to be more reliable, stable, and long-lasting than those intended for academic and research settings [12,25,26]. Industrial MV systems are frequently more affordable, offer good mechanical and thermal stability, high reliability and robustness, and sufficient accuracy. MV has exploded in popularity over the past few years, especially in fast-growing sectors like manufacturing and retail. These sectors use technology to enhance quality control, optimise resource use, and improve customer satisfaction [8,27,28].

One of the most often used tools in the sector is MV. It is used to perform an automated visual evaluation of the many objects generated every day. Algorithm development has advanced with advances in artificial intelligence. Deep learning-based inspection methods are often integrated with MV systems. These systems are easier to set up and train. Industrial MV systems are robust, stable, and trustworthy. They are reasonably priced, have exceptional mechanical and thermal stability, and are quite accurate [29]. MV is used in many different sectors [4,6,30–32]. The following are MV's key areas:

16.3.1 Measurement

As the name implies, measurement applications establish precise measurements of objects. This is accomplished by identifying specific points in a snapshot and deriving geometric measurements from them. For instance, MV may be used to analyse a 2D or 3D image from a camera to determine the inner diameter of an engine cylinder bore.

16.3.2 Parts counting

It requires little thinking and is a slow, tedious activity. However, manual labourers trying to accomplish their daily goals could need to be corrected when counting, which could cause major delays in part assembly. After identifying the pieces, MV can use its object recognition technique to quickly and accurately count them. In the manufacturing chain, for example, MV can efficiently count the piston rings in a stack.

16.3.3 Surface defect identification

Surface fault detection is another MV use that is essential to quality control. When balancing supply and demand, operators may overlook faults because manually identifying surface flaws is labour-intensive. MV can be used in a model that is easy to train to attain the accuracy and efficacy of surface inspection. In manufacturing, surface defect inspection is used to detect defects in bearings, castings, and other metal surfaces. Finding packing flaws, for example, can help find subpar packaging that will cause harm during shipment. One popular MV technique used in identification applications to recognise and categorise a variety of objects is reading barcodes and data matrix codes. Preventing errors in production and packaging processes depends on this. Compared to manual mistake verification, it is far quicker and more accurate. MV identification may also increase productivity by locating manufacturing pipeline bottlenecks.

16.3.4 Gauging

One technique utilised in production lines is MV gauging. In this application, a fixed-mount camera can identify two or more points on an object as it passes by on the assembly line. If the measured distances do not match the distances entered into the vision system, the part is removed from the production line. Compared to well-established techniques such as touch-based measuring, MV guiding is quicker and more precise.

16.3.5 Error detection

Error detection software detects surface faults, dents, and scratches on a product's surface. Thus, error-detection applications must be carefully designed to distinguish between unacceptable and acceptable errors. An MV-based on AI is ideal for use, as it trains the system through examples rather than ‘rules’.

16.3.6 Print defect identification

Finding anomalies in the prints, such as differences in colour, text, or pattern, is the first step in identifying print flaws. When performing manual inspections, human workers risk making mistakes that degrade the quality of the final product. MV may employ AI and deep learning to detect print flaws. Labels, packing, and printing are all handled. In many cases, manual print quality-checking techniques must catch up with what is required to identify errors. Sales, profitability, and the quality of final products all suffer from defects. By implementing automated platforms, such as an MV-based print-failure detection system, businesses can avoid it. The technique is beneficial for print, labels, and package inspection and has many applications. Ultimately, the solution eliminates the enormous burden of quality assurance on humans and assumes responsibility for flawless automatic detection.

16.3.7 Guidance

In the industrial sector, MV guidance has several useful applications. Most of the time, it involves locating a specific item and ensuring it is positioned correctly to minimise errors and downtime during production. MV approaches can also be used to define the orientation and position of a particular component. Information for production can be sent to a robot or machine controller. MV guidance enables more accurate, efficient output than manual positioning in assembly lines or while stacking items on pallets.

16.3.8 Object detection

When object detection, the algorithm looks for certain objects rather than scanning the entire image. In this instance, the algorithm successfully detects whether features are present or absent in the image rather than classifying the entire image. Several techniques are used to detect things effectively. Algorithms for object detection can be applied to a range of production processes, including sorting, assembly lines, inventory control, and quality control. An excellent example of this problem type is the engine assembly chain at a car manufacturer. Using its knowledge and experience, MV can assess the engine's condition and reliably verify whether specific parts are present in that engine block. On the machine side, component advancements produce much better raw materials, including cameras used to produce specialised image-capture solutions, new lenses, sophisticated robots, and more.

16.3.9 Localisation of object

MV is widely used in robotic navigation and other applications for object localisation. Finding a target object's position and coordinates is the goal of the MV system. Its data can do any task that requires this position, such as retrieving objects. Therefore, its application must teach the MV system component of interest so it can identify the part during production.

16.3.10 Barcode scanning

MV can read barcodes and data matrix codes. Consequently, by classifying the items AI detects, production and packing processes will be made error-proof. MV barcode scanners are more efficient than the potentially error-prone human classification process. Barcode scanners enable the separation of production products by their attributes.

16.3.11 Logistics

MV can be used to locate an item by providing its coordinates or its location relative to the searcher. The ability to locate an item at any time makes supply chain management and logistics more accessible. If something is lost, MV-based object detection can quickly find it. Since many parts are moved around during manufacturing, finding items is essential. MV can use various methods to track and identify these objects.

16.3.12 Inspection

MV inspection is used to identify faults and defects. In various industrial applications, MV inspection and defect detection provide greater flexibility for scanning items, including deterioration in agricultural goods, faults in textiles, branding markings on prescription pills, and more. MV inspection is faster and far more accurate than manual inspection techniques.

16.3.13 Classification

The classification process predicts an item's class or category. Certain classifiers produce binary classifications: yes or no. Some are multi-class and assign objects to one of multiple categories. Deep learning algorithms are widely used in production for classification, including image classification and identification. A complete image is frequently used as input for classification problems. After analysing the full image, the algorithm uses the training data to classify it further. One of the best examples of this is classifying products on a production line.

16.3.14 Print character reading

Print character identification is one MV application that uses OCR. MV may help trace a variety of items along the industrial supply chain by using print character reading. The name tag and other details of any item can be examined and modified as it moves through its life cycle. It is useful in logistics due to the serious risk of goods being missed.

16.3.15 Robotic guidance

To avoid production errors and downtime, this MV application requires locating a specific item and ensuring it is properly placed and oriented. Robotic guidance can perform tasks that require visual assistance via a robot or machine controller. When manual labourers are unable to perform repetitive, high-precision, high-accuracy work consistently to achieve maximum efficiency, robots may be used. For example, automatic placement and selection will quickly assemble the parts of any object.

16.4 Advantages of machine vision

The significant advantages of MV are as follows [33–36]:

- It saves time.
- Increases output and product quality, with better detection of defective products.
- Lowers the cost of manufacture.
- Increases the use of machines.
- Cutting down on the amount of labour required of testing and inspection staff.
- Lowers the quantity of products that are not qualified.
- Inventory control.

- Lower capital equipment costs.
- Higher precision.
- Useful for measurement.
- Assembly verification.
- Production flexibility.
- Reduced setup time.
- Scrap rate reduction.
- Reduced floor space requirements.
- Performs monotonous and repetitive tasks.

16.5 Roles of machine vision for Industry 4.0

For image-based automated inspection, MV is useful. Its applications have become critical in Industry 4.0 and are evolving in various industries. MV is now employed in various areas, including automobiles, pharmacies, food and beverage, electronics, etc. [37–39]. Fault detection, product and packaging inspection, part recognition and monitoring, and measurement are all examples of MV applications. This increases efficiency and productivity, reduces environmental impact, and reduces production costs. MV technology improves precision in critical activities, such as placing parts for assemblies (e.g. doors and panels), in the automotive sector [40–44]. The significant roles of MV for Industry 4.0 are as follows:

16.5.1 Improve the quality assurance process

The idea of automating quality testing using MV is gaining traction. It assists businesses in streamlining and improving the quality assurance process, particularly in modern manufacturing environments. From a cognitive standpoint, this technology handles high-priority tasks while workers increasingly have more complex abilities.

16.5.2 Streamlining particular operations

MV is a superior technology that helps producers streamline specific operations, such as testing multiple products on a production line. Because most manufacturers rely on high-quality output, it is critical to have a system that can automatically detect any differences. The most common use of MV systems is for positioning tools and details on production lines.

16.5.3 Part inspection

Several manufacturers employ MV for part inspection instead of human inspectors because it is well-suited to repetitive inspection tasks. It works faster, objectively, and continuously. MV systems can inspect thousands of items per minute and provide more accurate inspection findings than humans. MV may employ colour-detecting pixels, and pixel arrays are sometimes even more dominant.

16.5.4 Increase the efficiency of data collection

The intelligent plant is feasible by combining MV with automation systems, thereby boosting data-collection efficiency and volume. This is utilised to minimise waste and improve protection at higher output speeds. In the era of Industry 4.0, MV systems detect production changes in real time.

16.5.5 Accurate and useful information

Material processing equipment, production lines, and inspection stations benefit from MV's improved dissemination of accurate and helpful information. Equipment and workers function better in production, with fewer bottlenecks, overruns, and disruptions. It boosts productivity across all areas, such as installation, and ensures better output.

16.5.6 Industrial automation and intelligent manufacturing

Deep learning has revolutionised the role of MV in industrial automation and intelligent manufacturing. MV applications can now adjust to production variances independently through deep learning integration. Strengthened operational efficiency, faster inspection processes, and enhanced reliability have all improved the value proposition for the computer's view. MV integrates ML into several sensors, enabling it to monitor consumer and industrial hardware environments.

16.5.7 Eliminating operational losses

MV quality assurance ensures the product's precision and quality are consistently met, eliminating operational losses. Human involvement and evaluation of critical stages have long been integral to the production line. This provides smarter results with lower labour consumption and higher output.

16.5.8 Checking barcodes

Cross-referencing barcodes for thousands of products is labour-intensive, error-prone, and wasteful. With MV devices, barcodes can be quickly scanned and defective ones identified. This system can help with alerts, management automation, warehouse inventory tracking, and stock counting. MV systems can help prevent errors in inventory counting.

16.5.9 Supervision

The MV system will notify supervisors and others of the existence and severity of an ongoing mishap, allowing them to halt work in that field and ensure worker safety. The use of MV involves inspecting the proper delivery of products to the customers. This technology can automatically identify inadequate packing.

16.5.10 Satisfy customer needs

MV applications enable organisations to improve product quality, eliminate production errors, reduce production costs, and satisfy customer needs with high-quality products at a fair price. The principles of MV software and barcode decoding will make inspections more accessible, increase read speeds, and improve the accuracy and repeatability of the manufacturing line's output.

16.5.11 Identify faults and fractures

MV systems let a manufacturer detect faults and fractures in a physical product. Furthermore, these instruments can effectively verify precise, reliable measurements of materials and components. Producers can integrate computerised inventory-testing devices and precise quality control to satisfy customers.

16.5.12 Tracking and reporting

Tracking and reporting are also possible with this technology. The vision sensor's substantial dynamic range enables operation in challenging lighting conditions while providing significantly fewer data points than a traditional framework-based solution. MV uses images to autonomously analyse and inspect production lines, tracking operations, estimating, checking, guiding robots, and performing other duties.

16.5.13 Recognise images visually

MV enables recognising images, gathering information, and making judgments in the same manner as people do. This technology can automatically recognise specific patterns in distinct images and take the appropriate action. MV can automatically recognise specific patterns in distinct images and take the appropriate action.

16.5.14 Visual stock monitoring and control

In addition to content inspection, MV devices have a range of applications. These devices also track and regulate visual stock by scanning and counting barcodes and checking components at various manufacturing stages. A robot viewing device comprises many vital components, including a camera that captures an image and sends the information to the processing mechanism.

16.6 Roles of digital twin for machine vision

By providing a virtual replica of cameras, sensors, inspection stations, and production environments, digital twin technology significantly enhances MV systems in Industry 4.0. Before adjusting on the actual work floor, engineers can use this virtual model to simulate various lighting scenarios, camera angles, object placements, and defect types [44,45]. This enables the optimisation of processing algorithms, lens setups, and image capture settings without affecting the final output. Digital twins also help MV systems with predictive improvement and ongoing performance monitoring. This enables proactive modification of inspection criteria, updating AI models, and recalibrating cameras. By mimicking coordinated operations in a risk-free setting, digital twins also make it easier to integrate automated control systems, robotics, and MV [46–49]. The major roles of digital twin for MV are as follows:

- Industrial cameras are calibrated virtually before actual deployment.
- Lighting conditions are simulated to maximise the accuracy of image capture.
- Undertakes examination of focal settings and lens setups in a virtual production model.
- The creation of artificial fault datasets for deep learning vision model training.
- In automated inspection lines, camera positioning and viewing angles are optimised.
- Visual data is synchronised with the virtual production model in real time.
- Using 3D virtual environment analysis to identify blind spots in inspections.
- Benchmarking vision system performance at different production speeds.
- Product variation simulation for contexts involving bulk customisation.
- Evaluation of quality inspection false positive and false negative rates.
- Using performance drift analysis to predict camera recalibration.
- Keeping an eye on how temperature, vibration, and dust affect visual precision.
- Improving image processing parameters, including filtering and thresholding.
- Performance assessment of edge computing for real-time image analysis.
- Automated inspection stations can be virtually commissioned.
- Simulation of failure modes for situations involving camera or sensor malfunctions.
- Data-driven feedback loops for ongoing inspection strategy improvement.

16.7 Upcoming goals of machine vision

The upcoming goals of MV are as follows [50,51]:

- MV in manufacturing will become a diverse technical stack.
- This technology enhances automation, which will provide a strategic advantage for their competitors.
- This will ensure future economic growth and peace.
- This will introduce brand-new approaches for executing high-performance and low-waste manufacturing processes.
- It will have an impact on production, perforation, and mining.

16.8 Challenges with machine vision

The significant challenges with MV are as follows [5,52]:

- Different light sources will cause different imaging qualities and effects.
- It needs to rely on powerful computing capabilities.

- Image data transmission requires adjustments.
- Restricted by the cost-benefit economy.
- Necessity of specialists.
- When the machine or device fails, it does not announce or anticipate the problem.
- They were sometimes fail in image processing.

16.9 Conclusion

MV is used for automated visual examination, which can help solve various business issues. This technology has gradually replaced human quality inspections. It can identify items with varying degrees of accuracy and durability. Industry 4.0 benefits from MV, which improves accuracy, productivity, and overall efficiency while enhancing product traceability. Here, industrial equipment that uses high-resolution cameras in manufacturing can benefit from greater adaptability, fewer product defects, and increased technical efficiency. MV is useful for freight, supply chain, inventory processing, quality assurance, safety, and other vertical activities.

MV technologies capture images from various angles to create a 3D depiction. This technology promises to benefit the automotive, electrical, oil, gas, and energy industries. It provides engineers and operators with tools for better data collection, operator empowerment, and bug-resistant assembly. This technology is vital in product and component assembly in the manufacturing sector. When combined with machine learning, MV technology, and robotics, these technologies provide a slew of new possibilities for bringing cutting-edge concepts like the smart factory to life.

MV will soon be everywhere and fulfil various requirements of Industry 4.0. This technology is a set of advancements that allow machines to become aware of their surroundings. It makes it simpler to recognise visuals and make judgments based on them. Industrial equipment uses high-resolution cameras to take digital images of the surroundings and workpieces for MV applications. MV is relevant to manufacturers since it improves worker safety and protects the environment. MV devices would alert those concerned about the hazardous site, enabling them to implement the necessary safeguards promptly.

References

[1] V. Alonso, A. Dacal-Nieto, L. Barreto, A. Amaral, E. Rivero, Industry 4.0 implications in machine vision metrology: an overview, Procedia Manufacturing 41 (2019) 359–366. Available from https://doi.org/10.1016/j.promfg.2019.09.020.

[2] M. Javaid, A. Haleem, R.P. Singh, S. Rab, R. Suman, Exploring the impact and features of machine vision for a progressive Industry 4.0 culture, Sensors international 3 (2022) 100132.

[3] D.P. Penumuru, S. Muthuswamy, P. Karumbu, Identification and classification of materials using machine vision and machine learning in the context of Industry 4.0, Journal of Intelligent Manufacturing 31 (5) (2020) 1229–1241. Available from https://doi.org/10.1007/s10845-019-01508-6.

[4] M. Tzampazaki, C. Zografos, E. Vrochidou, G.A. Papakostas, Machine vision—moving from Industry 4.0 to Industry 5.0, Applied Sciences 14 (2024) 1471.

[5] P. Kumar, D. Singh, J. Bhamu, Machine vision in Industry 4.0: applications, challenges and future directions, Machine Vision for Industry 4.0, CRC Press, (2022), pp. 263–284.

[6] 9 Common Machine Vision Applications In Manufacturing. https://qualitastech.com/quality-control-insights/9-common-applications-of-machine-vision-in-manufacturing/#:~:text=Robotic%20Guidance,a%20machine%20controller%20or%20robot.

[7] R.L. Silva, O. Canciglieri Junior, M. Rudek, A road map for planning and deploying machine vision artefacts in the context of Industry 4.0, Journal of Industrial and Production Engineering 39 (2022) 167–180.

[8] A. Charan, C. Karthik Chowdary, P. Komal, The future of machine vision in industries-a systematic review, IOP Conference Series: Materials Science and Engineering 1224 (2022) 012027. Available from 10.1088/1757-899X/1224/1/012027.

[9] M. Hussain, Sustainable machine vision for Industry 4.0: a comprehensive review of convolutional neural networks and hardware accelerators in computer vision, AI 5 (3) (2024) 1324–1356. Available from https://doi.org/10.3390/ai5030064.

[10] J. Kaur, Machine vision working and its applications | A Quick Guide: https://www.xenonstack.com/blog/machine-vision-application, 2022.

[11] K. Palanikumar, E. Natarajan, A. Ponshanmugakumar, Application of machine vision technology in manufacturing industries—a study, Machine Intelligence in Mechanical Engineering, Academic Press, (2024), pp. 91–122.

[12] W. Jiang, A machine vision anomaly detection system for Industry 4.0 based on a variational fuzzy autoencoder, Computational Intelligence and Neuroscience 2022 (2022) 1945507.

[13] T. Benbarrad, M. Salhaoui, S.B. Kenitar, M. Arioua, Intelligent machine vision model for defective product inspection based on machine learning, Journal of Sensor and Actuator Networks 10 (2021) 7.

[14] A. Khang, V.A. Hajimahmud, R.N. Ali, V. Hahanov, Z. Avramovic, Role of machine vision in manufacturing and the Industrial Revolution 4.0, Machine Vision and Industrial Robotics in Manufacturing, CRC Press, (2024), pp. 1–13.

[15] M. Maślanka, D. Jancarczyk, J. Rysinski, Integration of machine vision and PLC-based control for scalable quality inspection in Industry 4.0, Sensors 25 (2025) 6383.

[16] P. Bedi, S.B. Goyal, A.S. Rajawat, P. Bhaladhare, A. Aggarwal, A. Prasad, Feature correlated auto encoder method for industrial 4.0 process inspection using computer vision and machine learning, Procedia Computer Science 218 (2023) 788–798.

[17] B. Batchelor, F. Waltz, Machine vision for industrial applications, Intelligent Machine Vision: Techniques, Implementations and Applications, Springer, London, (2001), pp. 1–29.

[18] M. Mirbod, A.R. Ghatari, S. Saati, M. Shoar, Industrial parts change recognition model using machine vision and image processing within the framework of industrial information integration, Journal of Industrial Information Integration 26 (2022) 100277. Available from https://doi.org/10.1016/j.jii.2021.100277.

[19] S.A. Hamza, A. Jesser, Advancing Industry 4.0 with real-time machine vision integration in cyber-physical systems, 2024 IEEE 3rd International Conference on Computing and Machine Intelligence (ICMI), IEEE, (2024), pp. 1–5. April.

[20] E. Zancul, H.O. Martins, F.P. Lopes, F.A. da Silva Neto, Machine vision applications in a learning factory, Procedia Manufacturing 45 (2020) 516–521.

[21] M. Umar, M. Gupta, R. Verma, N. Dhanda, Role of computer vision in manufacturing industry, Machine Vision and Industrial Robotics in Manufacturing, CRC Press, (2024), pp. 14–35.

[22] K.G. Suma, P. Patil, G. Sunitha, V.P. Mantri, N.D. Kale, Computer vision and its intelligence in Industry 4.0, Machine Learning Techniques and Industry Applications, IGI Global Scientific Publishing, (2024), pp. 119–142.

[23] R.L. Silva, M. Rudek, A.L. Szejka, O.C. Junior, Machine vision systems for industrial quality control inspections, IFIP International Conference on Product Lifecycle Management, Springer International Publishing, Cham, (2018), pp. 631–641, July.

[24] S.A. Singh, K.A. Desai, Automated surface defect detection framework using machine vision and convolutional neural networks, Journal of Intelligent Manufacturing 34 (2023) 1995–2011. Available from https://doi.org/10.1007/s10845-021-01878-w.

[25] V. Nandini, R.D. Vishal, C.A. Prakash, S. Aishwarya, A review on applications of machine vision systems in industries, Indian Journal of Science and Technology 9 (2016) 1–5.

[26] Y. Saif, Y. Yusof, K. Latif, A.Z. Abdul Kadir, M.B.I. Ahmad, A. Adam, et al., Development of a smart system based on STEP-NC for machine vision inspection with IoT environmental, The International Journal of Advanced Manufacturing Technology 118 (2022) 4055–4072.

[27] S. Sathiyamoorthy, Industrial application of machine vision, International Journal of Research in Engineering and Technology 3 (2014) 678–682.

[28] D. Guerra-Zubiaga, A. Franklin, D. Escobar-Escobar, T. Lemley, N. Hariri, J. Plattel, et al., Computer vision and machine learning to create an advanced pick-and-place robotic operation using Industry 4.0 trends, Proceedings of the ASME 2022 International Mechanical Engineering Congress and Exposition. Volume 2A: Advanced Manufacturing. Columbus, Ohio, USA. October 30–November 3, 2022. V02AT02A010, American Society of Mechanical Engineers, (2022). Available from https://doi.org/10.1115/IMECE2022-89743.

[29] 7 Applications of machine vision: https://qualitastech.com/quality-control-insights/7-applications-of-machine-vision/.

[30] P. Misra, N. Tiwari, The role of machine vision technology in the manufacturing of vehicles in industry, Turkish Journal of Computer and Mathematics Education 13 (03) (2022) 778–798.

[31] J. Werheid, H. Behnen, J.H. Woltersmann, S. He, T. Hamann, A. Abdelrazeq, et al., Machine vision in manufacturing SMEs: a review, Discover Applied Sciences 7 (2025) 371.

[32] K. Kozak, N.M. Kowalkowski, D. Perfido, Robust machine-vision models against adversarial attacks in Industry 4.0, Open Research Europe 6 (2026) 31.

[33] F. Frustaci, S. Perri, G. Cocorullo, P. Corsonello, An embedded machine vision system for an in-line quality check of assembly processes, Procedia Manufacturing 42 (2020) 211–218.

[34] C. Connolly, Machine vision advances and applications, Assembly Automation 29 (2009) 106–111.

[35] S. Răileanu, T. Borangiu, F. Anton, S. Anton, Open-source machine-vision platform for manufacturing and robotics, IFAC-PapersOnLine 54 (2021) 522–527.

[36] K. Dhibar, P. Maji, H.K. Mondal, S.K. Mondal, Accelerating Industry 4.0 using computer vision algorithms and analysis: analysing the significance and characteristics of machine vision in the advancing industry culture, Investigations in Pattern Recognition and Computer Vision for Industry 4.0, IGI Global Scientific Publishing, (2023), pp. 37–52.

[37] P. Pierleoni, A. Belli, L. Palma, M. Palmucci, L. Sabbatini, A machine vision system for manual assembly line monitoring, 2020 International Conference on Intelligent Engineering and Management (ICIEM), IEEE, (2020), pp. 33–38, June.

[38] A.O. Fernandes, L.F. Moreira, J.M. Mata, Machine vision applications and development aspects, 2011 9th IEEE International Conference on Control and Automation (ICCA), IEEE, (2011), pp. 1274–1278, December.

[39] D.K. Moru, D. Borro, A machine vision algorithm for quality control inspection of gears, The International Journal of Advanced Manufacturing Technology 106 (2020) 105–123.

[40] D. Mourtzis, J. Angelopoulos, A.N. Arvanitis, N. Panopoulos, Automating quality control based on machine vision towards automotive 4.0, IFIP International Conference on Advances in Production Management Systems, Springer Nature Switzerland, Cham, (2022), pp. 126–134, September.

[41] R. Packianathan, G. Arumugam, S.K. Natarajan, A. Malaiarasan, Machine vision and industrial robotics in manufacturing, Machine Vision and Industrial Robotics in Manufacturing, CRC Press, (2024), pp. 299–313.

[42] R. Gill, D. Srivastava, S. Hooda, C. Singla, R. Chaudhary, Unleashing sustainable efficiency: the integration of computer vision into Industry 4.0, Engineering Management Journal 37 (2025) 414–432.

[43] O. Döbrich, C. Brauner, Machine vision system for digital twin modeling of composite structures, Frontiers in Materials 10 (2023) 1154655.

[44] G.C. Deac, C.N. Deac, C.L. Popa, M. Ghinea, C.E. Cotet, Machine vision in manufacturing processes and the digital twin of manufacturing architectures, Proceedings of the 28th International DAAAM Symposium 2017, DAAAM International, (2017), pp. 733–736. Available from 10.2507/28th.daaam.proceedings.103.

[45] A.G. Bruzzone, M. Massei, M. Gotelli, R. Ferrari, A. De Paoli, A. Giovannetti, et al., Digital twin modeling for machine vision testing in autonomous systems, International Conference on Modelling and Simulation for Autonomous Systems, Springer Nature, Cham, (2023), pp. 221–231, October.

[46] A.G. Bruzzone, M. Massei, K. Sinelshchikov, M. Gotelli, A. Giovannetti, A. De Paoli, et al., Autonomous system digital twin to test machine vision, Proceedings of the International Conference on Modeling and Applied Simulation, MAS, ITA, (2023), pp. 1–5.

[47] M. Moretti, A. Rossi, N.J.A.M. Senin, In-process monitoring of part geometry in fused filament fabrication using computer vision and digital twins, Additive Manufacturing 37 (2021) 101609. Available from https://doi.org/10.1016/j.addma.2020.101609.

[48] C.A. de Souza Silva, E.P. Paladini, Smart machine vision system to improve decision-making on the assembly line, Machines 13 (2025) 98.

[49] A. Taatali, S.E. Sadaoui, M.A. Louar, B. Mahiddini, On-machine dimensional inspection: machine vision-based approach, The International Journal of Advanced Manufacturing Technology 131 (2024) 393–407.

[50] D.K. Moru, D. Agholor, F.A. Imouokhome, Machine vision and metrology systems: an overview, International Journal of Data Science 2 (2021) 77–84.

[51] Z. Ren, F. Fang, N. Yan, Y. Wu, State of the art in defect detection based on machine vision, International Journal of Precision Engineering and Manufacturing-Green Technology 9 (2022) 661–691.

[52] M. Nagy, G. Lăzăroiu, Computer vision algorithms, remote sensing data fusion techniques, and mapping and navigation tools in the Industry 4.0-based Slovak automotive sector, Mathematics 10 (2022) 3543.

17

Drones for Industry 4.0

17.1 Introduction to drone

A 'Drone' is a term used to describe any unpiloted aircraft. These machines, also known as 'unmanned aerial vehicles' (UAVs), can perform various tasks of Industry 4.0. Drones come in various sizes and are developed for the military and aerospace sectors. However, they have entered the public domain due to their improved safety and efficiency. These UAVs operate without a pilot and with varying autonomy [1–3].

The operational and financial advantages of drone-assisted work are beginning to be recognised by industries. Examples of implementable advanced technologies to reach Industry 4.0 include the Internet of Things, big data, drones, and other cutting-edge technologies. Drones are utilised in difficult or dangerous regions for inspections, mapping, and surveying. These are used worldwide in manufacturing, entertainment, and equipment. Drone-mounted thermal imaging equipment can detect temperature variations within a building, activate the sprinkler system, and alert emergency responders [4,5].

Autonomous drones significantly influence Industry 4.0 through their usefulness and versatility across energy, transportation, engineering, and more. Businesses are also increasingly using drones. These can do a wider range of Industry 4.0 tasks. Drones move finished goods from manufacturing to transportation and supplies from storage to the production floor. This technique results in a significant reduction in labour costs. The telecom sector widely uses drones to conduct site audits swiftly and efficiently, providing line-of-sight inspections and top-down views [6–8].

Drones can scan transmission lines, inspect boilers, monitor solar panels, and analyse and repair damage faster than humans. This technology offers a more cost-effective option for helicopter-based operations, particularly in acquisition and maintenance. It can help to reduce downtime, save money, and improve safety while increasing operational efficiency [9,10]. Rather than expensive tools, drones can scan industrial areas more precisely, enabling more accurate map construction, improved quality control, and cost reductions. Due to safety and compliance rules, drones will become an indispensable part of the building industry [11,12].

This technology changes the way business is conducted and prioritises employee safety. The risk of workplace accidents is reduced when self-driving drones are used to detect gas leaks and fires. Regular safety and inspection missions help raise industry safety standards. In a commercial drone, end-to-end communication is enabled by a wireless connection between the user and the controller [13,14]. This technology aims to automate the transportation of raw materials, components, and finished items. Today, the oil and gas industry's major players are employing drone technology to evaluate oil platforms, improving on-site performance and safety while reducing inspection times, surveillance [15–17].

Digital Twin and Industry 4.0. DOI: https://doi.org/10.1016/B978-0-443-54019-6.00017-0

17.2 Classification of unmanned aerial vehicles according to size

The classifications of UAVs according to size are as follows [7,18–21]:

Very small UAVs
The smallest UAVs range in length from 30 to 50 cm and are roughly the size of a huge bug. A typical micro design is an insect-like UAV with spinning or flapping wings. They can be utilised for biological warfare and espionage because of their small size and low weight. The typical aeroplane configuration is used in larger ones. Whether to use flapping or spinning wings depends on the necessary level of agility. Perching and landing on small surfaces are made possible by the designs of flapping-wing aircraft.
Small UAVs
Small UAVs, commonly called mini-UAVs, have at least one dimension more significant than 50 cm and a maximum size of 2 m. The fixed-wing model is the foundation for many designs in this category, and most are launched by hand, thrown into the air.
Medium UAVs
The medium UAV class includes UAVs that are too heavy for a single person to carry but are still smaller than light aircraft. They can carry 100–200 kg of cargo and typically have a wingspan of 5–10 m. There are a few more medium-sized UAVs. There are also many medium-sized UAVs with rotary motors.
Large UAVs
The large UAVs that the military mostly uses for combat missions are part of the big UAV class.

17.2.1 Classification of unmanned aerial vehicles, according to range and endurance

The classification of UAVs, according to range and endurance, is as follows [22–25]:

Very low-cost, close-range UAVs
UAVs in this category have a range of 5 km and an endurance time of 20 to 45 minutes. Model aeroplanes are very comparable to this class of UAVs.
Close-range UAVs
This class of UAVs has a 50 km range and a 1–6 hour endurance range. They are frequently used for reconnaissance and surveillance tasks.
Short-range UAVs
These UAVs have an endurance range of 8 to 12 hours and a minimum range of 150 km. Like close-range UAVs, they are primarily employed for surveillance and reconnaissance.
Mid-range UAVs
The mid-range class includes UAVs with a breakneck speed and an operational radius of 650 km. They are used for surveillance and reconnaissance, as well as for meteorological data collection.

Endurance UAVs

The endurance class of UAVs has a 36 hour endurance and a 300 km working radius. These UAVs have a maximum altitude of 30,000 feet. They are also used for reconnaissance and observation.

17.3 Different types of drones

Drones are a fascinating technology that might pique kids' interest in careers in the IT industry. Many small commercial drones can be quickly assembled and customised, bringing young people and college students into the field of aeronautical design. Additionally, drones may benefit businesses in several ways, as some need global positioning satellite (GPS) tracking [11,17,26–29]. Different types of drones are as follows [13]:

17.3.1 Single-rotor drones

Without a doubt, these are among the most basic types of drones. It has only one rotor, as the name implies, and is often able to generate thrust more efficiently than its multirotor rivals. For longer flights, they can be the best option. However, the design has problems. First off, because of their single-blade design and often gas-powered nature, single-rotor drones can sometimes be more expensive than their dual-rotor counterparts. Furthermore, there is a greater risk of accidental injury with larger rotor blades.

17.3.2 Multirotor drones

Unlike single-rotor drones, which resemble helicopters and can maintain flight with just one rotor, these drones have many rotors distributed throughout the craft. With these additional rotors, it might be easier for the drone to stay balanced and hover. Regardless of the many types of commercial drones, the longer the craft can remain in the air, the more rotors we install. Because of this, these gadgets often have a maximum flight duration of 30 minutes, despite their exceptional steadiness. The majority of multirotor drone designs cannot carry a significant payload because the offsetting rotors' ability to maintain balance limits payload capacity.

17.3.3 Fixed-wing drones

Because these drones lack the rotors of traditional drones and instead have fixed wings, they resemble controlled aeroplanes more than helicopters. They are ideal for long-range drones because their wings provide vertical lift rather than rotors, which require energy to keep moving forward. Gas engines might power some fixed-wing drones. A fixed-wing drone can fly continuously for up to 16 hours, while a multirotor aircraft cannot stay in the air for very long.

17.3.4 Fixed-wing hybrid drones

These drone models seek to build hybrid drones by combining the best aspects of fixed-wing and rotor-based technologies. On a hybrid drone with fixed wings, a few rotors are often installed at the wing tips. Aircraft designs from the 1950s and 1960s served as the basis for several of these drones. They were mostly abandoned before the development of drones because the technology needed to make them a reality was considered too complex. Since they are still in the experimental stage, they are much harder to find on the market than their single-rotor, multirotor, and fixed-wing counterparts. Given that numerous companies are working on them, they might become popular in the future.

17.3.5 Small drones

These drones typically measure between 20 and 80 inches long. These drones are only used for fun, and they frequently have to perform more business duties than other drones. For example, when fitted properly, cameras aboard stable multirotor aircraft can capture stunning aerial photos and films. However, small drones are sometimes too light and require more stability to achieve the picture-perfect balance needed for quality photography. However, these options can be a great, inexpensive introduction to the world of drones for children and amateurs.

17.3.6 Micro drones

Armies that utilise drones treat tiny drones strictly as business. Cameras might be overkill for small drones used for fun, but the unique micro cameras on these little drones might yield valuable intelligence. They range up to a mile and can fly for up to 25 minutes on a single battery charge.

17.3.7 Tactical drones

These drones are still too large to fit in a pocket, but they are significantly smaller than those utilised for more complex tasks and general combat. These drones are often used for surveillance purposes. These can be fitted with sophisticated infrared cameras that let soldiers see the area clearly even at night. GPS technology is integrated into the gadgets. Despite their simplicity and lack of bells and whistles, soldiers may easily operate them without specific training.

17.3.8 Large combat drones

These devices have a 14 hour runtime and a 1000 mile range. Drones like this have been employed in missions.

17.3.9 Photography drones

Taking pictures with drones can be profitable. Specifically, we should use a commercial drone to take professional photos. A weaker option can take a few pictures, and there are definitely plenty of drones that can do this for free. High-quality drones can produce high-quality photos. Additionally, some drones for photography use precise, automated flight modes to take pictures.

17.3.10 Noncombat large drones

Large drones, however, are designed for use outside of battle. These are utilised for larger-scale recon operations. These drones are inexpensive because they do not compete.

17.3.11 Global positioning satellite drones

These drones use a GPS link to connect to satellites. After that, they might utilise this to map out the rest of their trip, generating information that can be extracted and used to meet demands. This can be a great way to map out large topographies. It goes without saying that this drains the battery. The best GPS drones are therefore configured to automatically return to their base when their battery begins to run low.

17.4 Major parts of the drone

The significant parts of the drone are shown in Fig. 17.1 [4,30–34]:

17.4.1 Drone motor

To balance the turning power produced by the rotating propellers, drones are equipped with two clockwise and two anticlockwise motors. It claims that every action has an equal and opposite reaction. Therefore having an equal number of motors counteracting one another provides stability by balancing the spinning force. As a result, helicopters have a tail rotor to counteract the twisting force of the single main rotor.

17.4.2 Drone propellers

In addition to having two clockwise and two counterclockwise motors, quadcopters feature two different propellers, one for each motor orientation. The drone is propelled upward by the rotation of each propeller, which forces air downward along the airfoil's surface, generating areas of lower pressure above and higher pressure below each propeller.

17.4.3 Drone flight controller

The drone flight controller is called the brain of a drone. The flight controller converts information from the compass, GPS module, remote controller, and object-avoidance sensors into commands to move the motors. In the past, or if we had a cheap drone, it would just float around because there were no sensors sending information about the drone's location and how to adapt for changes. However, this drone's GPS and downward vision sensors allow it to pinpoint its exact location, so even in the presence of wind, it will remain there.

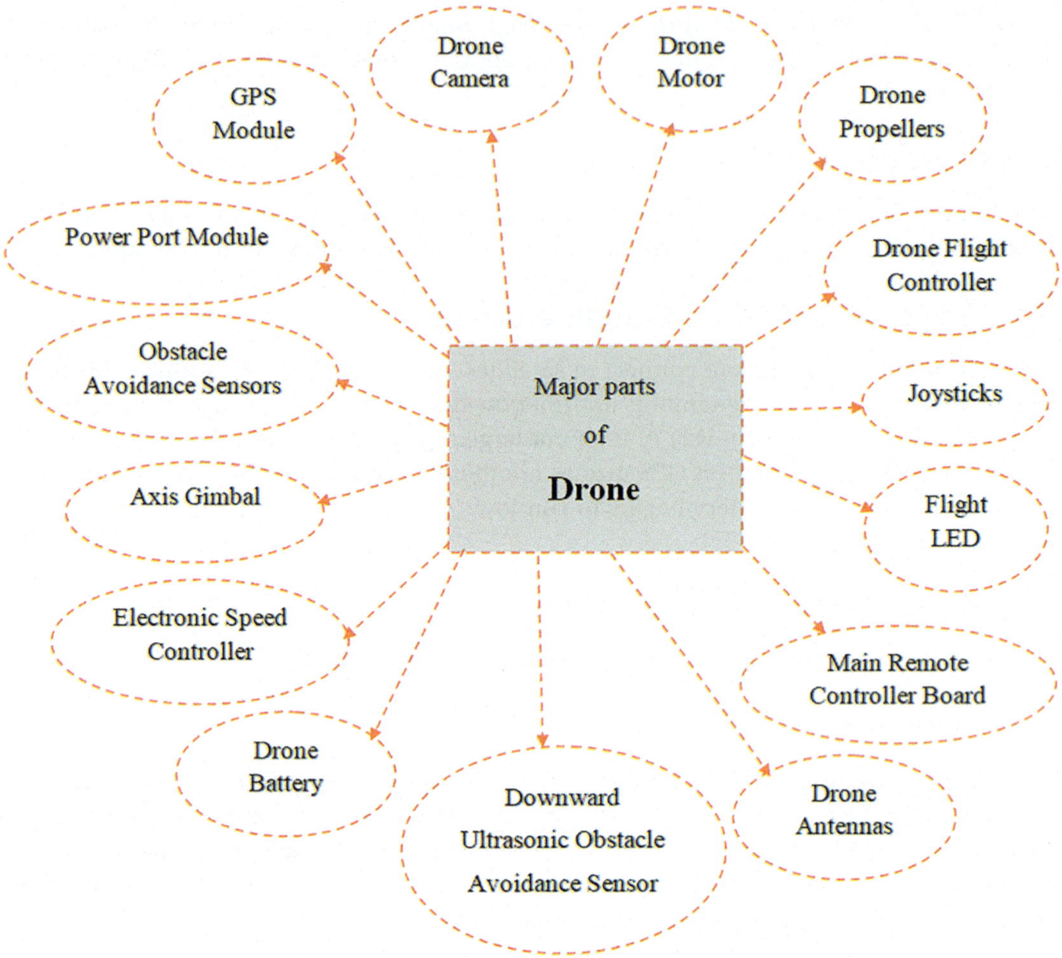

FIGURE 17.1 Major parts of a drone.

17.4.4 Global positioning satellite module

The GPS module uses information from two separate global positioning systems to pinpoint the drone's exact location. These satellites transmit position data to the planet's surface. These signals travel at the speed of light and are read by the drone's GPS module. After that, the drone calculates its geolocation by measuring the time it takes for signals from different satellites to reach it. The drone can locate itself on Earth and maintain its position using these GPS.

17.4.5 Electronic speed controller

The flight controller sends signals to the electronic speed controllers (ESCs), which adjust the power delivered to each motor. The ESCs are linked to the power distribution board and the flight controller.

17.4.6 Power port module

Power port modules control how much power is sent to the drone's ESCs and battle controller while monitoring the power flowing from the battery.

17.4.7 Obstacle avoidance sensors

This drone's front and bottom stereo vision sensors work in pairs, just like our eyes do. By identifying which image pixels from each sensor correspond to the same location, these sensors can determine depth. A drone's sensors allow it to measure the distance from an object in front of it.

17.4.8 Axis gimbal

In this, static drone footage is produced. A motor is mounted on each of the three axes surrounding the camera. The motors oppose the motion to cancel it when the sensors identify movement on any of these axes. As millions of computations are performed to ensure smooth footage, the result is virtually immediate.

17.4.9 Drone camera

A lens that opens out in front of the camera allows light to enter. An imaging sensor receives the incoming light rays and transforms them into a digital image.

17.4.10 Drone battery

These 'intelligent' batteries have temperature data, charge cycle history, overcharging protection, and power output communication with the drone. This ensures the battery is secure for frequent usage and prevents issues during flight.

17.4.11 Drone antennas

The transmission mechanism, housed inside the drone's legs, transmits data between the drone and the controller. Additionally, the drone has two compass sensors on its legs that tell the flight controller its location.

17.4.12 Downward ultrasonic obstacle avoidance sensor

One sensor transmits high-frequency sound pulses, and another receives them. By measuring the time between sending and receiving pulses, the drone calculates its altitude above the earth.

17.4.13 Flight LED

These flash various colours to indicate to the user which way the drone is pointed. The drone's front is visible using the two flashing red LEDs. Two green flashing LEDs indicate the drone's rear.

17.4.14 Joysticks

These convert the actual stick motion into data that the controller may utilise to interact with the drone. The drone may be raised, lowered, and panned right and left using the left joystick. The drone may go forward, backwards, up, and left with the right joystick.

17.4.15 Main remote controller board

This collects data on the drone's position, altitude, and the camera's capture. Inputs from the joysticks are also received, and directives are sent to the flying controller.

17.4.16 Main camera board

This processes data from the image sensor and gimbal motors to provide steady video. This board also writes the picture to the micro-SD card, which analyses camera data.

17.5 Application areas of drones

Drones are aircraft flown remotely by a person or automatically by an onboard computer. Drones are used across various industries for their cutting-edge technology and outstanding capabilities [10,35]. There are now more uses for drone technology due to people's rising interest in it. Drones are used in many applications, and as technology advances, these machines will become even more reliable and practical. Drones are currently used in every sector where humans once worked, including agriculture and the internet [14,36–40]. The significant areas of drones are as follows:

17.5.1 Aerial photography

Modern cameras mounted on drones are deployed to provide media coverage in complex regions. Before, only major news organisations that could afford helicopters could get crucial photos. However, today, practically all local journalists and small-scale media outlets can record crystal-clear air footage for compelling news coverage. Due to their tiny size, drones can also move about readily in confined spaces, making them helpful in filming in-person interviews.

17.5.2 Construction

Drones are becoming more common in the construction sector. Its key activities are surveying buildings and work sites, employee surveillance, security, surveying, and generating 2D and 3D data. Regular aerial monitoring might lead to better performance, more efficiency, and essential infrastructure improvements in the building industry. Drones conduct job-site observation and provide precise project estimates.

17.5.3 Fire fighting

Fire outbreaks can be controlled using this technology. Dangerous flames can be accessed and monitored by drones, eliminating the need for aerial helicopter monitoring. Thermal drones protect the public and can detect hotspots even in smoke and darkness. Because of this, it is usually preferable to deploy drones at the outset of a fire to assess the situation.

17.5.4 Wireless internet access

The internet is where drones are often used and deployed in sophisticated ways. It is particularly intriguing because the company will soon use drones to provide internet service in off-the-grid regions. Since the internet is one of the most critical technologies for people in the twenty-first century, any development in this area is greatly appreciated. Soon, a drone flying in the air will be able to pick up the signal for our mobile device.

17.5.5 Search and rescue operations

These days, drones are equipped with thermal sensors that help them locate lost people. Additionally, they can operate in dimly lit spaces and challenging terrain. Due to limited technology, search-and-rescue efforts have long suffered, but drones make everything feasible in minutes. They may be deployed and utilised for search-and-rescue operations during critical moments of conflict. They can drop goods in difficult-to-reach places.

17.5.6 Agriculture

The ability of drone technology to assist farmers in various ways is its most significant accomplishment. By monitoring unhealthy plants, drones can help farmers save money and increase their harvest. They may examine the sizable farmlands and monitor the effective irrigation system. Farmers can now anticipate rapid information updates and use drones to spray crops with fertiliser, insecticides, and water at the right times.

17.5.7 Shipping and delivery

Since modern drones can carry large payloads, they can now be used for delivery and shipping. It will help individuals receive their ordered items delivered to their doorsteps immediately. The most significant part is that it will only get stopped in traffic. Now, we may receive our meal order at home in a short time.

17.5.8 Research and natural science

Drone manoeuvrability is one of its essential characteristics. Since drones can be made incredibly compact while still powerful, they do not have a space problem; thus they can quickly reach confined or difficult-to-reach places.

17.5.9 Engineering applications

Nowadays, most engineering companies use drone technology to monitor intricate projects, such as transmission cables and oil pipelines, and to conduct maintenance checks. A drone may carry out a wide range of functions.

17.5.10 Safety surveillance

These little gadgets can help users achieve success in crowd control and public safety. Drones can detect criminal activity in crowded areas. They may also be used to carefully monitor border regions, enabling the quick apprehension of smugglers of migrants and illicit substances. Most nations today focus on constructing highly sophisticated drone units because drones act as an intelligent army that can safeguard a country from various issues.

17.6 Advantages of deploying drones in daily life

The significant advantages of drones when used in daily life are as follows [41,42]:

- To provide better public surveillance, security, and safety.
- With its excellent camera control, it can be well deployed for taking pictures, and it captures video while moving, and the same can be transmitted back.
- They can be developed with a high level of endurance and strength.
- It can have a flight period that lasts a long time.
- Collect a large amount of imaging data, depending on the support systems.
- Ability to work in a limited space.
- Helpful for remote locations with a large coverage area.
- Minimise health risks for operators.
- It has extensive military-based applications and is acting as a disruptive technology.
- Helpful for the protection of animals and other assets.
- Fixed wings can fly at a higher altitude.
- Helpful for collecting data on the weather and atmospheric conditions.
- With quick flying and better speed, it has good logistics applications.

17.7 Roles of drones for Industry 4.0

Drones are widely used in manufacturing for inventory management, intracompany material and part transfers, and asset inspection and maintenance. The most popular uses are maintenance and inspection. Industrial equipment anomalies or issues can be promptly identified by a drone equipped with a variety of sensors, including lidar, ultrasonic, laser, chemical, stability and orientation, and stereoscopic cameras. Because not every area can be examined, it is useful in refineries, pipelines, mines, and other large sites where human inspection is dangerous or pointless. Drones, therefore, help reduce the time, risk, and expense associated with human inspections [43–45]. Drones play the following important functions in Industry 4.0:

17.7.1 Record appropriate information

Drones can capture relevant information, such as infrared imagery and aerial photographs. This technology has proven a natural fit for the construction and insurance industries, as it allows them to inspect sites and structures without sending workers to the site. Drones may now be considered for recording ongoing information in the manufacturing industries.

17.7.2 Inspection

Drones with cameras can record images required for inspection and monitoring tasks. These sensor-based images can take on various shapes depending on the sensor used. The photographs may be viewed in real time or at any point in the past. They are used while visually examining specific structures, locations, and items.

17.7.3 Monitoring of the worksite's progress

Drones at worksites monitor progress, save time, and sort out numerous potential problems to improve safety. Drones are utilised at various stages of research and deployed across industries, as they greatly simplify monitoring, infrastructure evaluation, and even transportation. Drones will be used for agricultural applications, irrigation management, and mapping and surveying.

17.7.4 Analyse and eliminate the risk

The drone can help reduce the need for potentially dangerous in-person examinations. In industries, this can be monitored through ongoing production, which can be analysed and eliminated to mitigate any risk. Drones are attracting the attention of electricity providers because they promise to be faster, cheaper, and safer.

17.7.5 Leak detection in industries

This method may be used to detect leaks and possible actuation locations. It may be used in various applications, including electrical towers, wind turbines, moisture detection, solar panel inspection, and building energy efficiency inspection. Drones offer diverse options and configurations well-suited to various businesses.

17.7.6 Take care of an emergency

Drones can deliver critical goods in emergencies when other modes of transportation are unavailable. Drones are used in agriculture to monitor livestock and agricultural conditions, among other things. Consumer delivery services, currently under development, could benefit from drone deployment.

17.7.7 Lower maintenance costs

By utilising advanced drone technology, it is feasible to cut operating, maintenance, labour, and insurance costs. Drone technology has advanced, and its industrial and commercial applications have expanded as its reliability has increased. This helps maintain all industrial equipment.

17.7.8 Perform dangerous duties

Because of their abilities, they can perform more dangerous tasks, allowing people to focus on higher-value activities. They have been utilised under challenging conditions for decades, including subsea inspection, deep-water oil exploration, and hazardous plant zones. Drones use high-definition cameras, gas sensors, and infrared cameras to detect leaks.

17.7.9 Surveillance

Drones are used for various purposes, including aerial photography, surveillance, 3D mapping, shipping, and remote sensing. It is used in almost all forms of surveillance in industrial environments. Drones are used for asset management as part of their IoT adoption strategies.

17.7.10 Identify missing items

Drones can scan thousands of products in a warehouse while also looking for missing items. This technology helps conduct inspections quickly and safely, and at a lower cost. Drones are also frequently used by insurance companies for accident reconstruction.

17.7.11 Friendly to the environment

Drones are an environmentally friendly alternative to large delivery vans. Drones may refresh their navigational data in seconds, making it easier to reach their target safely. Operators may use advanced route-building simulators to establish optimal flight paths and timetables for a continuous drone fleet.

17.8 Roles of digital twin for drones

In Industry 4.0, digital twin technology is used to improve drone performance, safety, and operational effectiveness. Drone fleets can be deployed for industrial inspection, surveillance, mapping, and logistics may be monitored and optimised in real time with the help of digital twins. Operators can evaluate performance variances, optimise energy usage, and improve route efficiency by synchronising real-time drone data with the virtual model [46–49]. By modelling multidrone missions and airspace interactions in a safe virtual environment, digital twins facilitate coordinated fleet management in large-scale industrial operations. This facilitates data-driven decision-making, increases safety compliance, and improves operational reliability. Automation, accuracy, and intelligence are strengthened in smart industrial ecosystems through the integration of digital twin technology with drone systems [50–54]. The major roles of digital twins for drones are as follows:

- Aerodynamics and drone structure are virtually modelled to optimise performance.
- Flight path simulation prior to actual mission deployment.
- Testing collision avoidance systems in intricate industrial settings.
- Real-time power consumption and battery health monitoring.
- Payload capacity and weight distribution optimisation.
- Accurate navigation and GPS validation virtually.
- Performance evaluation of LiDAR sensors and onboard cameras.
- A simulation of fleet coordination for industrial missions involving several drones.
- Using mission data analytics to optimise energy efficiency.
- Simulation of failure modes for emergency landing situations.
- An evaluation of drone communication networks' cybersecurity risks.
- Optimisation of routes for delivery and inspection duties.
- Synchronisation between virtual models and telemetry data in real time.
- Evaluation of airspace safety as per the regulations and its compliance.
- Digital incident reconstruction for forensic analysis.
- Ongoing benchmarking of performance under different operating loads.

17.9 Upcoming goals of drones

The upcoming goals of drones are as follows [55,56]:

- Drones will soon be able to access regions that humans cannot.
- Drones with a large payload will be used to move products in warehouses and other locations and replenish inventories.
- Drones will be beneficial for inspecting high-risk areas.
- To address technological difficulties, drone designers and manufacturers will have much work ahead of them.
- This will carry out a variety of everyday chores as well as industrial work.

17.10 Challenges with drones

The significant challenges being faced with drones are as follows [9,12,19]:

- Higher levels of complexity.
- Payload capacity limitation.
- Specific knowledge and skills are required to operate drones properly.
- Data transfer speed is low.

- Its movement can be hacked, thus creating stress among people in a faraway place during hacking attempts.
- Launch and recovery require a large amount of space.
- It runs slower than aeroplanes.
- During the flight period, the battery is limited.

17.11 Conclusion

Drones are increasingly demonstrating their usefulness across industries and applications, and organisations worldwide are experimenting with incorporating them into their operations. Businesses want to automate their operations as drones become more common in order to profit from their success. Drones can finish on-site inspections in a fraction of the time required by human workers. Drones may be used to save money on a variety of things. These technologies are used to assess progress on construction projects and help analyse and remove dangers. It can provide information about the shop floor and other ongoing industrial operations and swiftly assess internal facilities.

This technology is eco-friendly and can help complete any work. Drones may be more effective with contemporary fleet management systems, enabling multiple drones to collaborate on a project site. The workplace is safer and more secure thanks to this technology, which also boosts productivity. This technology offers construction organisations real-time progress maps and surveys to plan projects and improve those already underway, resulting in significant cost savings. These technologies are essential to the manufacturing process. Manufacturers can also deploy drones specifically designed to work alongside machinery and workers.

References

[1] M. Javaid, I.H. Khan, R.P. Singh, S. Rab, R. Suman, Exploring contributions of drones towards Industry 4.0, Industrial Robot: The International Journal of Robotics Research and Application 49 (2022) 476–490.

[2] D. Moreno-Jacobo, G. Toledo-Nin, A. Ochoa-Zezzatti, V. Torres, F. Estrada-Otero, Evaluation of drones for inspection and control in Industry 4.0, Technological and Industrial Applications Associated with Intelligent Logistics, Springer International Publishing, Cham, (2021), pp. 579–595.

[3] K. Dwivedi, P. Govindarajan, D. Srinivasan, A. Keerthi Sanjana, R. Selvanambi, M. Karuppiah, Intelligent autonomous drones in Industry 4.0, Artificial Intelligence and Cyber Security in Industry 4.0, Springer Nature Singapore, Singapore, (2023), pp. 133–163.

[4] O. Maghazei, T. Netland, Drones in manufacturing: exploring opportunities for research and practice, Journal of Manufacturing Technology Management 31 (2020) 1237–1259.

[5] A. Arishi, K. Krishnan, M. Arishi, Machine learning approach to truck-drones-based last-mile delivery in the era of Industry 4.0, Engineering Applications of Artificial Intelligence 116 (2022) 105439.

[6] M.B. Hatoum, H. Nassereddine, The use of Drones in the construction industry: applications and implementation, ISARC. Proceedings of the International Symposium on Automation and Robotics in Construction, IAARC Publications, (2022), pp. 542–549.

[7] P. Zhang, C. Wang, Z. Qin, H. Cao, A multidomain virtual network embedding algorithm based on multiobjective optimisation for the Internet of Drones architecture in Industry 4.0, Software: Practice and Experience 52 (2022) 710–728.

[8] A.R. Svaigen, A. Boukerche, L.B. Ruiz, A.A. Loureiro, Security in the industrial internet of drones, IEEE Internet of Things Magazine 6 (2023) 110–116.

[9] J. Kuljanin, T. Zivojinović, N. Bojkovic, C. Barrado, Drone technology in Industry 4.0: challenges and obstacles in urban environments, Intelligent and Sustainable Engineering Systems for Industry 4.0 and Beyond, CRC Press, (2025), pp. 180–220.

[10] M. Wales, Drone applications at present and in the future. Available from https://filmora.wondershare.com/drones/drone-applications-and-uses-in-future.html, 2022.

[11] D. Stroud, V. Timperley, M. Weinel, Digitalised drones in the steel industry: the social shaping of technology, Relations Industrielles/Industrial Relations 75 (2020) 730–750.

[12] D. Bansal, N. Bhattacharya, P. Shandilya, Exploring applications: intelligent drones and robots in industrial settings, Building Embodied AI Systems: The Agents, the Architecture Principles, Challenges, and Application Domains, Springer Nature Switzerland, Cham, (2025), pp. 159–180.

[13] List of 14 different types of drones explained with photos. Available from https://aerocorner.com/blog/types-of-drones/.

[14] C. Bai, P. Dallasega, G. Orzes, J. Sarkis, Industry 4.0 technologies assessment: a sustainability perspective, International Journal of Production Economics 229 (2020) 107776.

[15] F. Elghaish, S. Matarneh, S. Talebi, M. Kagioglou, M.R. Hosseini, S. Abrishami, Toward digitalisation in the construction industry with immersive and drone technologies: a critical literature review, Smart and Sustainable Built Environment 10 (2021) 345–363.

[16] K. Tantawi, I. Fidan, O. Huseynov, Y. Musa, A. Tantawy, Advances in Industry 4.0: from intelligentization to the industrial metaverse, International Journal on Interactive Design and Manufacturing (IJIDeM) 19 (2025) 1461–1472.

[17] M.M. Gunal, Data collection inside industrial facilities with autonomous drones, Simulation for Industry 4.0: Past, Present, and Future, Springer International Publishing, Cham, (2019), pp. 141–151.

[18] Classification of the Unmanned Aerial Systems. Available from https://courses.ems.psu.edu/geog892/node/642.

[19] T.M. Fernández-Caramés, O. Blanco-Novoa, I. Froiz-Míguez, P. Fraga-Lamas, Towards an autonomous industry 4.0 warehouse: a UAV and blockchain-based system for inventory and traceability applications in big data-driven supply chain management, Sensors 19 (2019) 2394.

[20] D. Mourtzis, J. Angelopoulos, N. Panopoulos, Uavs for industrial applications: identifying challenges and opportunities from the implementation point of view, Procedia Manufacturing 55 (2021) 183–190.

[21] W. Pensec, D. Espes, C. Dezan, Smart Anomaly Detection and Monitoring of Industry 4.0 by Drones, 2022 International Conference on Unmanned Aircraft Systems (ICUAS), Dubrovnik, Croatia, IEEE, (2022), pp. 705–713. Available from https://doi.org/10.1109/ICUAS54217.2022.9836057.

[22] Classification of the Unmanned Aerial Systems. Available from https://www.e-education.psu.edu/geog892/node/5.

[23] D. Mourtzis, J. Angelopoulos, N. Panopoulos, Unmanned Aerial Vehicle (UAV) path planning and control assisted by Augmented Reality (AR): the case of indoor drones, International Journal of Production Research 62 (2024) 3361–3382.

[24] Sustainable product design engineering in Industry 4.0: Civilian and military drones vis-à-vis digital transformation, AIP Conference Proceedings 2485 (2023) 070008. Available from https://doi.org/10.1063/5.0106297.

[25] A.H. Hussain, A. Ummihusna, S. Eni, W.N.W. Mohamad, N.C. Musa, Integrating Industry 4.0 into higher education: a case study of a drone course for multidisciplinary learning course, International Journal of Academic Research in Business and Social Sciences 14 (2024).

[26] S.H. Alsamhi, A.V. Shvetsov, S. Kumar, J. Hassan, M.A. Alhartomi, S.V. Shvetsova, et al., Computing in the sky: a survey on intelligent ubiquitous computing for uav-assisted 6g networks and Industry 4.0/5.0, Drones 6 (2022) 177.

[27] F. Conceição, C. Guimarães, L. Cominardi, S.T. Talat, M.F. Ardiansvah, C. Zhang, et al., Empowering Industry 4.0 and autonomous drone scouting use cases through 5G-DIVE solution, 2021 Joint European Conference on Networks and Communications & 6G Summit (EuCNC/6G Summit), Porto, Portugal, IEEE, (2021), pp. 265–270. Available from https://doi.org/10.1109/EuCNC/6GSummit51104.2021.9482590.

[28] D. Palaniappan, T. Premavathi, R. Jain, K.J. Parmar, Drones in manufacturing and industrial applications, Computer Vision and Edge Computing Technologies for the Drone Industry, IGI Global Scientific Publishing, (2025), pp. 61–88.

[29] A. Rejeb, K. Rejeb, S.J. Simske, H. Treiblmaier, Drones for supply chain management and logistics: a review and research agenda, International Journal of Logistics Research and Applications 26 (2023) 708–731.

[30] Anatomy of A Drone - What's inside a DJI Phantom Drone. Available from https://www.dronefly.com/the-anatomy-of-a-drone.

[31] J. Shahmoradi, E. Talebi, P. Roghanchi, M. Hassanalian, A comprehensive review of applications of drone technology in the mining industry, Drones 4 (2020) 34.

[32] L.F. Luque Vega, E. Lopez-Neri, C.A. Arellano-Muro, L.E. González-Jiménez, J. Ghommam, R. Carrasco-Navarro, UAV Flight Instructional Design for Industry 4.0 based on the Framework of Educational Mechatronics, IECON 2020 The 46th Annual Conference of the IEEE Industrial Electronics Society, Singapore, IEEE, (2020), pp. 2313–2318. Available from https://doi.org/10.1109/IECON43393.2020.9255295.

[33] A. Kumar, B. Muhammad, On how internet of drones is going to revolutionise the technology application and business paradigms, 2018 21st International Symposium on Wireless Personal Multimedia Communications (WPMC), IEEE, (2018), pp. 405–410.

[34] O. Bongomin, G. Gilibrays Ocen, E. Oyondi Nganyi, A. Musinguzi, T. Omara, Exponential disruptive technologies and the required skills of Industry 4.0, Journal of Engineering 2020 (2020) 4280156.

[35] Drone applications in the Future. Available from https://www.advanceecomsolutions.com/blog/drone-applications-in-the-future.

[36] T.M. Fernández-Caramés, O. Blanco-Novoa, M. Suárez-Albela, P.A. Fraga-Lamas, A UAV and Blockchain-Based System for Industry 4.0 Inventory and Traceability Applications, Proceedings 4 (1) (2019) 26. Available from https://doi.org/10.3390/ecsa-5-05758.

[37] T. Alladi, V. Chamola, R.M. Parizi, K.K.R. Choo, Blockchain applications for Industry 4.0 and industrial IoT: a review, IEEE Access 7 (2019) 176935–176951.

[38] S.S. Ali, S. Khan, N. Fatma, C. Ozel, A. Hussain, Utilisation of drones in achieving various applications in smart warehouse management, Benchmarking: An International Journal 31 (2024) 920–954.

[39] A. Gohari, A.B. Ahmad, R.B.A. Rahim, N.I.M. Elamin, M.S.M. Gismalla, O.O. Oluwatosin, et al., Drones for road accident management: a systematic review, IEEE Access 11 (2023) 109247–109256.

[40] M. Ridolfi, N. Macoir, J. V. -V. Gerwen, J. Rossey, J. Hoebeke, E. de Poorter, Testbed for warehouse automation experiments using mobile AGVs and drones, IEEE INFOCOM 2019 - IEEE Conference on Computer Communications Workshops (INFOCOM WKSHPS), Paris, France, IEEE, (2019), pp. 919–920. Available from https://doi.org/10.1109/INFCOMW.2019.8845218.

[41] O. Maghazei, M.A. Lewis, T.H. Netland, Emerging technologies and the use case: a multi-year study of drone adoption, Journal of Operations Management 68 (2022) 560–591.

[42] L. Wawrla, O. Maghazei, T. Netland, Applications of drones in warehouse operations, Whitepaper. ETH Zurich, D-MTEC, 221, (2019), p. 1.

[43] N. Grigoropoulos, S. Lalis, Simulation and Digital Twin Support for Managed Drone Applications, 2020 IEEE/ACM 24th International Symposium on Distributed Simulation and Real Time Applications (DS-RT), Prague, Czech Republic, IEEE, (2020), pp. 1–8. Available from https://doi.org/10.1109/DS-RT50469.2020.9213676.

[44] J. Arora, M. Tushir, Digital twin: towards internet of drones, Digital Twin Technology (2021) 23–45.

[45] S.H. Lim, K.M. Choi, G.S. Cho, A study on 3D model building of drones-based urban digital twin, Journal of Cadastre & Land Informatix 50 (2020) 163–180.

[46] A. Islam, S.Y. Shin, A digital twin-based drone-assisted secure data aggregation scheme with federated learning in artificial intelligence of things, IEEE Network 37 (2023) 278–285.

[47] D. Carramiñana, L. Rey Braga, A.M. Bernardos, A.D. Dobrzycki, L. Bergesio, J.A. Besada, et al., A digital twin concept to prototype multi-drone based applications, IoT '24: Proceedings of the 14th International Conference on the Internet of Things, Association for Computing Machinery, New York, NY, (2024), pp. 150–154. Available from https://doi.org/10.1145/3703790.3703808.

[48] A. Hamzah, F. Aqlan, S. Baidya, Drone-based digital twins for water quality monitoring: a systematic review, Digital Twins and Applications 1 (2024) 131–160.

[49] A. To, M. Liu, M. Hazeeq Bin Muhammad Hairul, J.G. Davis, J.S. Lee, H Hesse, et al., Drone-based AI and 3D reconstruction for digital twin augmentation, International Conference on Human-Computer Interaction, Springer International Publishing, Cham, (2021, July), pp. 511–529.

[50] J. Wu, J. Guo, Z Lv, Deep learning driven security in digital twins of drone network, ICC 2022-IEEE International Conference on Communications, IEEE, (2022, May), pp. 1–6.

[51] O. Maghazei, T.H. Netland, D. Frauenberger, T. Thalmann, Automatic drones for factory inspection: The role of virtual simulation, IFIP International Conference on Advances in Production Management Systems, Springer International Publishing, Cham, (2021, August), pp. 457–464.

[52] A. Ossamah, Blockchain as a solution to drone cybersecurity, 2020 IEEE 6th World Forum on Internet of Things (WF-IoT), IEEE, (2020), pp. 1–9.

[53] H. Jahani, Y. Khosravi, B. Kargar, K.L. Ong, S. Arisian, Exploring the role of drones and UAVs in logistics and supply chain management: a novel text-based literature review, International Journal of Production Research 63 (2025) 1873–1897.

[54] P. Boccadoro, A. Santorsola, L.A. Grieco, A dual-stack communication system for the internet of drones, Ad-Hoc, Mobile, and Wireless Networks. ADHOC-NOW 2020. Lecture Notes in Computer Science, vol 12338, Springer, Cham, (2020), pp. 71–83. Available from https://doi.org/10.1007/978-3-030-61746-2_6.

[55] N. Singh, V. Niranjan, N.R. Chauhan, M. Saraswat, Impact of drones on various industries, AIP Conference Proceedings 3330 (2025) 020010. Available from https://doi.org/10.1063/5.0292945.

[56] S.S. Ali, R. Kaur, H. Abbas, Feasibility of drones as the next step in solution for emerging society, Artificial Intelligence in Industry 4.0 and 5G Technology, Wiley, (2022), pp. 179–208. Available from https://doi.org/10.1002/9781119798798.ch9.

18

Cyber-physical systems for Industry 4.0

18.1 Introduction to cyber-physical systems

Cyber-physical systems (CPS) play an essential role in Industry 4.0. CPS, also known as cyber manufacturing, is an Industry 4.0-enabled manufacturing environment that provides real-time data collection, transparency, and analysis across all stages of a manufacturing process. Executives can make data-driven decisions using analytics, thereby increasing productivity [1,2]. The two components of CPS are physical and cyber. CPS is being incorporated into various products, including cars and other equipment, to perform specific tasks. Due to its ease of use, CPS can be used in any business, including engineering, manufacturing, transportation, and even healthcare. Collaborative systems for the CPS network are designed for communication. CPS also uses cutting-edge network technologies, such as cloud solutions [3–5].

A new generation of physical processes that integrate with computer and networking activities is referred to as CPS. It is the merging of the internet with the physical world. The computing and communication infrastructure of the widely used, networked, and computerised digital technology known as 'cyberspace' defines it. Manufacturing, energy, infrastructure, consumer products, the military, robotics, intelligent buildings, communications, healthcare, transportation, and other sectors are employing CPS [6–8]. CPS integrates monitoring and management of physical components with a computer, storage, and communication capabilities. Giving physical items 'intelligence' will enable them to interact with one another. As a result of their interconnectedness, they employ global digital networks to learn, work together, and grow, as well as to monitor, manage, and use the information available in the virtual world [9,10].

Industry 4.0 relies heavily on semiautonomous and autonomous decision-making, supported by CPS, Internet of Things (IoT), artificial intelligence (AI), integrated systems, sophisticated analytics, and other technologies. Within Industry 4.0, CPS is the next evolution of continuous improvement and functional integration. Manufacturing equipment is outfitted with IP-addressable sensors to communicate with other web-enabled devices. One of the key advantages of this revolution is the increased efficiency and adaptability of personalised goods [11,12]. Flexibility might enable companies to create unique, customised, or one-of-a-kind products and to make last-minute changes across different manufacturing stages using CPS. Because certain environments are more suitable or cost-effective for specific workloads, they may optimise their workloads across all of their clouds [13,14].

Depending on the network connection, CPS could take a while for data to be sent from the corporate cloud to the production floor. This also keeps data close to its source and lowers security risks. The Industrial Internet of Things leverages advanced control technology, machine learning

Digital Twin and Industry 4.0. DOI: https://doi.org/10.1016/B978-0-443-54019-6.00010-8

(ML), and big data to enable smart machines and, eventually, smart factories and warehouses by linking sensor data, machine communications, and automation systems that are becoming more prevalent in industrial settings [15–17]. To protect critical industrial systems and production lines from cyberattacks, Industry 4.0 technologies require greater connectivity. System capabilities, adaptability, scalability, resilience, safety, security, and usability will increase significantly using CPS developments compared to the present generation of simple embedded systems [18–20].

A CPS combines physical components, transducers, and information technology systems in Industry 4.0. CPS is a system that connects the physical and digital worlds. A CPS may be an open-loop or closed-loop system, meaning it can either recognise and control the physical system's real-world features or only make that data available for analysis. Industry 4.0 enables cyber manufacturing, which provides real-time data collection, analysis, and transparency across all aspects of the manufacturing process [21,22]. CPS must provide the necessary networking, integration processes, and tools. CPS enables us to create, develop, enhance, and sustain intelligent systems that benefit organisations, communities, and people. CPS has the potential to impact technology across several areas of Industry 4.0. The use of CPS to promote sustainability is expanding [23,24].

18.2 Design of cyber-physical systems

Cosimulation has recently been utilised across disciplines to interact without imposing new design processes, as there is no conventional approach to design practice that incorporates a wide variety of technical and creative disciplines in CPS [4,12,15,25–27]. The 5C architecture can be used to develop and implement a cyber-physical production system, as illustrated in Fig. 18.1.

18.2.1 Connection

Devices that self-connect and self-sense their activities may be created.

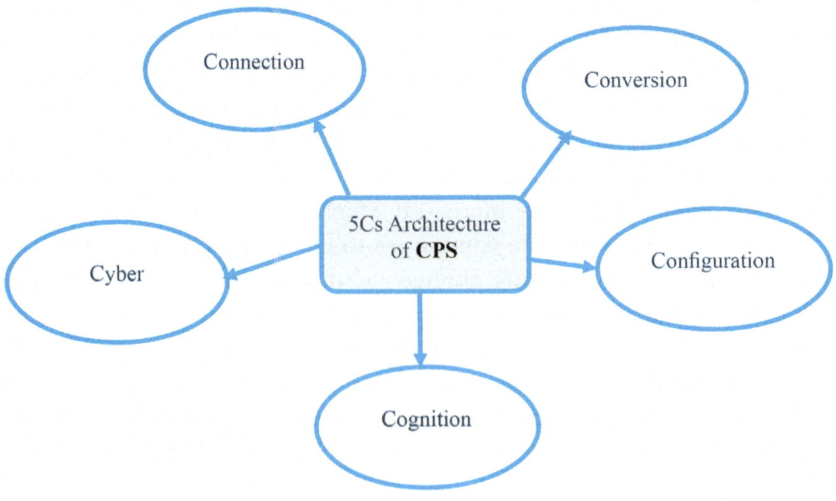

FIGURE 18.1 5Cs architecture of cyber-physical systems.

18.2.2 Conversion

Machines may use self-awareness to predict potential problems. Self-connected gadgets and sensors analyse the features of significant issues using self-aware capabilities.

18.2.3 Cyber

Each machine creates its own 'clone' by employing instrumented attributes and a few methods to specify the machine pattern more precisely. For additional synthesis, the established 'clone' might conduct a self-comparison to assess peer-to-peer performance.

18.2.4 Cognition

Users will see the outcomes of the self-assessment and self-evaluation in an 'infographic' that depicts the content and context of possible issues.

18.2.5 Configuration

The equipment or production system may be modified based on priority and risk criteria to ensure reliable performance.

18.3 Advantages of cyber-physical systems

The significant advantages of CPS are as follows [28–30]:

- Management of smart cities.
- Improve the experience of businesses.
- Simultaneously improving the quality of life.
- Ability to perform countless calculations instantly.
- Potential to bring about a positive revolution in the world.
- Protection from viruses and hacks.
- Provides better privacy.
- Everyone can use CPS with ease in any industry.
- Provides robust functionality in a variety of contexts.
- CPS is based on network cooperation and is designed for communication.
- CPS implements embedded network solutions with sturdiness and intelligence.

18.4 Application areas of cyber-physical systems

Data security, especially the protection of personal data, is included in CPS as a crucial component of industrial assets and infrastructure. Manufacturers can use this to optimise logistics, manage inventory, and maintain quality control [31–35]. Global asset intelligence may be easier for factory workers to obtain. Standard asset management procedures, such as transfers, sales, reclassifications, and revisions, can be accelerated, centrally administered, and controlled in real-time [12,20,36–39]. The following are the main areas where CPS is being used:

18.4.1 Manufacturing

The industry primarily uses CPS to monitor its production processes and operations. Production is significantly increased by the information shared between machines, corporate systems, supply chains, suppliers, and customers. This increases product security and traceability by improving supply chain visibility and control. In the industrial sector, sensors are also used to diagnose machine and equipment flaws and forecast when equipment will wear out. Appropriate data analysis improves equipment performance and reduces maintenance costs.

18.4.2 Water distribution

Because of initiatives such as the CPS, water distribution systems are increasingly being automated. Pipelines, wells, pumps, tanks, and reservoirs make up the water supply systems in our homes. Systems use devices to monitor several aspects of water distribution system operations. For example, a sensor may measure the amount of water that overflows from a pressure pipe or tank. Additionally, the systems automatically open valves using programmable control circuitry.

18.4.3 Agriculture

CPS-related technology, also known as digital farming or smart agriculture, has enabled innovations that boost farm productivity. These include drones and satellites that provide photos linked to plant health, as well as smart sensors on tractors or harvesters that provide information on soil type and condition. CPS significantly affects agriculture, boosting productivity and decreasing hunger. The input could be handled remotely via network services. The designs also include station sensors for humidity, temperature, soil moisture, and light. Sensors for temperature and humidity control are placed, enabling fans and sprinklers to be turned on and off to change the temperature. The CPS provides significant benefits to the agriculture sector through smart greenhouses.

18.4.4 Sustainability

Advancements like accessible solar power, affordable energy storage, public electric transportation, and clean recycling programmes are often made possible by CPS technology. The enormous need for sustainable practices in business, healthcare, and a host of other sectors remains a priority for society.

18.4.5 Smart grids

To increase the electrical efficiency, reliability, and security of the grid, the smart grid is a CPS that collects and analyses data from power plants, energy transmission, distribution networks, and customers.

18.4.6 Aircraft systems

CPS is used in the sensors, control algorithms, and communication systems of aerospace systems such as aircraft, satellites, and missiles to enhance their security and effectiveness.

18.4.7 Building management

CPS controls buildings' lighting, ventilation, air conditioning, and heating. They increase the effectiveness of these systems using sensors, actuators, and control algorithms.

18.4.8 Security

The advancement of smart technology has, in many ways, increased security measures. From the creation of mobile apps for real-time remote monitoring to fully functional intelligent surveillance systems, CPS technology has enabled smart security to advance and improve.

18.4.9 Healthcare

Numerous advancements in medicine have been made possible by the development of CPS technology. From a smart monitoring system that tracks cancer patients' responses to therapy to a continuous smart device that sends glucose data to the wearer's smartphone, CPS research and technology have revolutionised healthcare. In the medical field, CPS is frequently utilised for sensor control and real-time monitoring of patient status. This reduces the need for hospitalisation for patients while improving care for the elderly and disabled.

18.4.10 Smart structures

Buildings' energy efficiency increases while energy use and greenhouse gas emissions are reduced through CPS. The system is designed to sense and detect temperature, humidity, and other actuators, such as fans, heating, ventilation, and air conditioning (HVAC) systems, and water heaters.

18.4.11 Transportation

Cars can now communicate with one another via CPS to share information about the road, including traffic, position, and issues, thereby improving safety and preventing collisions. CPS is advanced enough to use networked systems to merge hardware and software. It successfully blends the real and virtual worlds to boost productivity across sectors such as engineering, healthcare, transportation, smart greenhouses, and smart buildings.

18.5 Roles of cyber-physical systems for Industry 4.0

CPS significantly influences Industry 4.0, increasing productivity. Because a CPS comprises several heterogeneous components, it is necessary to use complex models to characterise each subsystem and its behaviour. To ensure the behaviour of each subdeterministic system and, later, the dynamic interactions between the subsystems, the overarching model serves as a control entity [40–42]. This also concentrates on controllable elements, including humidity, light intensity, temperature, and watering. These are made to respond to specific computer programmes to assist growth. Additionally, the gadget continuously updates the user on the greenhouse's status. Industry 4.0 changes how businesses produce, improve, and distribute their products through CPS [43–47]. The significant roles of CPS for Industry 4.0 are as follows:

18.5.1 Link and interact with one another

CPS enables industrial systems to link and interact with one another, thereby improving current production capacities. Networked, communicative cyber-physical modules and systems make connected or smart factories, smart logistics, smart health, smart cities, and other realities possible. The end-to-end digital supply chain, suppliers, the sources of the materials and components needed for different types of smart manufacturing, and the final consumer must all be taken into account, regardless of the number of intermediate steps and participants. This means reducing the costs and inefficiencies of intermediaries in a digital supply chain paradigm while enabling direct methods for customised production, services, and customer engagement. The main feature of Industry 4.0, a comprehensive vision with distinctive frameworks and reference architectures, is the Integration of digital technologies and industrial assets in CPS.

18.5.2 Address crucial issues

A skilled workforce is needed to support and sustain the expansion of innovation in the CPS, as it is built on contributions from many other sectors. It may transform and progress the industry while addressing several crucial issues. The IoT has improved economic benefits, productivity, and efficiency while transforming and advancing CPS's role. It makes backing up data, accessing data from local and mobile devices, and enhancing cybersecurity easier in some instances. Real-time data management, improved procurement and supply chain efficiency and transparency, and automation-driven process improvement throughout the organisation's business divisions are all included in the CPS. This gives a virtual representation of anything with a tangible counterpart in the real world, such as hardware or software. They are used to provide testing data for new version development, predictive maintenance, and to improve the performance of their physical equivalent.

18.5.3 Automation

The usage of automation in water distribution systems employing CPS has grown during the last several years. Our homes' water-delivery systems include pipes, wells, pumps, tanks, and reservoirs. Devices in the system monitor various functions of the water distribution system. The systems also have programmable control logic for automated valve openings. CPS are electronic systems that interact with the outside environment. They are found in many services and applications and are widely used in modern industrial processes.

When integrating IoT and CPS devices into business networks, it is essential to ensure that existing systems are modified to realise the expected benefits and that the technology does not pose a significant liability risk. A monitoring mechanism for the ecosystem might be implemented to ensure optimal CPS performance.

18.5.4 Improve the production system

CPS proves improvement in the production system. Industry 4.0 transforms how businesses create, improve, and distribute their products. Manufacturers use this technology for their manufacturing facilities and operations. These intelligent factories use modern sensors, embedded software, and robotics to collect data, analyse it, and aid in decision-making. Intelligent robots with ML capabilities might be developed as a result of CPS breakthroughs, helping modern civilisation become more individualistic. Developing and maintaining a CPS roadmap that starts with simple implementations and advances to increasingly complex, AI-driven systems is necessary. The interactions among the many components of a CPS and the ability to perform multiple activities simultaneously are substantially affected by the network's latency, bandwidth, and stability.

18.5.5 Interact with the physical process

A CPS combines many systems, with the primary goal of manipulating and interacting with a physical process while adapting in real time to changing circumstances via feedback. Global digital networks that link CPS to the virtual world often connect them. They arise from the merger of computers, networks, and physical processes. A central organisation directs and controls interactive systems supported by intelligent devices using CPS. CPS conveys operational data to competent employees. CPS can rethink processes, optimise machines, and identify general improvements to increase income and productivity by using data from connected smart factories. Networked industrial systems maximise efficiency and effectiveness, which ultimately boosts profit margins.

18.5.6 Efficiently meet consumer demand

Future factories will have autonomy, human oversight, and the ability to respond quickly and effectively to changing customer and market demands through CPS, which orchestrates networking, computer, and physical processes as a unified unit. Since they enable the fusion of machines, products, people, and the environment into a single intelligent production system, connectivity and interoperability are crucial in the digitised industry. This helps provide tangible value, while information technologies assist with data storage, collection, and analysis. Real-time data analytics streamlines business intelligence and financial planning processes, while automation and machine learning enhance supply chain, vendor relationship management, and manufacturing operations. Digital twins are virtual representations of a CPS's discrete components, such as machines. They are necessary for providing information that might benefit a company and are considered essential parts of CPS.

18.5.7 Improve production quality

CPS is used in smart industries to improve production quality. By substituting AI-powered visual insights for manual inspection, business models reduce manufacturing errors and save money and time. Professionals in quality control can quickly set up a smartphone linked to the cloud to monitor production from anywhere. Manufacturers may find that repair work is more costly when using ML algorithms. Technology is integrated into or connected to the objects that make up the IoT, enabling them to detect, collect, and share data for specific purposes. Depending on the purpose and objective, this may include collecting data on movement, location, gas presence, temperature, and other factors.

18.5.8 Improve information and communication

CPS improves the networking of processes and intelligent industrial machines, as well as information and communication. It alludes to an idea that was developed by the industrial sector. The goal is to enable autonomous decision-making processes, real-time asset and process monitoring, and equally real-time linked value-generating networks via early stakeholder engagement and vertical and horizontal Integration. Industry 4.0 is a continually evolving vision, policy, and concept that includes reference designs, standardisation, and definitions. Computers and networks in Industry 4.0 monitor and control all physical processes via feedback loops. The system uses software to interpret actions and monitor results, much as a physical system would.

18.5.9 Autonomous decision-making

Transferring autonomy and autonomous decision-making to CPS and machines is at the heart of Industry 4.0. With the generation, leveraging, and use of actionable data and information to realise smart industries, industrial innovation, and collaboration ecosystems, Industry 4.0 offers an information-intensive transformation of manufacturing in a connected environment of big data, people, processes, systems, services, and IoT-enabled industrial assets. The fields of engineering, manufacturing, and mechanics all see CPS as the next evolution. Data may be shared across CPS. They can link to other CPS in a network, which gives them advanced control systems, embedded software, and communication capabilities. CPS may be recognised explicitly. They have an IP address, which allows them to be directly addressed, and they are part of the Internet of Everything.

18.5.10 Product development and design

The CPS serves as the foundation for new capabilities in remote control, prototyping, product design and development, services and diagnosis, condition monitoring, proactive and predictive maintenance, planning, innovation capability, track and trace, agility, real-time applications, and others. The 'smart' phenomenon, enabled by Industry 4.0's enhanced capabilities, typically attracts the most attention. Industry 4.0, which includes CPS, the IoT, cloud computing, cognitive computing, and the development of the smart factory, is the current trend of automation and data interchange in manufacturing technology. From a benefits perspective, Industry 4.0 is mainly concerned with increasing productivity, automating and optimising business processes and operations, and the most important IoT use case.

18.5.11 Integrated computing and storage capabilities

CPS comprises intelligent physical objects, systems, and components with integrated computing and storage capabilities linked through networks in an IoT, data, and services environment that emphasises processes. From an Industry 4.0 perspective, CPS is built on the latest embedded software platforms, control systems, and IP. The main goal of Industry 4.0 is to speed up production and related sectors, such as logistics, while also going beyond automation and efficiency to identify new business opportunities and models. Businesses that embrace comprehensive Industry 4.0 strategies are much more successful in generating revenue, achieving social impact, developing human capital, and investing in technology. With opportunities for value creation at each level and an emphasis on improving total value throughout the journey's stages, preparations, implementations, and actions, the Industry 4.0 strategy is the best approach.

18.5.12 Track, trace, and remote control

CPS offers new manufacturing capabilities for Industry 4.0, including track and trace and remote control. Energy, oil and gas, utilities, smart manufacturing, and other smart application areas are all made possible by CPS. Sensors, actuators, and other IoT components are also included in CPS. Both the IoT and CPS need a connection. The specific networking technology used depends on the circumstances. Networked machinery and embedded sensors provide industrial organisations with crucial data. Using data analytics, organisations can analyse historical trends, identify patterns, and make better decisions. Manufacturers may use sales, warehousing, and human resources data to analyse manufacturing choices based on labour and profit margins.

18.5.13 Increases information transparency

The interconnectedness of Industry 4.0 increases information transparency and empowers businesses to collect massive amounts of data across the whole production process through CPS. Using aggregate data, operators may identify problems, inefficiencies, and previously undiscovered opportunities. Additionally, manufacturers of IoT devices can collect data long after the product has left the factory, enhancing their ability to optimise, modify, and correct products. This may also help small and medium-sized manufacturers avoid upfront costs by allowing them to gauge demand and expand as their businesses grow. CPS can deliver dependable, intelligent embedded network systems. Manufacturing businesses often use CPS to self-monitor their operations and production methods. The information transferred between tools, business systems, supply chains, suppliers, and customers significantly improves production.

18.5.14 Utilise production resources effectively

By automatically responding to delays, supply chain changes, and other variables that may affect each site differently, CPS can utilise the organisation's production resources more efficiently. CPS integrates robots, IoT, and ML. A CPS is any mechanical process automatically controlled by software. Using sensors and other inputs from mechanical components, the software executes algorithms that determine how the machinery, equipment, or infrastructure should be operated. A CPS can be easily modified to meet a manufacturer's evolving needs, as it can adapt to environmental changes and operate across various settings and configurations. As Industry 4.0 takes shape, industrial organisations will increasingly rely on CPS principles to achieve operational excellence through productivity gains, efficient resource allocation, customer satisfaction, and increased shareholder value.

18.5.15 New possibilities

As CPS gains popularity, new possibilities emerge in the form of new products and business/service models. It will also automate the manual tasks currently carried out. This strikes a balance between humans selecting and rewarding value-added activity and CPS executing monotonous jobs. Using AI and ML, manufacturers can efficiently leverage the vast amounts of data generated on the production line and across all their business divisions, as well as from partners and external sources. Operations and business processes may become visible, predictable, and automated as a result of insights generated by AI and ML. Businesses can use the data from these assets and ML algorithms to do predictive maintenance, increasing uptime and efficiency. CPS may help manufacturers improve operations, create innovative products, and operate more efficiently. Manufacturers, for instance, might evaluate improvements to boost capacity or reduce downtime by simulating a manufacturing process.

18.5.16 Prevention of future problems

CPS helps prevent future problems. Operations in business and operations may profit from insights, predictability, visibility, and automation. For instance, industrial equipment is susceptible to failure throughout the manufacturing process. Thus to boost productivity and uptime, businesses may use ML algorithms and asset data to perform predictive maintenance. Manufacturers may create virtual clones of processes, factories, supply networks, and production lines. Every operation, process, worker, division, and piece of equipment involved in production in Industry 4.0 requires communication. There must be transparency and coordination across the whole research and development process, the supply chain, the production floor, customer service, marketing, and sales. Regardless of the applications and devices, they must work seamlessly and intelligently together.

18.5.17 Understanding of the supply chain and manufacturing process

Using CPS, manufacturers have improved their understanding of the whole supply chain and manufacturing process. Supply chain optimisation is now easily achievable in real time thanks to this level of visibility and AI and ML capabilities. Building networks, automating everything, placing sensors everywhere, and analysing everything with the IoT, sophisticated robots, and cutting-edge big data analytics may significantly enhance supply chain management. To guarantee CPS security and integrity, combining traditional cybersecurity safeguards inside the programme with authentication, advanced security engineering, encryption, cryptography, and physical security measures is essential. The management of CPS development requires the use of new-generation process solutions that provide hybridised artefacts, management, orchestration, automation, enforcement, traceability, and visibility. These solutions must also be extensible and integrable with a wide range of tools and technologies.

18.5.18 Design and modelling options

Modelling, design, and analysis may benefit from the integrated abstractions and alternatives offered by CPS. This relatively new idea and solution are given much thought in the context of enterprise digital transformation. However, this technology, which is being researched and enhanced worldwide, still holds much promise. A fundamental CPS concern is to determine and support new cooperative engineering paradigms that enable the fusion of mechanical and software design and development. It is essential to present relevant data across several domains in comprehensible ways, using different techniques for each domain. Progress and outcomes must also be gathered and connected. Monitoring the consequences of a change requires predictive analytics, engineering accountability, and audibility through cross-domain traceability.

18.5.19 Reduce waste

Lean manufacturing places a high premium on cutting waste and boosting output. CPS supports efficiency initiatives at every stage of the manufacturing process by combining Industry 4.0 technologies with lean manufacturing production methodologies. Time, money, energy, material resources, and human resources are all advantageous to lean manufacturing. Because of their networks, sensors, numerous processors, and embedded systems, modern cars are systems. In manufacturing, data streams connect people and machines. For instance, a sensor and its digital equivalent may be linked to a company's systems. The physical sensor provides the system with real-time data. At the same time, the digital twin is tested across different use cases to gain insight into potential process improvements, hardware upgrades, and other performance characteristics before those changes are applied in the real world.

18.5.20 Manage energy use

The CPS regulates the quantity needed for each operation and recycles unused energy to limit the energy utilised throughout manufacturing. CPS comprises software programmes that allow computerised control of industrial operations. Instead of individual control through an onboard controller, this permits communication between each station and a centralised platform, enabling convergence across management, production planning, and logistics.

Using sensors and other IoT measurement tools, the system may self-monitor and identify critical issues to boost productivity. Businesses can make quick decisions on any issue or new need with improved automation. It encourages manufacturing adaptation and agility while increasing production. Producers are urged by Industry 4.0 to make better use of resources at every stage of the supply chain and to keep a closer eye on improving quality, inventory, and logistics opportunities.

18.5.21 Software management

Software management has seen tremendous advances enabled by CPS technologies. By using the same design, we can deploy and run conventional embedded apps using cloud-native tools and infrastructure. Previously dominated by CPS, software documentation regimes now need to be managed and coordinated across the development lifecycle and multiple domain lifecycles. CPS adds new abstractions and artefacts. Numerous artefacts, abstractions, and models relating to various design characteristics typically need to be synthesised into forms and formats that have significance in other relevant domains. Software in a CPS requires considerably more quality assurance, as it often implements life-critical functions and must ensure human safety. The interconnection and traceability of diverse quality processes are necessary for successful CPS innovation.

18.5.22 Connect and monitor the complete production system

Data can be utilised to inform automated decisions because the CPS can connect to and monitor the entire production floor as a single system. In these CPS, machines communicate with each other and with people. Measurements and data on machine usage, component counts, and quality are collected from the shop floor using smart sensors. Because they only pay for the services they use, manufacturers can swiftly scale their IT infrastructure up and down using cloud computing. This enables companies to keep IT personnel close to clients and to provide temporary test environments. CPS is a technical systems that use robots and networked computers to communicate with the real world. It is essential to consider CPS's influence in this area, as it is used across many goods and services and has been rapidly incorporated into modern industrial processes.

18.5.23 Collect and evaluate data

Data is critical to Industry 4.0, especially the ability to collect and analyse information from sensors and intelligent devices integrated into machines or processes. Preventative maintenance, analytics-driven production optimisation, and predictive diagnostics are just a few industrial applications that employ the information derived from this data. As organisations transition from early phases to creative methods, Industry 4.0 is moving toward an end-to-end value chain, and decentralised intelligence in production is the next step toward CPS. Industry 4.0 is built on data modelling and mapping that covers the end-to-end product lifecycle and value stream.

18.5.24 Management of resources

The construction of CPS optimises resource management, user input, and consumer interactions with the product. The actual revolution links the mechanised and computerised manufacturing floor via sensors, computing modules, and processing centres. CPS significantly alters the communication between machines. The cloud also expands the possibility of machine-to-machine communication. It may provide many possibilities for Integration. The cooperation and networking of machines, data, and people in production are aided by IoT. Consequently, it leverages seamlessly interconnected and interfaced IoT devices, sensors, and data. Every part of manufacturing processes may be networked and linked in the IoT, and the data produced can be utilised to maximise efficiency across the whole production process.

18.6 Roles of digital twin for cyber-physical systems

In Industry 4.0, digital twin technology is strategically important for improving the dependability and performance of CPS. A digital twin enables continuous synchronisation between the digital and physical layers of CPS by creating a real-time virtual counterpart of physical assets, embedded controllers, sensors, and communication networks. By using intelligent optimisation and predictive analytics, digital twins improve CPS [48–52]. The digital twin can identify irregularities, predict equipment failures, and suggest remedial measures in advance by analysing real-time sensor data and system behaviour. By giving autonomous control units precise system insights, it also facilitates decentralised decision-making. When new technologies or components are added to the CPS framework, digital twins enable integration validation, scalability testing, and performance benchmarking [53–57]. The major roles of digital twin for CPS are as follows:

- Synchronisation of physical assets and virtual CPS models in real time.
- Embedded control algorithm simulation before actual implementation.
- Verification of the correctness of sensor-actuator communication.
- Evaluation of decentralised decision-making systems in a digital setting.
- Using anomaly pattern analysis to identify faults early.
- CPS component predictive maintenance based on real-time operational data.
- To evaluate system resilience, cyber-physical assault scenarios are modelled.
- Optimising feedback control loop performance.
- The integration of CPS production lines through virtual commissioning.
- Analysis and optimisation of energy use in CPS networks.
- To increase system dependability, use failure mode and effects simulation.
- Monitoring latency and network communication delays in real time.
- Verification of safety mechanisms in unusual operating circumstances.
- Reconstructing system occurrences digitally for root cause analysis.
- Production scheduling optimisation using CPS frameworks.
- Evaluation of the compatibility of various CPS components.
- Assistance for self-healing and adaptive system behaviour on its own.

18.7 Disadvantages of cyber-physical systems

The significant disadvantages of CPS are as follows [18,58–60]:

- The need for protection of both physical and digital components
- Legacy system vulnerabilities.
- Catastrophic failure potential, as with attacks on any such infrastructure, will cause huge damage.
- Privacy concerns.
- Socio-economic disadvantages, with workforce displacement, job losses in manufacturing and logistics.
- Skills gap, with high initial and maintenance costs.
- System abstraction complexity.
- The need for defence against both deliberate and accidental assaults.
- The need to defend against dangers like viruses, worms, or denial-of-service assaults.
- The need for attack-resistant gadget design.
- Required management of networks.
- It is not easy to develop a data architecture that meets the stringent performance, scalability, and availability requirements of industrial contexts.
- Adopting new data-flow technologies and integrating them with older systems takes significant effort.

18.8 Upcoming goals of cyber-physical systems

The primary upcoming goals of CPS are as follows [1,61]:

- Future industry standards for production procedures will be introduced through CPSs.
- Production environments that self-configure, self-tune, and self-optimise will become more agile, flexible, and cost-effective.
- This will profoundly change the industrial environment, as by fusing the physical and digital worlds, CPS has ushered in a new age of productivity, growth, and information.
- By including intelligence in their products, manufacturers will be able to perform predictive maintenance.
- Different businesses will adopt this digital transformation at different speeds and in different ways.
- This will pave the way for Industry 4.0 ideas as well as a new generation of highly profitable manufacturing methods.

18.9 Conclusion

CPS are systems that combine physical and digital components, and the embedded software communicates with the network to manage actual actions and collect data from sensors. With embedded software, modern control systems, and an Internet address, they are connected to and addressed via IoT. For developing several cutting-edge CPS applications, a well-designed communication network with dependable quality of service assurances is a crucial foundation. A CPS is a group of related systems that work together to control a physical process and, via feedback, adjust to changing conditions. They combine computer, networking, and physical operations in real-time. CPS changed how people interact with planned systems, just as the internet changed how people engage with information.

It supports the production process and fills a crucial function. This enables automatic model rebuilding to address challenging issues. Smart, linked goods and tech-enabled systems are powered by robust, fail-safe technology and data analytics solutions at the heart of Industry 4.0. The CPS stands out for its integration of operational technology, smart cities, IoT, and IT infrastructure. The CPS innovation is the labour of many other professions, and to support and maintain its expansion, it requires skilled staff. It may revolutionise the industry and progress it while resolving various urgent problems. The expansion of the number of IoT-connected devices has changed and enhanced the role of CPS, thereby improving productivity, efficiency, and financial outcomes.

References

[1] J. Lee, B. Bagheri, H.A. Kao, A cyber-physical systems architecture for Industry 4.0-based manufacturing systems, Manufacturing letters 3 (2015) 18–23.

[2] Y. Lu, Cyber-physical system (CPS)-based Industry 4.0: a survey, Journal of Industrial Integration and Management 2 (3) (2017) 1750014.

[3] M. Javaid, A. Haleem, R.P. Singh, R. Suman, An integrated outlook of Cyber–Physical Systems for Industry 4.0: topical practices, architecture, and applications, Green Technologies and Sustainability 1 (2023) 100001.

[4] L. Wang, G. Wang, Big data in cyber-physical systems, digital manufacturing and Industry 4.0, International Journal of Engineering and Manufacturing (IJEM 6 (2016) 1–8.

[5] P.J. Mosterman, J. Zander, Industry 4.0 as a cyber-physical system study, Software & Systems Modelling 15 (2016) 17–29.

[6] S.J. Oks, M. Jalowski, M. Lechner, S. Mirschberger, M. Merklein, B. Vogel-Heuser, et al., Cyber-physical systems in the context of Industry 4.0: a review, categorisation and outlook, Information Systems Frontiers 26 (2024) 1731–1772.

[7] L.D. Xu, L. Duan, Big data for cyber physical systems in Industry 4.0: a survey, Enterprise Information Systems 13 (2019) 148–169.

[8] J.H. Kim, A review of cyber-physical system research relevant to the emerging IT trends: industry 4.0, IoT, big data, and cloud computing, Journal of Industrial Integration and Management 2 (2017) 1750011.

[9] Cyber Physical Systems Research Challenges. Available from https://www.tutorsindia.com/blog/cyber-physical-systems-research-challenges/.

[10] M.M. Aslam, W. Shafik, A.F. Hidayatullah, K. Kalinaki, H. Gul, R.Y. Zakari, et al., Intelligent transportation systems: a critical review of integration of cyber-physical systems (CPS) and Industry 4.0, Digital Communications and Networks (2025).

[11] B. Bagheri, S. Yang, H.A. Kao, J. Lee, Cyber-physical systems architecture for self-aware machines in Industry 4.0 environment, IFAC-PapersOnLine 48 (2015) 1622–1627.

[12] What are Cyber Physical Systems? (Design Elements & Factors). Available from https://www.erp-information.com/cyber-physical-systems.

[13] R.S. Ahmed, E.S.A. Ahmed, R.A. Saeed, Machine learning in cyber-physical systems in Industry 4.0, Artificial Intelligence Paradigms for Smart Cyber-Physical Systems, IGI Global Scientific Publishing, (2021), pp. 20–41.

[14] Afrizal, A., Mulyanti, B., Widiaty, I., 2020. Development of cyber-physical system (CPS) implementation in Industry 4.0. In: IOP Conference Series: Materials Science and Engineering, vol. 830, No. 4, p. 42090. IOP Publishing.

[15] Cyber Physical Systems: Advantages and Disadvantages. Available from http://cyberphysicalsystems.blogspot.com/2011/07/cyber-physical-systems-advantages-and.html.

[16] K. Zhou, T. Liu, L. Liang, From cyber-physical systems to Industry 4.0: make future manufacturing possible, International Journal of Manufacturing Research 11 (2016) 167–188.

[17] B. Dafflon, N. Moalla, Y. Ouzrout, The challenges, approaches, and used techniques of CPS for manufacturing in Industry 4.0: a literature review, The International Journal of Advanced Manufacturing Technology 113 (2021) 2395–2412.

[18] Cyber-physical systems. Available from https://www.luos.io/blog/cyber-physical-systems.

[19] E. Mutua, Cyber-physical systems and their role in Industry 4.0, Journal of Technology and Systems 6 (2024) 57–69.

[20] N.H. Tran, H.S. Park, Q.V. Nguyen, T.D. Hoang, Development of a smart cyber-physical manufacturing system in the Industry 4.0 context, Applied Sciences 9 (2019) 3325.

[21] J.R. Jiang, An improved cyber-physical systems architecture for Industry 4.0 smart factories, Advances in Mechanical Engineering 10 (6) (2018) 168781401878419. Available from https://doi.org/10.1177/1687814018784192.

[22] M. Krugh, L. Mears, A complementary cyber-human systems framework for Industry 4.0 cyber-physical systems, Manufacturing Letters 15 (2018) 89–92.

[23] D.G. Pivoto, L.F. De Almeida, R. da Rosa Righi, J.J. Rodrigues, A.B. Lugli, A.M. Alberti, Cyber-physical systems architectures for industrial internet of things applications in Industry 4.0: a literature review, Journal of Manufacturing Systems 58 (2021) 176–192.

[24] D. Sinha, R. Roy, Reviewing cyber-physical system as a part of smart factory in industry 4.0, IEEE Engineering Management Review 48 (2020) 103–117.

[25] O.C. Abikoye, A.O. Bajeh, J.B. Awotunde, A.O. Ameen, H.A. Mojeed, M. Abdulraheem, et al., Application of internet of thing and cyber physical system in Industry 4.0 smart manufacturing, Emergence of Cyber Physical System and IoT in Smart Automation and Robotics: Computer Engineering in Automation, Springer International Publishing, Cham, (2021), pp. 203–217.

[26] C. Serôdio, P. Mestre, J. Cabral, M. Gomes, F. Branco, Software and architecture orchestration for process control in Industry 4.0 enabled by cyber-physical systems technologies, Applied Sciences 14 (2024) 2160.

[27] J. Yan, M. Zhang, Z. Fu, An intralogistics-oriented Cyber-Physical System for workshop in the context of Industry 4.0, Procedia manufacturing 35 (2019) 1178–1183.

[28] Cogliati, D., Falchetto, M., Pau, D., Roveri, M., Viscardi, G., 2018. Intelligent cyber-physical systems for Industry 4. 0. In: 2018 First International Conference on Artificial Intelligence for Industries (AI4I), IEEE, pp. 19-22.

[29] P. Fantini, M. Pinzone, M. Taisch, Placing the operator at the centre of Industry 4.0 design: modelling and assessing human activities within cyber-physical systems, Computers & Industrial Engineering 139 (2020) 105058.

[30] M. Sony, Design of cyber physical system architecture for Industry 4.0 through lean six sigma: conceptual foundations and research issues, Production & Manufacturing Research 8 (2020) 158–181.

[31] What are Cyber-Physical Systems?: https://www.analyticssteps.com/blogs/what-are-cyber-physical-systems.

[32] H. Singh, Big data, Industry 4.0 and cyber-physical systems integration: a smart industry context, Materials Today: Proceedings 46 (2021) 157–162.

[33] Sbaglia, L., Giberti, H., Silvestri, M., 2018. The cyber-physical systems within the Industry 4.0 framework. In: The International Conference of IFToMM ITALY, pp. 415-423. Cham: Springer International Publishing.

[34] S. Karnouskos, P. Leitao, L. Ribeiro, A.W. Colombo, Industrial agents as a key enabler for realising industrial cyber-physical systems: multiagent systems entering Industry 4.0, IEEE Industrial Electronics Magazine 14 (2020) 18–32.

[35] T. Ji, X. Xu, Exploring the integration of cloud manufacturing and cyber-physical systems in the era of Industry 4.0–an OPC UA approach, Robotics and Computer-Integrated Manufacturing 93 (2025) 102927.

[36] Savtschenko, M., Schulte, F., Voß, S., 2017. IT governance for cyber-physical systems: the case of Industry 4.0. In: International Conference of Design, User Experience, and Usability, pp. 667–676. Cham: Springer International Publishing.

[37] G. Matana, A. Tadeu Simon, D.R. Tase Velazquez, L. Hernández Mastrapa, A. Luís Helleno, Cyber-physical systems as key element to Industry 4.0: characteristics, applications and related technologies, Engineering Management Journal 35 (2023) 377–404.

[38] J. Lee, M. Azamfar, J. Singh, A blockchain-enabled Cyber-Physical System architecture for Industry 4.0 manufacturing systems, Manufacturing letters 20 (2019) 34–39.

[39] Hofer, F., 2018, October. Architecture, technologies and challenges for cyber-physical systems in Industry 4.0: a systematic mapping study. In: Proceedings of the 12th ACM/IEEE International Symposium on Empirical Software Engineering and Measurement, pp. 1-10.

[39] Frontoni, E., Loncarski, J., Pierdicca, R., Bernardini, M., Sasso, M., 2018, June. Cyber physical systems for industry 4.0: towards real time virtual reality in smart manufacturing. In: International Conference on Augmented Reality, Virtual Reality and Computer Graphics, pp. 422–434. Cham: Springer International Publishing.

[41] M.H. Abidi, H. Alkhalefah, U. Umer, Fuzzy harmony search based optimal control strategy for wireless cyber physical system with Industry 4.0, Journal of Intelligent Manufacturing 33 (2022) 1795–1812.

[42] K. Chukalov, Horizontal and vertical Integration, as a requirement for cyber-physical systems in the context of Industry 4.0, Industry 4.0 2 (2017) 155–157.

[43] T.M. Fernández-Caramés, P. Fraga-Lamas, M. Suárez-Albela, M.A. Díaz-Bouza, A fog computing based cyber-physical system for the automation of pipe-related tasks in the Industry 4.0 shipyard, Sensors 18 (2018) 1961.

[44] F. Matsunaga, V. Zytkowski, P. Valle, F. Deschamps, Optimisation of energy efficiency in smart manufacturing through the application of cyber–physical systems and Industry 4.0 technologies, Journal of Energy Resources Technology 144 (2022) 102104.

[45] K.M. Alam, A. El Saddik, C2PS: a digital twin architecture reference model for the cloud-based cyber-physical systems, IEEE Access 5 (2017) 2050–2062.

[46] Xu, Q., Ali, S., Yue, T., 2021, April. Digital twin-based anomaly detection in cyber-physical systems. In: 2021 14th IEEE Conference on Software Testing, Verification and Validation (ICST), pp. 205-216, IEEE.

[47] K. Josifovska, E. Yigitbas, G. Engels, Reference framework for digital twins within cyber-physical systems, 2019, IEEE/ACM 5th International Workshop on Software Engineering for Smart Cyber-Physical Systems (SEsCPS), IEEE, (2019), pp. 25–31.

[48] F. Tao, Q. Qi, L. Wang, A.Y.C. Nee, Digital twins and cyber–physical systems toward smart manufacturing and Industry 4.0: correlation and comparison, Engineering 5 (2019) 653–661.

[49] Eckhart, M., Ekelhart, A., 2019. Digital twins for cyber-physical systems security: state of the art and outlook. In: Security and Quality in Cyber-Physical Systems Engineering: With Forewords by Robert M. Lee and Tom Gilb, pp. 383-412.

[50] J. Leng, H. Zhang, D. Yan, Q. Liu, X. Chen, D. Zhang, Digital twin-driven manufacturing cyber-physical system for parallel controlling of smart workshop, Journal of Ambient Intelligence and Humanised Computing 10 (2019) 1155–1166.

[51] J. Lee, M. Azamfar, J. Singh, S. Siahpour, Integration of digital twin and deep learning in cyber-physical systems: towards smart manufacturing, IET Collaborative Intelligent Manufacturing 2 (2020) 34–36.

[52] M. Groshev, C. Guimarães, J. Martín-Pérez, A. de la Oliva, Toward intelligent cyber-physical systems: digital twin meets artificial intelligence, IEEE Communications Magazine 59 (2021) 14–20.

[53] F. Biesinger, D. Meike, B. Kraß, M. Weyrich, A digital twin for production planning based on cyber-physical systems: a case study for a cyber-physical system-based creation of a digital twin, Procedia CIRP 79 (2019) 355–360.

[54] Q. Xu, S. Ali, T. Yue, Digital twin-based anomaly detection with curriculum learning in cyber-physical systems, ACM Transactions on Software Engineering and Methodology 32 (2023) 1–32.

[55] B. Bordel Sánchez, R.P. Alcarria Garrido, D. Sánchez de Rivera Córdoba, Á. Sánchez Picot, Enhancing process control in Industry 4.0 scenarios using cyber-physical systems, Journal of Wireless Mobile Networks, Ubiquitous Computing, and Dependable Applications 7 (2016) 41–64.

[56] G.C. da Silva, P.C. Kaminski, From embedded systems (ES) to cyber-physical systems (CPS): an analysis of transitory stage of automotive manufacturing in the Industry 4.0 scenario, 25th SAE BRASIL International Congress and Display, SAE Technical Paper, (2016).

[57] M.A. Alohali, F.N. Al-Wesabi, A.M. Hilal, S. Goel, D. Gupta, A. Khanna, Artificial intelligence enabled intrusion detection systems for cognitive cyber-physical systems in Industry 4.0 environment, Cognitive Neurodynamics 16 (2022) 1045–1057.

[58] Jamaludin, J., Rohani, J.M., 2018, November. Cyber-physical system (CPS): state of the art. In: 2018 International Conference on Computing, Electronic and Electrical Engineering (ICE Cube), pp. 1-5. IEEE.

[59] H.M. Kang, K.T. Hwang, Analysis of research trends of cyber physical system (CPS) in the manufacturing industry, Informatisation Policy 25 (2018) 3–28.

[60] R.J. Eleftheriadis, O. Myklebust, Industry 4.0 and cyber-physical systems in a Norwegian industrial context, International workshop of advanced manufacturing and automation, Springer, Singapore, (2017), pp. 491–499.

[61] M. Ryalat, H. ElMoaqet, M. AlFaouri, Design of a smart factory based on cyber-physical systems and Internet of Things towards Industry 4.0, Applied Sciences 13 (2023) 2156.

Index

Note: Page numbers followed by "*f*" refer to figures.

D

E

F

Printed and bound by CPI Group (UK) Ltd, Croydon, CR0 4YY
14/09/2026
02216408-0012